Lecture Notes in Computer Science 16147

The series Lecture Notes in Computer Science (LNCS), including its subseries Lecture Notes in Artificial Intelligence (LNAI) and Lecture Notes in Bioinformatics (LNBI), has established itself as a medium for the publication of new developments in computer science and information technology research, teaching, and education.

LNCS enjoys close cooperation with the computer science R & D community, the series counts many renowned academics among its volume editors and paper authors, and collaborates with prestigious societies. Its mission is to serve this international community by providing an invaluable service, mainly focused on the publication of conference and workshop proceedings and postproceedings. LNCS commenced publication in 1973.

Jianing Qiu · Jinlin Wu · Curtis Langlotz ·
Baoru Huang · Zhen Lei · Honghan Wu ·
Hongbin Liu · Weidi Xie
Editors

AI for Clinical Applications

First International Workshops
Agentic AI 2025, CREATE 2025, and Clinical MLLMs 2025
Held in Conjunction with MICCAI 2025
Daejeon, South Korea, September 23 and 27, 2025
Proceedings

 Springer

Editors
Jianing Qiu
The Chinese University of Hong Kong
Sha Tin, Hong Kong

Curtis Langlotz
Stanford University
Palo Alto, CA, USA

Zhen Lei
Centre for Artificial Intelligence and Robotics
(CAIR)
Pak Shek Kok, New Territories, Hong Kong

Hongbin Liu
Centre for Artificial Intelligence and Robotics
(CAIR)
Pak Shek Kok, New Territories, Hong Kong

Jinlin Wu
Centre for Artificial Intelligence and Robotics
(CAIR)
Pak Shek Kok, Hong Kong

Baoru Huang
The University of Liverpool
Liverpool, UK

Honghan Wu
University of Glasgow
Glasgow, UK

Weidi Xie
Shanghai Jiao Tong University
Shanghai, China

ISSN 0302-9743   ISSN 1611-3349 (electronic)
Lecture Notes in Computer Science
ISBN 978-3-032-06003-7   ISBN 978-3-032-06004-4 (eBook)
https://doi.org/10.1007/978-3-032-06004-4

This Springer imprint is published by the registered company Springer Nature Switzerland AG
The registered company address is: Gewerbestrasse 11, 6330 Cham, Switzerland

If disposing of this product, please recycle the paper.

Agentic AI for Medicine Preface

The field of medical AI has evolved from predictive AI to generative AI and is about to enter the era of agentic AI, poised for rapid growth in the upcoming years. Unlike previous generations of AI, which passively execute human instructions and are limited to specific tasks, AI agents are more proactive and autonomous. Utilizing large AI models as their digital brain, these agents can perceive information, plan and make decisions, recall and reflect, interact and collaborate, leverage various tools, and take decisive action. As a result, AI agents exhibit a higher level of autonomy and intelligence in addressing medical and clinical challenges, operating within an open-ended action space that enables them to tackle a wide array of tasks across diverse data modalities.

AI agents have shown great potential and value in science domains. They have recently been used in novel nanobody design, chemical synthesis, and clinical diagnosis. However, the development of AI agents in the medical AI field is still in its nascent stages, with numerous challenges and risks remaining unsolved such as security and safety of AI agents, and the lack of benchmarks and evaluation methods.

Despite all these challenges and risks, AI agents represent a research opportunity that can drive new biomedical discoveries and transform the landscape of current healthcare systems and clinical practices.

The first Agentic AI for Medicine workshop, held in conjunction with MICCAI 2025, received 16 valid submissions, among which 13 were accepted for oral and poster presentations after double-blind peer review via the OpenReview platform, and 12 of them are published in this volume of the MICCAI joint proceedings. Each submitted paper received an average of 2.25 reviews and was handled by a designated area chair.

Topics of accepted work include addressing the safe development of agentic AI in mental health, new theories and principles of medical AI agents, and clinical applications of agentic systems such as large language model (LLM)-based agents for Alzheimer's disease analysis and multi-agent systems for automated machine learning in medical imaging.

By bringing together the latest findings and innovations, as well as open discussions, we hope this workshop can contribute to future research and development of medical AI agents and drive the responsible use of AI agents in medicine and healthcare.

July 2025

Jianing Qiu
Baoru Huang

Organization

General Chair

Wu Yuan	Chinese University of Hong Kong, China

Program Committee Chairs

Jianing Qiu	Chinese University of Hong Kong, China
Baoru Huang	University of Liverpool, UK

Steering Committee

Anh Nguyen	University of Liverpool, UK
Weidi Xie	Shanghai Jiao Tong University, China
Ehsan Adeli	Stanford University, USA
Daniel Rueckert	Technical University of Munich, Germany

Program Committee

Lin Li	University of Oxford, UK
Jiankai Sun	Stanford University, USA
Haibao Yu	University of Hong Kong, China
Sheng Liu	Stanford University, USA
Linyuan Li	Chinese University of Hong Kong, China
Jiawei Ma	City University of Hong Kong, China
Emma Anran Ran	Chinese University of Hong Kong, China
Bang Zheng	Peking University, China
Zhe Xu	Chinese University of Hong Kong, China
Minqing Zhang	Chinese University of Hong Kong, China
Kyle Lam	Imperial College London, UK
Artur Krzysztof Banach	EPFL, Switzerland
Xiao Gu	University of Oxford, UK
Frank P.-W. Lo	Imperial College London, UK

Additional Reviewers

Wei Zhang

Lujun Gui

Jinghan Sun

Yixin Wang

Jinghao Feng

Wenting Chen

Zirui Zhou

Zhe Min

Junhong Chen

Qiaoyu Zheng

CREATE 2025 Preface

The CREATE 2025 Workshop (Clinical-driven Robotics and Embodied AI TEchnology) was held in Daejeon, South Korea, on September 27, 2025, in conjunction with the 28th International Conference on Medical Image Computing and Computer-Assisted Intervention (MICCAI).

The integration of artificial intelligence (AI) and robotics is fundamentally reshaping clinical intervention and patient care. The CREATE workshop was established to advance this progress by emphasizing the synergy between embodied AI—including recent advances in large language models (LLMs)—and robotics for healthcare applications. Embodied AI systems demonstrate advanced reasoning, adaptability, and the ability to interact intelligently within dynamic clinical environments, which are essential for developing next-generation surgical robots and intelligent clinical assistants. By bringing together researchers, engineers, and clinicians, CREATE fosters multidisciplinary collaboration and drives innovations that address real-world clinical challenges.

Authors were invited to submit original research on topics such as clinical-driven surgical robotics, intelligent surgical assistants, human-robot collaboration, AI-augmented operating rooms, and the application of foundation models in surgery. The workshop received 12 high-quality submissions from international research teams. The double-blind review process was organized by a Program Committee of 9 experts, supported by 10 additional reviewers from 14 research institutions worldwide. Each submission was assigned to three reviewers to ensure fairness and consistency, with papers scored from 1 (lowest) to 9 (highest), and an average score above 4.5 considered acceptable. Final decisions were made after thorough Program Committee discussion. Based on this process, 10 outstanding papers were selected for inclusion in these proceedings.

The accepted papers cover a broad spectrum of topics, including clinical-driven challenges in surgery, AI-powered robotic systems, intelligent assistance in the operating room, foundation models for surgical applications, and innovations in human-robot interaction. These contributions reflect the momentum toward intelligent, adaptive, and clinically relevant technologies in surgical care, highlighting the importance of cross-disciplinary efforts for real clinical impact.

We thank all authors for their excellent submissions, our reviewers for their valuable feedback, our invited speakers for sharing their expertise, and all participants for their engagement and support. We hope that the CREATE 2025 Workshop and these proceedings will inspire further interdisciplinary collaboration and innovation at the intersection of embodied AI, robotics, and medicine, advancing the future of healthcare.

October 2025

Jinlin Wu
Zhen Lei
Hongbin Liu

Organization

Workshop Chairs

Nassir Navab	Technische Universität München, Germany
Hongbin Liu	CAIR, HKISI-CAS, China
Zhen Lei	CAIR, HKISI-CAS, China
Gaofeng Meng	CAIR, HKISI-CAS, China
Hongliang Ren	Chinese University of Hong Kong, China
Danny T. M. Chan	Prince of Wales Hospital, China
Jinlin Wu	CAIR, HKISI-CAS, China

Program Committee

Long Bai (Program Chair)	Chinese University of Hong Kong, China
Kun Yuan	University of Strasbourg, France
Han Li	Technische Universität München, Germany
Tong Chen	University of Sydney, Australia
Mingcong Chen	City University of Hong Kong, China
Mingyang Zhao	Academy of Mathematics and Systems Science, CAS, China
Zelin Zang	CAIR, HKISI-CAS, China
Hongyuan Zhang	CAIR, HKISI-CAS, China
Boqiang Xu	CAIR, HKISI-CAS, China

Additional Reviewers

Qingyao Tian	Guankun Wang
Siqi Fan	Wei Zhang
Xusheng Liang	Hongqiu Wang
Xingjian Luo	Jingsong Liu
Miao Xu	Shi Li

MLLM Preface

The First Workshop on Multimodal Large Language Models (MLLMs) in Clinical Practice was held on September 23, 2025, in Daejeon, South Korea, in conjunction with the 28th International Conference on Medical Image Computing and Computer Assisted Intervention (MICCAI 2025).

Recent breakthroughs in medical MLLMs—such as MedGemini, which can process 2D and 3D medical images as well as genomic sequences—have ushered in a transformative era for clinical AI. These models integrate diverse data modalities, including text, medical images, and genomic data, enabling more comprehensive and precise analyses than were previously possible with unimodal approaches. Unifying multimodal data sources such as clinical notes, diagnostic imaging, and graph-based representations within a single framework has shown promise in improving diagnostics, enhancing workflow efficiency, and advancing personalized medicine.

Despite these advancements, several critical challenges remain before MLLMs can be fully integrated into clinical practice. Data scarcity, the fragmented nature of healthcare records, and concerns over patient privacy hinder the development of the large, representative datasets necessary for robust model generalization across different patient populations. Furthermore, current evaluation practices often emphasize accuracy alone, overlooking essential dimensions like fairness, robustness, and generalization across clinical tasks and data types. Developing comprehensive benchmarks and evaluation protocols is essential to ensure MLLMs can be safely, effectively, and ethically deployed in real-world clinical practice.

Simultaneously, MLLMs open exciting new possibilities for human-AI collaboration in healthcare. By incorporating clinician-driven inputs, such as gaze patterns or contextual prompts, these systems have the potential to enhance diagnostic precision and tailor care to individual patient needs. As these tools evolve, they may help streamline clinical workflows and support more data-driven decision-making. Realizing this potential, however, will require continued progress in developing models capable of handling diverse modalities, alongside thoughtful deployment strategies that address data limitations, safety, and evaluation standards.

Recognizing the growing importance of this field, we established this inaugural workshop to catalyze dialogue, interdisciplinary collaboration, and focused research on the development and clinical translation of MLLMs. Our goal was to spotlight key technical and practical challenges, showcase novel research, and foster a community dedicated to advancing multimodal AI in clinical care.

The workshop received 19 submissions. One was desk-rejected for not meeting submission guidelines, while the remaining 18 underwent rigorous double-blind peer review, with each paper evaluated by three independent reviewers. Based on the assessments of 21 expert reviewers, 4 papers were selected for oral presentation and 7 for poster presentation. All 11 accepted papers are included in this volume, published in the Springer

Lecture Notes in Computer Science (LNCS) series as part of the joint MICCAI 2025 proceedings.

We extend our heartfelt thanks to all contributing authors for their innovative work, to the reviewers for their thoughtful and thorough evaluations, and to the participants for their lively engagement and insightful discussions. We hope this volume serves as a valuable resource and inspiration for the continued advancement of multimodal large language models in clinical settings.

July 2025

Yunsoo Kim
Chaoyi Wu
Justin Xu
Hyewon Jeong
Sophie Ostmeier
Michelle Li
Zhihong Chen
Xiaoqing Guo
Yuyin Zhou
Weidi Xie
Honghan Wu
Curtis Langlotz

Organization

General Chairs

Curtis Langlotz	Stanford University, USA
Honghan Wu	University of Glasgow, UK
Weidi Xie	Shanghai Jiao Tong University, China

General Student Chairs

Yunsoo Kim	University College London, UK
Chaoyi Wu	Shanghai Jiao Tong University, China
Justin Xu	University of Oxford, UK

Program Committee Chairs

Zhihong Chen	Stanford University, USA
Xiaoqing Guo	Hong Kong Baptist University, China
Hyewon Jeong	Massachusetts Institute of Technology, USA
Yunsoo Kim	University College London, UK
Curtis Langlotz	Stanford University, USA
Michelle Li	Harvard Medical School, USA
Sophie Ostmeier	Stanford University, USA
Chaoyi Wu	Shanghai Jiao Tong University, China
Honghan Wu	University of Glasgow, UK
Weidi Xie	Shanghai Jiao Tong University, China
Justin Xu	University of Oxford, UK
Yuyin Zhou	Stanford University, USA

Keynote Chairs

Michelle Li	Harvard Medical School, USA
Yuyin Zhou	Stanford University, USA

Panel Chairs

Zhihong Chen	Stanford University, USA
Xiaoqing Guo	Hong Kong Baptist University, China

Publicity Chair

Hyewon Jeong	Massachusetts Institute of Technology, USA

Program Committee

Mansu Kim	Gwangju Institute of Science and Technology, South Korea
Jianan Chen	University College London, UK
Xiaoman Zhang	Harvard Medical School, USA
Xiao Zhou	Shanghai AI Laboratory, China
Abul Hasan	University of Oxford, UK
Hyeryun Park	Seoul National University, South Korea
Weike Zhao	Shanghai Jiao Tong University, China
Yusuf Abdulle	King's College London, UK
Haoning Wu	Shanghai Jiao Tong University, China
Jie Liu	City University of Hong Kong, China
Wenting Chen	City University of Hong Kong, China
Qiushi Yang	City University of Hong Kong, China
Pengcheng Qiu	Shanghai Jiao Tong University, China
Kevin Yuan	University of Oxford, UK
Jinge Wu	University College London, UK
Quang N. Nguyen	University College London, UK
Teya Bergamaschi	Massachusetts Institute of Technology, USA
Clemence Mottez	Stanford University, USA

Additional Reviewers

Venkat Nilesh Dadi	Yao Zhang

Contents

The Clinical-Driven Robotics and Embodied AI Technology (CREATE 2025)

The First Workshop on Multimodal Language Models (MLLMs 2025)

The MICCAI Workshop on Agentic AI 2025

An Explainable Multimodal Framework with LLM Agents for Intracranial Hemorrhage Detection

Shashwath Punneshetty[1], Dhyey Italiya[1], Vinti Agarwal[1(✉)],
Chandresh Maurya[2], and Amit Agrawal[3]

[1] Birla Institute of Technology and Science Pilani, Pilani, India
`{shashwath.p,f20211463,vinti.agarwal}@pilani.bits-pilani.ac.in`
[2] Indian Institute of Technology Indore, Indore, India
`chandresh@iiti.ac.in`
[3] All India Institute of Medical Sciences Bhopal, Bhopal, India

Abstract. Explainability in intracranial hemorrhage (ICH) diagnosis is essential for timely and accurate clinical decisions, especially in life–threatening situations. We propose a framework that generates explainable, clinically relevant text from 2D CT scans using two cooperative GPT-4o agents: a Multi-modal User Agent (MUA) and a Planner Agent. The MUA interprets scans with YOLOv10 (mosaic augmentation), SAM2, and clustering; the Planner selects tools and outputs key imaging parameters: bleed location, midline shift, calvarial fracture, and mass effect crucial for urgent interventions. Explainability is enforced via chain-of-thought prompting to ensure transparent decision-making. Experiments show YOLOv10 with mosaic improves mAP@0.5:0.95 by 4.1% over existing methods, and the LLM agents extract clinical parameters with 78.1% accuracy (Our code is available at https://github.com/Shashwathp/Explainable-ICH-Detection-with-LLM-Agents/tree/main). These results underscore the potential of explainable AI to enhance trust and reliability in critical healthcare applications.

Keywords: Intracranial Hemorrhage · Explainable AI · LLM Agents

1 Introduction

Intracranial hemorrhage (ICH) requires rapid diagnosis to prevent irreversible damage. Radiologists evaluate bleed type, location, fractures, mass effect, and midline shift for treatment urgency. While recent explainable AI using self-attention [4] and CAM [5] provides visual explanations, comprehensive diagnostic summaries remain challenging.

We investigate generating explainable ICH diagnostic summaries by integrating LLMs with visual tools. Our system prioritizes critical slices, segments

© The Author(s), under exclusive license to Springer Nature Switzerland AG 2026
J. Qiu et al. (Eds.): Agentic AI 2025/CMLLMs 2025/CREATE 2025, LNCS 16147, pp. 3–12, 2026.
https://doi.org/10.1007/978-3-032-06004-4_1

4   S. Punneshetty et al.

hemorrhages, and generates interpretable summaries. We adopt Hillal et al.'s [8] method for slice selection and combine LLM agents with YOLOv10 [6] and SAM2 [7] (Fig. 1). Key contributions:

- **Multi-modal Integration**: First multimodal ICH approach using LLM agents with visual encoders for automated diagnostic reporting.
- **Explainable Summaries**: Unlike prior works, we provide diagnostic summaries beyond detection.
- **Critical Slice Prioritization**: Identifies most informative slices for effective decision-making.

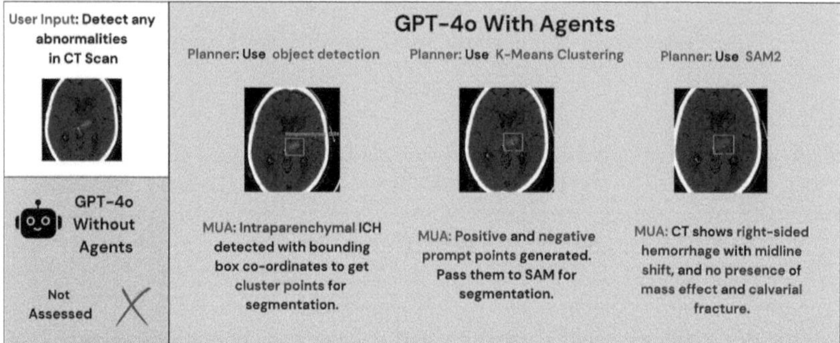

Fig. 1. Vanilla GPT-4o failed to provide diagnostic summaries (Output: 'Not Assessed') when presented with CT scans shown in the left. In contrast, agent-equipped GPT-4o utilized vision tools: object detection, K-means clustering, and SAM2 to generate intermediate reasoning steps, enabling detailed and explainable hemorrhage analysis.

2   Related Work

Traditional ICH diagnosis relies on manual CT examination. Recent developments focus on semi-supervised approaches [9], weak supervision using class activation maps (CAM), and binary CNNs for ICH location detection [10]. Recent innovations include Head-Wise Gradient-Infused Self-Attention Maps (Swin-HGI-SAM) [11] and weakly supervised segmentation combining YOLO and uncertainty-rectified SAM [12], demonstrating improved detection while maintaining explainability.

Foundation models have emerged for medical imaging, including 3D models for head CT disease detection [1], vision-language models for 3D CT [2], and multimodal medical capabilities [3]. However, these require extensive computational resources and massive datasets, limiting clinical accessibility. The Sketchpad framework [13] enables multimodal language models to utilize specialist vision models interpretably, though medical applications remain unexplored. Our approach leverages pre-trained components with targeted fine-tuning for practical clinical deployment while maintaining explainability.

3   Proposed Methodology

Our multi-agent framework integrates YOLOv10 and SAM2 vision tools with LLM agents (MUA and Planner) for hemorrhage analysis, using clustering for SAM2 prompting and selecting key slices based on clinical significance (Fig. 2).

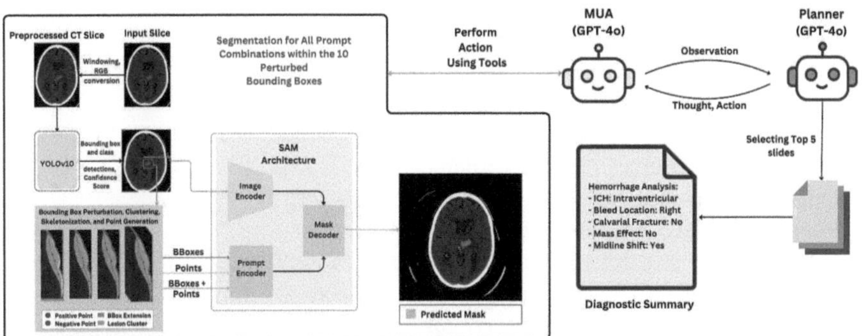

Fig. 2. CT slices are preprocessed and lesions detected via YOLOv10, followed by bounding box perturbation and point generation. SAM performs segmentation based on these inputs. MUA and Planner agents interact through observation–thought–action cycles. The top 5 slices are selected using weighted criteria for final diagnostic summarization.

3.1   Bounding Box Detection, Perturbations and Segmentation

We utilize a YOLOv10-SAM2 framework for ICH segmentation, integrating a fine-tuned YOLOv10 model [6] for lesion detection with SAM2 [7] for segmentation. First, YOLOv10 detects lesions and generates bounding boxes. When coordinate errors occur during the detection phase, our multi-agent system performs automatic validation and correction—the MUA agent detects invalid bounding box dimensions (e.g., negative height/width) and applies corrective transformations before proceeding. The validated boxes are then perturbed with random directional shifts to create multiple bounding variations. These perturbed bounding box prompts are passed to SAM2 for segmentation. To further refine the prompts, K-means clustering is applied to CT slices, producing four clusters: residual skull, ICH, healthy brain tissue, and background. The brightest cluster, assumed to represent the bleed, provides the positive prompt, while the remaining clusters serve as negative prompts. SAM2 uses these refined prompts along with the agent-corrected bounding boxes to create segmentation masks for each perturbed input, ensuring robust segmentation even when initial detections contain errors.

Input : CT scan slice - *"Detect any abnormalities"*

Output: Explainable diagnostic report with reasoning trace

With Agents - Multi-step Reasoning:

Step 1: Object Detection Planning

   Planner: *"Use object detection for hemorrhage identification"*;

   MUA Action: Execute YOLO detection on CT slice;

   Thought: "Analyzing CT for hemorrhage patterns...";

   Result: Intraparenchymal ICH detected, **confidence 93.2%**;

   Output: Bounding box coordinates for hemorrhage region;

Step 2: K-Means Clustering

   Planner: *"Use K-Means clustering for density analysis"*;

   MUA Action: Generate positive/negative prompt points;

   Thought: "Brightest cluster likely represents hemorrhage core";

   Result: 4 clusters identified, hemorrhage region isolated;

   Output: Prompt points for SAM segmentation;

Step 3: SAM2 Segmentation

   Planner: *"Use SAM2 for precise hemorrhage boundary"*;

   MUA Action: Execute SAM2 with generated prompts;

   Thought: "Segmenting hemorrhage for volume calculation";

   Result: Binary mask generated, **area = 0.9%** of image;

Step 4: Clinical Analysis

   MUA Reasoning: Extract clinical parameters from results;

   Location: x-coordinate 0.506 > 0.5 → **"Right hemisphere"**;

   Mass Effect: size 0.9% < 2% threshold → "No";

   Midline Shift: L/R volume ratio 1.88 > 1.1 → **"Yes"**;

   Fracture: No discontinuity in skull boundary → "No";

Final Report:

   Findings: Right-sided intraparenchymal hemorrhage;

   Critical Features: *Midline shift present, no mass effect*;

  Algorithm 1: GPT-4o with Agents: Chain-of-Thought CT Analysis

3.2   Agents

The framework employs two specialized agents *Multi-modal User Agent (MUA)* and a *Planner Agent*, powered by GPT-4o, each configured with distinct roles and responsibilities within the architecture. For this work, we relied on two frameworks: Visprog [14] and ViperGPT [15]. These frameworks utilize large language models (LLMs) to generate Python code, which sequentially invokes specialized vision tools to solve subproblems effectively.

Building on this concept, we wrap our modules into Python functions that GPT-4o agents can call during runtime. These agents collaborate dynamically, leveraging the decompositional capabilities of LLMs to handle intricate clinical imaging tasks.

3.3 Multi-modal User Agent and Planner Collaboration

The Multi-modal User Agent (MUA) and Planner work together in a step-by-step, interactive process to analyze CT scan slices as shown in Algorithm 1. This collaboration involves several key stages.

Initial Detection: The MUA begins by analyzing a single slice using a fine-tuned YOLOv10 model. This model detects potential hemorrhagic areas and provides bounding boxes around regions of interest for the Planner to evaluate.

Refinement Process: If the Planner's confidence level is insufficient, it requests the MUA to perform a more detailed analysis of the identified areas. The MUA then employs K-means clustering within the bounding boxes to more precisely define hemorrhage boundaries by grouping similar points.

Detailed Segmentation: Using the clustered data, SAM2 performs advanced segmentation to accurately delineate the fine-grained boundaries of the lesion. This segmented output is then passed back to the Planner for assessment.

Iterative Assessment: The Planner continuously directs actions based on the MUA's observations. This iterative collaboration enables real-time adjustments and deeper analysis of hemorrhage characteristics.

Clinical Insights: The combined analysis from both components enables a comprehensive clinical assessment.

3.4 Selection of Key Slices

To reduce GPT-4o inference costs, we select five critical slices using weighted factors: *confidence, size, centrality,* and *cluster* scores [8]. This reduces cost to $0.70 per patient while ensuring clinical relevance.

Confidence reflects the YOLOV10 model's certainty regarding each detected hemorrhage, with values ranging from 0 to 1. Higher confidence scores indicate greater certainty that the detected region is indeed hemorrhagic.

Size measures the relative area of the hemorrhage in proportion to the image size, favoring larger hemorrhages (up to a threshold) that could signal more substantial findings.

Centrality assesses how close the hemorrhage is to the image center, as centrally located abnormalities are often more clinically significant.

Cluster consistency considers the number of nearby hemorrhages across adjacent slices, rewarding clusters of detections to highlight patterns that may suggest progressing or widespread pathology.

These four scores are aggregated for each detection using the following formula:

$$\text{Total Score} = (w_{\text{confidence}} \times \text{Confidence}) + (w_{\text{size}} \times \text{Size})$$
$$+ (w_{\text{centrality}} \times \text{Centrality}) + (w_{\text{cluster}} \times \text{Cluster consistency}) \tag{1}$$

where: $w_{\text{confidence}} = 0.3$, $w_{\text{size}} = 0.3$, $w_{\text{centrality}} = 0.2$, $w_{\text{cluster}} = 0.2$.

3.5   Diagnostic Summary

The top five slices with the highest total scores are selected for detailed analysis. After each slice is annotated using the proposed methodology for Bleed Location (left or right), Calvarial Fracture (yes or no), Mass Effect (yes or no), and Midline Shift (yes or no), these slice-level observations are aggregated to form a single, cohesive diagnostic summary. The generated summary is compared against ground truth data, ensuring consistency and accuracy in the final diagnosis.

4   Experiments

4.1   Datasets

We utilize two publicly available datasets to train, validate, and test our framework. The BHX dataset [16] contains 15,979 expert-annotated bounding boxes covering six ICH categories (Intraventricular, Intraparenchymal, Subarachnoid, Chronic, Subdural, and Epidural). We adopt an 80-10-10 split for training (12,783), validation (1,599), and testing (1,597) annotations. The Seg-CQ500 dataset [17] serves as an independent test set with 51 scans providing comprehensive labels for ICH subtypes and clinical parameters including midline shift, bleed location, fracture, and mass effect, ensuring refined slice-level annotations across varying injury severities.

4.2   Windowing and Preprocessing

CT slices are processed using three windows (brain, subdural, bone) stacked as RGB images and normalized to [0,1], ensuring consistency for YOLOv10 training.

4.3   Baseline ICH Detector Training

The YOLOv10 model is trained on resized 640×640 input images for 100 epochs with a batch size of 16 using an NVIDIA A100 GPU. Additional training details including hyperparameters, hardware specifications, and optimization settings are publicly available at Github[1]. To enhance detection performance, we apply mosaic augmentation (factor 1.0), which merges four brain images into one composite sample, simulating varied hemorrhage patterns and improving the model's ability to detect diverse hemorrhage types.

4.4   Agent-Based System Integration

The fine-tuned YOLOv10 model is integrated into the multi-agent system built on the AutoGen framework for automated ICH analysis. Each vision expert (YOLOv10 for ICH detection, SAM2 for segmentation, and the clustering module for point generation) is implemented as a server using Gradio, allowing them to be accessed by the agent system. The agents use these vision capabilities in a coordinated pipeline: first detecting potential ICH regions, then using clustering to determine optimal points for guiding the segmentation process, and finally generating precise masks using SAM2.

5   Results

5.1   ICH Detection Performance

Table 1 presents a comparative analysis of our model against existing state-of-the-art approaches for ICH detection. Our model demonstrates competitive performance across all evaluation metrics, particularly in maintaining balanced precision and recall scores. Additionally, our model shows superior performance in mAP@.5:.95, suggesting better robustness across various intersections over union (IoU) thresholds.

Table 1. Performance comparison of hemorrhage detection models.

Model	Precision	Recall	F1-Score	mAP@.5:.95
2D Faster RCNN [18]	0.897	–	0.908	–
Faster R-CNN [19]	0.857	0.844	0.85	–
R-FCN [19]	0.905	0.826	0.864	–
YOLOv5s-CAM [20]	0.935	0.908	0.921	0.650
GA (TL-LFF Net) [21]	0.935	0.945	0.940	0.728
Ours	**0.938**	**0.946**	**0.941**	**0.758**

5.2   Diagnostic Summary Generation

Unlike traditional medical report generation baselines that produce free-text narratives, our framework generates structured clinical parameters essential for ICH triage. Since the SegCQ500 ground truth consists of discrete labels rather than full reports, we convert our generated diagnostic summaries into binary labels for each parameter to enable direct comparison. This structured approach prioritizes actionable clinical decisions over narrative generation, making metrics like BLEU scores (used for text generation) inappropriate for our task.

Table 2 presents our model's performance across four critical aspects of clinical parameters. The results demonstrate particular strength in mass effect identification and midline shift detection, which are key diagnostic elements in ICH assessment. The high precision in mass effect detection indicates reliability in identifying this important clinical finding.

Table 2. Per-class and overall performance comparison of generated clinical parameters from the proposed pipeline on SegCQ500.

Class	Accuracy	Precision	Recall
Bleed Location	0.765	0.720	0.683
Calvarial Fracture	0.798	0.675	0.742
Mass Effect	0.812	0.892	0.754
Midline Shift	0.749	0.831	0.796
Average	**0.781**	**0.780**	**0.744**

5.3 Discussion

Our model demonstrates robust performance in both bounding box detection and clinical report generation for ICH analysis. When YOLOv10 predictions are highly confident, the model directly generates clinical reports. When confidence is lower, the model employs K-means clustering and SAM2 to refine predictions, ensuring precise and clinically meaningful reports. When uncertain, it generates "not assessed" entries, adding reliability and transparency. Comparing clinical parameters predicted by LLM agents against Seg-CQ500 ground truth, our model achieves 78.1% accuracy, 78.0% precision, and 74.4% recall overall (Table 2), establishing it as a reliable decision-support tool for automated ICH analysis from CT scans.

6   Conclusion

This study introduces a dual-agent framework for the detection of intracranial hemorrhage (ICH) and automated generation of diagnostic summaries, guided by chain-of-thought prompting to promote interpretability and clinical transparency. By integrating YOLOv10, SAM2, and reasoning-based prompts, the proposed system enhances diagnostic performance while enabling traceable clinical insights. The cooperative agent architecture facilitates efficient processing of CT scans and structured output generation. Future work will focus on validating the framework across larger and more diverse clinical datasets. Furthermore, the adoption of a cost-effective slice selection strategy enables inference at an estimated cost of \$0.70 per patient, demonstrating the framework's feasibility for deployment in both resource-rich and resource-constrained healthcare environments.

Acknowledgments. We acknowledge financial support received from the Department of Computer Science and Information Systems (CSIS), BITS Pilani. All contributors are listed as co-authors.

Disclosure of Interests. The authors have no competing interests to declare that are relevant to the content of this article.

References

1. Zhu, W., et al.: 3D foundation AI model for generalizable disease detection in head computed tomography. arXiv preprint arXiv:2502.02779 (2025)
2. Blankemeier, L., et al.: Merlin: a vision language foundation model for 3D computed tomography. arXiv preprint arXiv:2406.06512 (2024)
3. Yang, L., et al.: Advancing multimodal medical capabilities of Gemini. arXiv preprint arXiv:2405.03162 (2024)
4. Prakasam, P., et al.: Spatial attention-based CSR-Unet framework for subdural and epidural hemorrhage segmentation and classification using CT images. BMC Med. Imaging **24**, 285 (2024)
5. Salehinejad, H., et al.: A real-world demonstration of machine learning generalizability in the detection of intracranial hemorrhage on head computerized tomography. Sci. Rep. **11**(1), 17051 (2021)
6. Wang, A., et al.: YOLOv10: real-time end-to-end object detection. arXiv preprint arXiv:2405.14458 (2024)
7. Ravi, N., et al.: SAM 2: segment anything in images and videos. arXiv preprint arXiv:2408.00714 (2024)
8. Hillal, A., et al.: Computed tomography in acute intracerebral hemorrhage: neuroimaging predictors of hematoma expansion and outcome. Insights Imaging **13**(1), 180 (2022)
9. Wang, J.L., et al.: Segmentation of intracranial hemorrhage using semi-supervised multi-task attention-based U-Net. Appl. Sci. **10**(9), 3297 (2020)
10. Nemcek, J., et al.: Localization and classification of intracranial hemorrhages in CT data. In: 8th European Medical and Biological Engineering Conference, pp. 767–773. Springer (2021)
11. Rasoulian, A., et al.: Weakly supervised intracranial hemorrhage segmentation using head-wise gradient-infused self-attention maps from a swin transformer in categorical learning. arXiv preprint arXiv:2304.04902 (2023)
12. Spiegler, P., et al.: Weakly supervised intracranial hemorrhage segmentation with YOLO and an uncertainty rectified segment anything model. arXiv preprint arXiv:2407.20461 (2024)
13. Hu, Y., et al.: Visual Sketchpad: sketching as a visual chain of thought for multimodal language models. arXiv preprint arXiv:2406.09403 (2024)
14. Gupta, T., Kembhavi, A.: Visual programming: compositional visual reasoning without training. In: Proceedings of the IEEE/CVF Conference on Computer Vision and Pattern Recognition, pp. 14953–14962 (2022)
15. Surís, D., et al.: ViperGPT: visual inference via python execution for reasoning. In: Proceedings of the IEEE/CVF International Conference on Computer Vision, pp. 12123–12133 (2023)
16. Reis, L.M., et al.: Brain hemorrhage extended (BHX) dataset. arXiv preprint arXiv:2012.02783 (2020)
17. Spahr, N., et al.: Label-efficient segmentation of acute intracranial hemorrhages on non-contrast CT. Med. Image Anal. **88**, 102865 (2023)
18. Myung, M.J., et al.: Novel approaches to detection of cerebral microbleeds: single deep learning model to achieve a balanced performance. J. Stroke Cerebrovasc. Dis. **30**(9), 105886 (2021)
19. Le, T.N., et al.: Automatic identification of intracranial hemorrhage on CT/MRI image using meta-architectures improved from region-based CNN. In: Optimization of Complex Systems: Theory, Models, Algorithms and Applications, pp. 740–750. Springer (2020)

20. Vidhya, V., et al.: YOLOv5s-CAM: a deep learning model for automated detection and classification for types of intracranial hematoma in CT images. IEEE Access (2023)
21. Kothala, S., et al.: TL-LFF Net: transfer learning based lighter, faster, and frozen network for the detection of multi-scale mixed intracranial hemorrhages through genetic optimization algorithm. Int. J. Mach. Learn. Cybern., 1–27 (2024)

Agentic Surgical AI: Surgeon Style Fingerprinting and Privacy Risk Quantification via Discrete Diffusion in a Vision-Language-Action Framework

Huixin Zhan[1,2] and Jason H. Moore[1(✉)]

[1] Department of Computational Biomedicine, Cedars-Sinai Medical Center, Los Angeles, CA 90048, USA
huixin.zhan@cshs.org, jason.moore@csmc.edu
[2] Department of Computer Science and Engineering, New Mexico Institute of Mining and Technology, Socorro, NM 87801, USA

Abstract. Surgeons exhibit distinct operating styles shaped by training, experience, and motor behavior—yet most surgical AI systems overlook this personalization signal. We propose a novel agentic modeling approach for surgeon-specific behavior prediction in robotic surgery, combining a discrete diffusion framework with a vision-language-action (VLA) pipeline. Gesture prediction is framed as a structured sequence denoising task, conditioned on multimodal inputs including surgical video, intent language, and personalized embeddings of surgeon identity and skill. These embeddings are encoded through natural language prompts using third-party language models, allowing the model to retain individual behavioral style without exposing explicit identity. We evaluate our method on the JIGSAWS dataset and demonstrate that it accurately reconstructs gesture sequences while learning meaningful motion fingerprints unique to each surgeon. To quantify the privacy implications of personalization, we perform membership inference attacks and find that more expressive embeddings improve task performance but simultaneously increase susceptibility to identity leakage. These findings demonstrate that while personalized embeddings improve performance, they also increase vulnerability to identity leakage, revealing the importance of balancing personalization with privacy risk in surgical modeling. Code is available at: https://github.com/huixin-zhanai/Surgeon_style_fingerprinting.

Keywords: Agentic AI · Diffusion model · Vision-language-action model

1 Introduction

Personalized modeling of surgical behavior has the potential to improve intraoperative decision support, skill assessment, and robot-assisted training. In particular, fine-grained prediction of surgical gestures—low-level action primitives

J. Qiu et al. (Eds.): Agentic AI 2025/CMLLMs 2025/CREATE 2025, LNCS 16147, pp. 13–22, 2026.
https://doi.org/10.1007/978-3-032-06004-4_2

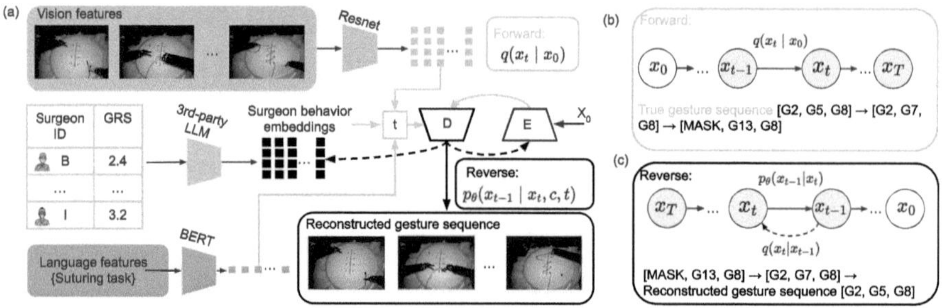

Fig. 1. Multimodal diffusion for personalized gesture prediction. (a) ResNet, BERT, and a third-party LLM extract vision, language, and surgeon embeddings (ID + GRS). (b) Discrete gesture sequences are noised over time via forward diffusion. (c) The model denoises gestures conditioned on multimodal context.

such as "grasp needle" or "position needle"—can reveal patterns in technique that vary across surgeons, tasks, and experience levels. Recent work [18] has leveraged vision and language models for surgical understanding, but most existing systems remain surgeon-agnostic, failing to capture individual variation critical for personalization and performance benchmarking.

Modeling this variation, however, raises unique challenges: How can models account for stylistic behavioral differences between surgeons? Can we learn surgeon-specific representations without exposing identity information? And how can we build robust predictive systems that generalize across users while still capturing meaningful intra-surgeon variability?

To address these challenges, we introduce a discrete diffusion-based framework for personalized gesture sequence modeling in robotic surgery, grounded in a vision-language-action (VLA) paradigm. We formulate gesture prediction as a structured denoising process over discrete tokens, where the reverse diffusion model is *agentically* conditioned on multimodal inputs: (i) visual context from surgical video, (ii) task-level language cues representing surgical intent, and (iii) surgeon-specific behavioral fingerprints, encoded via natural language prompts. These prompts integrate both the surgeon's ID and their averaged Global Rating Score (GRS [10]), and are embedded using frozen third-party large language models (LLMs). The resulting embeddings are trained end-to-end with the VLA-conditioned diffusion model, enabling adaptive and personalized gesture generation. Although identity and skill cues are provided during prompt construction, they are abstracted through frozen third-party LLMs and not directly exposed to the gesture prediction module, thereby offering a degree of privacy while supporting rich personalization (Fig. 1).

We evaluate our method on the JIGSAWS dataset [9] and show that surgeon-specific embeddings improve structured sequence denoising performance. To assess potential privacy risks introduced by personalization, we conduct membership inference attacks targeting the learned embeddings. Our results reveal that while LLM+GRS embeddings enhance personalization, they also increase susceptibility to re-identification, highlighting a trade-off between behavioral fidelity and privacy. This work underscores the need for careful evaluation of privacy leakage in personalized surgical AI systems.

Our key contributions are: (1) a diffusion-based formulation for personalized discrete gesture prediction in robotic surgery, (2) a language-model-based embedding scheme that encodes surgeon identity and skill cues via natural language prompts via third-party large language models, (3) a quantitative privacy analysis using membership inference attacks to evaluate identity leakage, and (4) comprehensive experiments regarding both gesture prediction accuracy and stylistic coherence on the JIGSAWS dataset demonstrating both improved personalization and the trade-offs between performance and privacy. Together, these components define an agentic modeling framework that captures context- and identity-aware behavior representation, enabling personalized gesture sequence generation and advancing the development of adaptive, privacy-conscious AI systems for surgical robotics.

2 Method

We formulate surgeon gesture prediction as a discrete denoising diffusion process. Each gesture sequence $x_0 \in \mathcal{G}^T$, where $\mathcal{G} = \{1, \ldots, K\}$ is the gesture vocabulary, is corrupted over time via a multinomial noise process. The goal is to learn a conditional model that reconstructs x_0 from a noisy sequence $x_t \sim q(x_t \mid x_0)$, given multimodal context.

2.1 Forward Process: Discrete Multinomial Corruption

We define a time-indexed transition matrix $Q_t \in \mathbb{R}^{K \times K}$ that governs the noise schedule. Each diagonal entry $Q_t(i, i)$ decays linearly as:

$$Q_t(i, i) = 1 - \frac{t+1}{T}, \quad Q_t(i, j \neq i) = \frac{1 - Q_t(i, i)}{K - 1}$$

This process flips gesture tokens at increasing rates as t progresses, simulating categorical corruption.

2.2 Reverse Process: Personalized Gesture Denoising

We train a transformer-based model [19] to approximate the reverse conditional distribution $p_\theta(x_{t-1} \mid x_t, c, t)$, where the conditioning context c includes:

- Vision features extracted from endoscopic video using a ResNet encoder [11],
- Language features derived from a BERT encoding [3] of the task prompt (e.g., "suturing task"),
- Surgeon embedding $\mathbf{s}_i \in \mathbb{R}^d$ that encodes stylistic behavioral information,
- Timestep embedding t. These components are fused and injected into the denoising model at each timestep.

2.3   Surgeon Embeddings and Personalization

To model inter-surgeon variation, each surgeon is assigned an embedding vector s_i. These embeddings are either learned directly through a trainable embedding layer or derived from natural language prompts—such as "Surgeon ID: i, GRS: 3.4"—using a frozen third-party language model (e.g., Sentence-BERT [15] or MiniLM [20]). By conditioning on these embeddings, the model learns to denoise gesture sequences in a surgeon-specific manner, enabling stylistic profiling and personalized gesture generation. After training, the resulting embeddings can be used for downstream applications such as clustering, retrieval, or surgeon re-identification.

2.4   Training Objective

We sample a timestep $t \sim \mathcal{U}(0, T)$ uniformly and compute a noisy gesture sequence $x_t \sim q(x_t \mid x_0)$. The model is trained to predict the original gesture x_0 using a cross-entropy loss:

$$\mathcal{L}_{\text{CE}} = \mathbb{E}_{x_0, t, x_t} \left[-\log p_\theta(x_0 \mid x_t, c, t) \right]$$

The loss is computed over all gesture positions, and gradients are backpropagated through the entire diffusion trajectory.

2.5   Inference

At test time, we formulate gesture prediction as a generative inference task over discrete sequences. Starting from a fully corrupted gesture sequence $x_T \sim q(x_T \mid x_0)$, which represents near-uniform categorical noise, the model applies the learned reverse process iteratively from $t = T$ to 0. This produces a denoised sequence $\hat{x}_0$ that reconstructs the likely gesture trajectory under the current visual, linguistic, and behavioral context.

This formulation departs from traditional gesture recognition methods, which typically rely on direct sequence classification or frame-wise decoding. Instead, we cast inference as a structured sequence generation problem conditioned on multimodal and surgeon-specific embeddings. The process is inherently probabilistic and allows for uncertainty modeling and sample diversity, enabling the system to generate plausible personalized behaviors.

We evaluate inference performance along two key axes: (i) *gesture prediction accuracy*, which quantifies the model's ability to reconstruct the ground truth gesture sequence, and (ii) *stylistic coherence*, which assesses how closely the generated trajectory reflects the unique behavioral fingerprint of the target surgeon, as captured by their personalized embedding.

2.6   Privacy Risk Analysis via Membership Inference

To quantify the privacy risk associated with personalized embeddings, we conduct a membership inference attack [17]. Given the learned surgeon embeddings

$\{\mathbf{s}_i\}$, we simulate an adversary that trains a binary classifier (e.g., XGBoost [2]) to distinguish between in-training (member) embeddings and synthetic out-of-distribution (non-member) ones. High classifier performance—measured by AUC, accuracy, and F1-score—indicates a greater risk of identity leakage.

We compare privacy vulnerability across three embedding strategies. As reported in Sect. 3, the LLM (ID + GRS) embedding achieves the best gesture prediction performance, but also exhibits the highest susceptibility to membership inference attacks (AUC = 1.000). This reveals a key privacy-performance trade-off: richer personalization yields more discriminative embeddings, which are easier to link back to training data. In contrast, the non-private baseline produces less structured embeddings that offer weaker personalization but lower risk under attack.

These findings emphasize the need for explicit privacy evaluation when deploying personalized models in clinical settings, especially when embedding behavioral signals like skill.

3   Experiments

To evaluate the effectiveness of our proposed framework, we design a series of experiments aimed at answering the following key scientific questions:

- **Q1:** How do different surgeon representation strategies (e.g., learnable ID embeddings, third-party LLMs (ID only), and third-party LLMs (ID + GRS)) impact personalized gesture sequence modeling?
- **Q2:** Do the learned surgeon embeddings capture meaningful and structured variation in skill or behavior across different individuals?
- **Q3:** To what extent do different personalization methods expose the model to privacy leakage under adversarial conditions such as membership inference attacks?

We train for 20 epochs using Adam (lr=1e−3, batch size=32). All components use 512-d hidden states. ResNet (1000-d) and BERT (768-d) features are projected into a shared space. Inputs are 5-token gesture sequences from a 16-token vocabulary (G1–G15 + [MASK]). Corruption spans 10 timesteps with categorical noise. Ablations examine embedding variants, GRS fingerprinting, and privacy via membership inference. The results and insights are presented in the following sections.

3.1   Ablation on Surgeon Embedding Strategies

To assess the impact of surgeon-specific conditioning on gesture prediction, we compare three distinct embedding strategies:

1. **Non-private baseline:** We use a learnable embedding layer that maps each surgeon ID to a unique vector. This approach directly exposes identity and serves as an upper-bound reference in terms of capacity and specificity.

Table 1. Gesture prediction under different surgeon representations. LLM (ID + GRS) encodes identity and skill via natural language prompts; **LLM (ID only)** uses ID prompts alone; **Baseline** relies on learned embeddings.

Task	Surgeon Representation	Top-1 Accuracy	Top-5 Accuracy	Weighted F1-Score
Suturing	Third-party LLM (ID + GRS)	**0.8389**↑	0.9984↓	**0.8447**↑
	Third-party LLM (ID only)	0.8327	**0.9988**	0.8324
	Non-private baseline	0.8240	0.9968	0.8237

2. **Third-party LLM (ID only):** Surgeon IDs are formatted into short natural language prompts (e.g., "Surgeon ID: 3") and encoded using a publicly available sentence-level language model (e.g., MiniLM). This embedding is kept fixed and projected to the model's hidden dimension via a learnable linear layer.
3. **Third-party LLM (ID + GRS):** To incorporate behavioral priors, we construct natural language prompts that include both the surgeon ID and their averaged GRS, e.g., "Surgeon ID: 3, average skill score: 3.75". These prompts are encoded using a frozen third-party language model (e.g., Sentence-BERT), and projected into the model's hidden space. This strategy captures personalized style in a clinically grounded manner, while avoiding direct integration of raw identifiers into the gesture prediction model.

We evaluate all strategies within the same diffusion-based gesture prediction framework, under identical training and testing conditions. As shown in Table 1, the third-party LLM (ID + GRS) method achieves the highest Top-1 accuracy (83.89%) and weighted F1-score (0.8447), with near-perfect Top-5 accuracy (99.84%), outperforming the ID-only and non-private baselines. These results reflect the benefit of incorporating clinically grounded behavioral cues like GRS into surgeon representations.

However, our privacy analysis reveals that this gain in personalization comes at the cost of increased identity leakage under membership inference attacks. This finding reveals a core tension in surgical modeling: richer behavioral embeddings improve prediction, but also make models more susceptible to privacy breaches. Therefore, while third-party LLM embeddings with GRS offer strong personalization capabilities, they also demand careful consideration of privacy risk.

3.2   Surgeon Fingerprinting and GRS Association

To explore whether our model captures meaningful variations in surgical behavior, we visualize surgeon embeddings using t-SNE. Each point in Fig. 2 represents an individual surgeon, and colors indicate their average GRS performance score.

The left panel shows embeddings learned when the model is only provided surgeon IDs. In this setting, the embedding space primarily reflects identity but

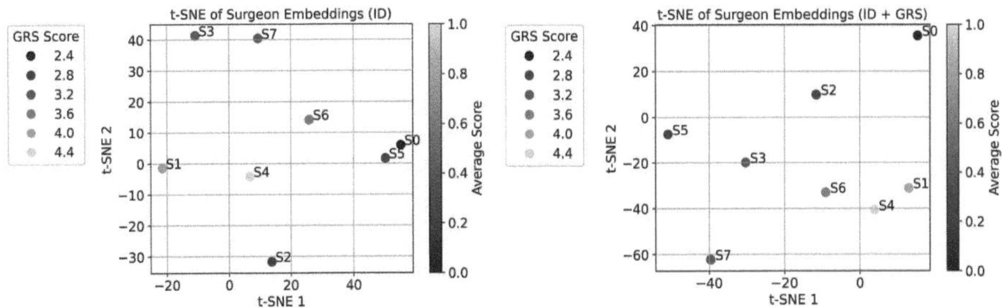

Fig. 2. t-SNE of surgeon embeddings. (**Left**) Trained using surgeon ID only. (**Right**) Trained with both ID and GRS supervision. Colors indicate average skill score (GRS).

does not exhibit clear separation by skill level. In contrast, the right panel shows embeddings learned with both surgeon ID and GRS supervision. Here, surgeons with similar skill levels are more coherently grouped, and the embeddings demonstrate smoother gradients along the GRS axis.

This suggests that our personalized representation—guided by visual, linguistic, and behavioral cues—can encode latent skill information in an unsupervised or weakly-supervised fashion. Embedding models that integrate clinical supervision like GRS may enhance both interpretability and downstream personalization.

Table 2. Privacy risk via membership inference. We compare embedding strategies by training a classifier to distinguish training vs. non-training samples. Higher AUC and precision signal greater identity leakage. **LLM (ID + GRS)** offers stronger personalization but increases re-identification risk.

Task	Surgeon Representation	Accuracy	Precision	Recall	F1 Score	AUC
Suturing	Third-party LLM (ID + GRS)	**0.997**↑	**0.998**↑	**1.000**↑	**0.988**↑	**1.000**↑
	Third-party LLM (ID only)	0.889	0.500	1.000	0.667	1.000
	Non-private baseline	0.889	0.000	0.000	0.000	0.469

3.3 Privacy Evaluation via Membership Inference

We assess privacy leakage from different surgeon embedding strategies using a membership inference attack, in which an adversary attempts to determine whether a given embedding originated from the model's training data. Results are shown in Table 2.

The **third-party LLM (ID + GRS)** representation is the most expressive, but also the most vulnerable. It enables highly personalized gesture modeling, but the resulting embeddings are easily distinguishable from synthetic (non-member) embeddings, with an AUC of 1.000, precision of 0.998, and F1-score of 0.988. This indicates that the embedding space reveals strong identity-linked signals, raising concerns about privacy leakage.

The **third-party LLM (ID only)** strategy also exhibits high recall and AUC, but its lower precision (0.500) and F1-score (0.667) suggest a less precise separation between members and non-members. This reflects a weaker entanglement of identity information compared to the GRS-conditioned variant.

The **non-private baseline** shows low attack performance, with zero precision and recall and low AUC (0.469). While this may seem desirable from a privacy standpoint, it more likely reflects noisy or unstructured embeddings rather than true robustness.

Overall, these results highlight a trade-off: more expressive, personalized embeddings improve task performance but increase susceptibility to identity inference attacks. This underscores the importance of quantifying privacy risks when designing surgeon-specific models in surgical AI.

4   Discussion

Recent advances in surgical AI have explored multimodal learning through vision-language-action (VLA) frameworks [4,21], where visual and linguistic signals are fused to model procedural context. Despite promising progress in gesture segmentation and phase prediction [5,13,14], most existing systems adopt a one-size-fits-all approach that neglects surgeon-specific behavioral differences. This limits their ability to support personalized training or adaptive guidance—use cases where modeling individual style is critical. Building on this gap, we introduce a generative approach that embeds per-surgeon behavioral fingerprints and predicts gesture sequences using a discrete diffusion model. Our method complements prior studies on skill assessment and motion analysis [8,22] and extends gesture forecasting frameworks [12,16] with personalized generation capabilities. By conditioning on structured style embeddings derived from natural language prompts, our model learns to capture surgical behavior at both the gesture and phase level. While prior privacy studies have primarily focused on image reconstruction and membership inference in classification models [6,7,17], we explore privacy risks in behavioral sequence embeddings. Our analysis reveals a trade-off between expressiveness and identity leakage, consistent with recent findings on embedding-based leakage [1].

Collectively, our findings underscore the need for models that are not only multimodal but also personalized, privacy-aware, and behaviorally grounded. Our framework advances the direction of agentic AI by enabling systems to adapt to user identity, intent, and context. By capturing surgeon-specific behavioral fingerprints and reasoning over structured gesture sequences, we take a step toward building surgical agents capable of individualized learning, proactive assistance, and trust-aware interaction in clinical environments.

References

1. Carlini, N., et al.: Extracting training data from large language models. In: USENIX Security Symposium (2021)
2. Chen, T., Guestrin, C.: XGBoost: a scalable tree boosting system. In: Proceedings of the 22nd ACM SIGKDD International Conference on Knowledge Discovery and Data Mining, pp. 785–794, 2016
3. Devlin, J., Chang, M.-W., Lee, K., Toutanova, K.: BERT: pre-training of deep bidirectional transformers for language understanding. In: Proceedings of the 2019 Conference of the North American Chapter of the Association for Computational Linguistics: Human Language Technologies, vol. 1 (long and short papers), pp. 4171–4186 (2019)
4. Ding, D., Yao, T., Luo, R., Sun, X.: Visual question answering in robotic surgery: a comprehensive review. IEEE Access 2025
5. DiPietro, R., Lea, C., Malpani, A., Hager, G.D.: Segmental spatiotemporal CNNs for fine-grained action segmentation. In: European Conference on Computer Vision (ECCV), pp. 36–52. Springer (2019)
6. Dwork, C., Roth, A.: The algorithmic foundations of differential privacy. Found. Trends Theor. Comput. Sci. $9(3\text{–}4)$, 211–407 (2014)
7. Fredrikson, M., Jha, S., Ristenpart, T.: Model inversion attacks that exploit confidence information and basic countermeasures. In: Proceedings of the 22nd ACM SIGSAC Conference on Computer and Communications Security (CCS), pp. 1322–1333 (2015)
8. Funke, I., Mees, S.T., Weitz, J., Speidel, S.: Video-based surgical skill assessment using 3D convolutional neural networks. In: International Conference on Medical Image Computing and Computer-Assisted Intervention (MICCAI), pp. 101–108. Springer (2019)
9. Gao, Y., et al.: JHU-ISI gesture and skill assessment working set (JIGSAWS): a surgical activity dataset for human motion modeling. In: MICCAI Workshop: M2cai, vol. 3, p. 3 (2014)
10. Gray, J.D.: Global rating scales in residency education. Acad. Med. $71(1)$, S55-63 (1996)
11. He, K., Zhang, X., Ren, S., Sun, J.: Deep residual learning for image recognition. In: Proceedings of the IEEE Conference on Computer Vision and Pattern Recognition, pp. 770–778 (2016)
12. Honarmand, M., Jamal, M.A., Mohareri, O.: VidLPRO: a video-language pre-training framework for robotic and laparoscopic surgery. Explainability, Robustness, Security, and Beyond, In Advancements In Medical Foundation Models (2024)
13. Kiyasseh, D., et al.: A vision transformer for decoding surgeon activity from surgical videos. Nat. Biomed. Eng. $7(6)$, 780–796 (2023)
14. Ma, L., H. Kang, N. Magnenat-Thalmann, and K. Wac. TransSG: a spatial-temporal transformer for surgical gesture recognition. In: Computer Graphics International Conference, pp. 151–165. Springer (2024)
15. Reimers, N., Gurevych, I.: Sentence-BERT: sentence embeddings using Siamese BERT-networks. arXiv preprint arXiv:1908.10084 (2019)
16. Shi, C., Zheng, Y., Fey, A.M.: Recognition and prediction of surgical gestures and trajectories using transformer models in robot-assisted surgery. In: 2022 IEEE/RSJ International Conference on Intelligent Robots and Systems (IROS), pp. 8017–8024. IEEE (2022)

17. Shokri, R., Stronati, M., Song, C., Shmatikov, V.: Membership inference attacks against machine learning models. In: 2017 IEEE Symposium on Security and Privacy (SP), pp. 3–18. IEEE (2017)
18. van Amsterdam, B., Clarkson, M.J., Stoyanov, D.: Gesture recognition in robotic surgery: a review. IEEE Trans. Biomed. Eng. **68**(6), (2021)
19. Vaswani, A., et al.: Attention is all you need. In: Advances in Neural Information Processing Systems, vol. 30 (2017)
20. Wang, W., Wei, F., Dong, L., Bao, H., Yang, N., Zhou, M.: MiniLM: deep self-attention distillation for task-agnostic compression of pre-trained transformers. Adv. Neural. Inf. Process. Syst. **33**, 5776–5788 (2020)
21. Zargarzadeh, S., Mirzaei, M., Ou, Y., Tavakoli, M.: From decision to action in surgical autonomy: Multi-modal large language models for robot-assisted blood suction. IEEE Robot. Autom. Lett. (2025)
22. Zia, A., Essa, I.: Automated assessment of surgical skills using frequency analysis. IEEE Trans. Biomed. Eng. **65**(9), 2155–2166 (2018)

ADAgent: LLM Agent for Alzheimer's Disease Analysis with Collaborative Coordinator

Wenlong Hou[1], Guangqian Yang[1], Ye Du[1], Yeung Lau[1], Lihao Liu[2], Junjun He[3], Ling Long[4], and Shujun Wang[1,5,6(✉)]

[1] Department of Biomedical Engineering, The Hong Kong Polytechnic University, Hong Kong SAR, China
`shu-jun.wang@polyu.edu.hk`
[2] Amazon, Seattle, USA
[3] Shanghai Artificial Intelligence Laboratory, Shanghai, China
[4] Department of Geriatrics, Third Affiliated Hospital of Sun Yat-Sen University, Guangzhou, China
[5] Research Institute for Smart Ageing, The Hong Kong Polytechnic University, Hong Kong SAR, China
[6] Research Institute for Artificial Intelligence of Things, The Hong Kong Polytechnic University, Hong Kong SAR, China

Abstract. Alzheimer's disease (AD) is a progressive and irreversible neurodegenerative disease. Early and precise diagnosis of AD is crucial for timely intervention and treatment planning to alleviate the progressive neurodegeneration. However, most existing methods rely on single-modality data, which contrasts with the multifaceted approach used by medical experts. While some deep learning approaches process multi-modal data, they are limited to specific tasks with a small set of input modalities and cannot handle arbitrary combinations. This highlights the need for a system that can address diverse AD-related tasks, process multi-modal or missing input, and integrate multiple advanced methods for improved performance. In this paper, we propose ADAgent, the first specialized AI agent for AD analysis, built on a large language model (LLM) to address user queries and support decision-making. ADAgent integrates a reasoning engine, specialized medical tools, and a collaborative outcome coordinator to facilitate multi-modal diagnosis and prognosis tasks in AD. Extensive experiments demonstrate that ADAgent outperforms SOTA methods, achieving significant improvements in accuracy, including a 2.7% increase in multi-modal diagnosis, a 0.7% improvement in multi-modal prognosis, and enhancements in MRI and PET diagnosis tasks. The code is available at https://github.com/willenhou/ADAgent.

Keywords: Alzheimer's Disease · AI Agent · Multi-modality

1 Introduction

Alzheimer's disease (AD) is a neurodegenerative disorder characterized by the progressive decline in cognitive function, memory, and behavior [11,17]. It affects

J. Qiu et al. (Eds.): Agentic AI 2025/CMLLMs 2025/CREATE 2025, LNCS 16147, pp. 23–32, 2026.
https://doi.org/10.1007/978-3-032-06004-4_3

millions of people worldwide, leading to a significant burden on patients, families, and healthcare systems. The impact of AD is not only detrimental to the patient's quality of life but also has profound social and economic consequences [2,16]. Early diagnosis plays a crucial role in slowing down the disease's progression and providing patients and families with opportunities for better care planning, treatment, and support [1]. Traditional methods for diagnosing AD include cognitive tests, clinical assessments, and neuroimaging techniques such as magnetic resonance imaging (MRI) and positron emission tomography (PET) scans [5]. In the early stage of AD, subtle changes in the brain and mild cognitive decline would occur, which are also common in the normal ageing cases, leading difficulty to directly judge from any traditional methods mentioned above. In addition, the heterogeneity of data from different methods and the subjectivity in doctors' assessments increase the difficulty of extracting multi-modal information to identify early-stage AD patients.

With the recent advancements in deep learning (DL), new DL algorithms have emerged, offering more objective and efficient ways to automatically detect early AD instead of subjective assessments based on humans. Most methods [4,9, 10,13–15,18,24,25] in the AD domain rely on data from a single modality, which is contrary to the multifaceted approach that medical experts use to detect the disease. Despite several DL methods attempting to process multi-modal data [19, 21,22], they remain limited in only handling a specific range of tasks with a small set of input modalities, and cannot tackle multiple tasks with arbitrary input modalities. In addition, even if multiple existing methods are designed for the same task, they cannot be collaboratively integrated to fully utilize their advantages for performance improvement. This highlights the need for a system with these three properties: (1) can address a wide range of AD-related tasks, (2) can process multi-modal input or missing modality input, and (3) can achieve objective assessments by collaborative coordinator of existing multiple methods.

The emergence of AI Agents driven by LLMs has fundamentally changed how we approach autonomous reasoning, planning, and tool-using [6]. This paradigm shift has enabled AI Agents to surpass traditional task-specific models by dynamically adapting to diverse applications without additional training. There are several medical agents for medical reasoning on chest and abdomen X-rays [6,12], showing notable performance on a wide range of tasks. Motivated by these, we aim to develop a solution via building an LLM-driven Agent integrated with multiple tools.

In this paper, we propose the Alzheimer's Disease Analysis Agent (ADAgent), the first specialized AI agent for AD analysis. Specifically, it is a large language model (LLM)-based system that could solve user (*i.e.* doctors) queries (*e.g.*, determining the disease stage) by incorporating reasoning, planning, and tool-using capabilities to facilitate task execution and support decision-making. ADAgent consists of three primary components: (1) a reasoning engine driven by the LLM, which serves as a tool action planner; (2) a set of specialized medical tools, each targeting specific tasks in AD detection; and (3) a collaborative outcome coordinator, which leverages the LLM's reasoning capabilities to make informed decisions based on aggregated results. ADAgent is integrated with four

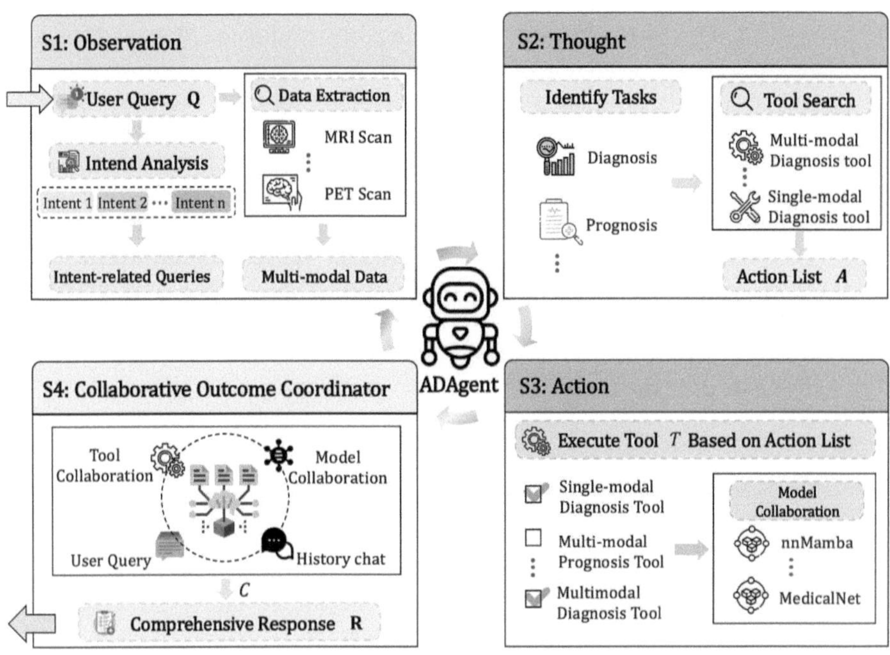

Fig. 1. Overall workflow of ADAgent. Given a use query, the framework performs reasoning in four stages: (i) observation, (ii) thought, (iii) action, and (iv) collaborative outcome coordinator.

tools capable of processing multi-modal input data for both the diagnosis and prognosis of AD. ADAgent can be extended to more tasks by accessing more tools/methods seamlessly without extra training. To access the performance of the proposed ADAgent, we extensively compared the result of ADAgent against multiple baseline methods on multi-modal tasks and missing modality diagnosis tasks. The results show that with the help of the collaborative outcome coordinator of ADAgent, the accuracy performance improves 2.7%, 0.7%, 0.7%, and 4.4% on multi-modal diagnosis task, multi-modal prognosis task, MRI diagnosis task, and PET diagnosis task, respectively.

2 ADAgent

In this section, we first present the overall workflow of ADAgent in Sect. 2.1, followed by a detail description of all four modules in Sect. 2.2. The extensibility analysis and the interaction example of ADAgent are described in Sects. 2.3 and 2.4, respectively.

2.1 Overall Workflow of ADAgent

Our developed ADAgent processes multi-modal use queries, including MRI scan and/or PET scan, along with text-based intents from users, to perform advanced analysis and generate diagnostic and prognostic results. ADAgent employs a large language model (LLM), denoted by $\mathcal{L}$, as the core to drive a reasoning and acting loop, which decomposes complex medical queries into sequential analytical steps. The loop is formalized as $\mathcal{R}$ for reasoning and $\mathcal{A}$ for acting, inspired by MedRAX approach [6], which is a medical reasoning agent for chest X-ray. To address a user query $\mathbf{Q}$, ADAgent follows a four-step process, as shown in Fig. 1: (1) **Observation** – The system observes the current state $s \in \mathcal{S}$ and interprets the user's query $\mathbf{Q}$, where $\mathcal{S}$ represents the state space. (2) **Thought** – The system determines the necessary actions $\mathbf{A}$ to proceed, where $\mathbf{A}$ is the action set. (3) **Action** – The system executes relevant tools $\mathcal{T}_i$ to gather outcomes $\mathbf{O}_i$ from each tool, where $\mathcal{T}_i \in \mathcal{T}$ are the tools in the tool set $\mathcal{T}$. (4) **Collaborative Outcome Coordinator** – The system synthesizes the findings from the previous steps to generate a comprehensive response $\mathbf{R}$ for the user.

2.2 Observation, Thought, Action, and Collaborative Outcome Coordinator

Observation Module - The Observation module divides the user query into more refined sub-queries based on the user's intentions and then gathers the necessary text and image for the next Thought module.

Thought Module with Tool Integration - ADAgent integrates a diverse set of tools $\mathcal{T} = \{\mathcal{T}_1, \mathcal{T}_2, \ldots, \mathcal{T}_n\}$ designed to address various AD-related analysis tasks. For each tool $\mathcal{T}_i$, we attempt to find multiple public models, denoted by $\mathcal{M}_i = \{\mathcal{M}_{i1}, \mathcal{M}_{i2}, \ldots, \mathcal{M}_{ik}\}$, for collaboration to enhance task execution. ADAgent primarily focuses on two core tasks: diagnosis and prognosis.

(i) **Diagnosis task** aims to assess the current stage of AD based on user input, such as MRI or PET scan data. Formally, the diagnosis task can be expressed as

$$\hat{y}_{\mathrm{diag}} = \mathcal{D}(\mathcal{X}, \mathcal{M}), \tag{1}$$

where $\hat{y}_{\mathrm{diag}}$ represents the predicted stage of AD, $\mathcal{X} = \{\mathbf{X}_1, \mathbf{X}_2\}$ denotes the multi-modal inputs (e.g., MRI and PET scans), and $\mathcal{M}$ is the set of models used for diagnosis (e.g., CMViM, MCAD, etc.).

(ii) **Prognosis task** focuses on predicting the future progression of Alzheimer's disease. The prognosis task can be modeled as

$$\hat{y}_{\mathrm{prog}} = \mathcal{P}(\mathcal{X}, \mathcal{M}), \tag{2}$$

where $\hat{y}_{\mathrm{prog}}$ is the predicted progression, and $\mathcal{X}$ and $\mathcal{M}$ are the same as in the diagnosis task, but with models specialized for prognosis.

(iii) **Missing modality scenario** is common in clinical practice, where certain modalities may be unavailable or incomplete. In such cases, ADAgent integrates single-modal tools. If only one modality, say MRI $\mathbf{X}_{\mathrm{MRI}}$, is available, the system uses the MRI Diagnosis Tool $\mathcal{T}_{\mathrm{MRI}}$, and the prediction is made as

$$\hat{y}_{\mathrm{diag}} = \mathcal{T}_{\mathrm{MRI}}(\mathbf{X}_{\mathrm{MRI}}). \tag{3}$$

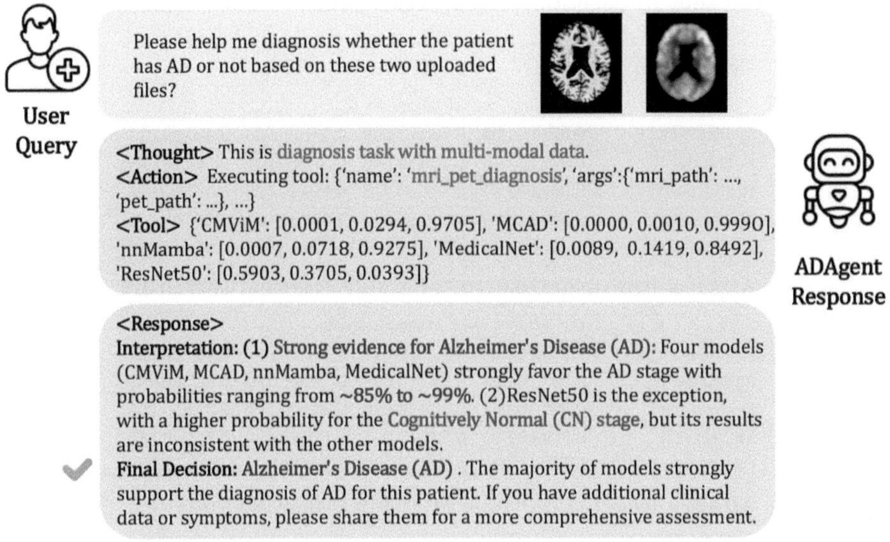

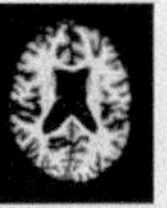
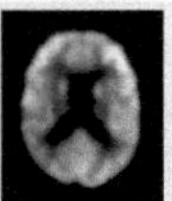

User Query

ADAgent Response

Fig. 2. ADAgent Interaction Example. An example of how ADAgent handles a conversation through its workflow (<thought>, <action>, <Tool>) along with tool outputs and final response.

Similarly, if only the PET modality is available, the PET Diagnosis Tool $\mathcal{T}_{\text{PET}}$ is used to process $\mathbf{X}_{\text{PET}}$ and make predictions.

Although ADAgent currently integrates existing tools for AD diagnosis and prognosis, its architecture is highly extensible to integrate new tools. Integrating a new tool requires only a class definition that specifies the tool's input/output formats without any extra training, which are mentioned in detail in Sect. 2.3.

Action Module - The Action module follows the planning and executes the task by utilizing relevant tools to get necessary outcomes for further analysis.

Collaborative Outcome Coordinator Module - The Collaborative Outcome Coordinator is responsible for synthesizing the response from the various specialized tools in ADAgent. Each tool $\mathcal{T}_i$ is designed to handle a distinct AD-related analysis task, and the tool collaborates with multiple models to ensure accurate and efficient task execution. The collaborative outcome coordinator leverages the reasoning capabilities of the LLM $\mathcal{L}$ to make informed decisions based on the aggregated results from the different tools. Mathematically, this process can be expressed as

$$\mathbf{R} = \mathcal{C}\left(\mathcal{O}_1(\mathcal{M}_1), \mathcal{O}_2(\mathcal{M}_2), \ldots, \mathcal{O}_n(\mathcal{M}_n)\right), \tag{4}$$

where $\mathcal{C}$ represents the coordinator function that combines the results from selected tools, and $\mathcal{O}_i(\mathcal{M}_i)$ denotes the output of tool $\mathcal{T}_i$ based on $\mathcal{M}_i$.

Table 1. Quantitative comparison on multi-modal tasks

Method	ACC	SPE	SEN	F1
Multi-modal diagnosis task				
MedicalNet [3]	0.571±0.009	0.768±0.009	0.591±0.030	0.586±0.015
nnMamba [7]	0.527±0.016	0.753±0.008	0.572±0.021	0.541±0.028
ResNet50 [8]	0.567±0.039	0.771±0.018	0.599±0.036	0.581±0.043
MCAD [23]	0.548±0.029	0.766±0.014	0.591±0.027	0.556±0.026
CMViM [18]	0.617±0.014	0.659±0.003	0.630±0.013	0.633±0.010
ADAgent (Ours)	**0.644±0.014**	**0.794±0.010**	**0.644±0.021**	**0.647±0.018**
Multi-modal prognosis task				
MedicalNet [3]	0.763±0.026	0.902±0.017	0.333±0.052	0.407±0.064
nnMamba [7]	0.733±0.038	0.843±0.068	0.394±0.189	0.405±0.147
ResNet50 [8]	0.726±0.034	0.853±0.029	0.333±0.052	0.374±0.065
MCAD [23]	0.719±0.026	0.892±0.061	0.182±0.091	0.231±0.080
CMViM [18]	0.815±0.046	0.912±0.029	0.515±0.105	**0.575±0.109**
ADAgent (Ours)	**0.822±0.010**	**0.941±0.019**	**0.545±0.043**	0.561±0.030

2.3 Extensibility

ADAgent is built using the LangChain and LangGraph frameworks. Its reasoning engine can be powered by any LLM, supporting both text-only and multi-modal models, whether open-source or proprietary. This flexibility allows ADAgent to be deployed in various environments, from local installations to cloud-based solutions, while accommodating diverse healthcare privacy requirements [6]. Each tool in ADAgent functions as an independent module, with clearly defined input, output, and specific capability. The modular design allows tools to be modified, replaced, or repurposed without affecting the overall system. Integrating a new tool requires only a class definition that specifies the tool's input/output formats and capabilities, enabling the LLM to learn its usage without extra training.

2.4 Interface and Interaction Example

ADAgent includes a production-ready interface built with Gradio, facilitating seamless deployment in clinical settings. The interface supports the uploading of MRI and PET scans in NIFTI format and maintains an interactive chat session for natural interactions. The interface also provides transparency by tracking and displaying intermediate outputs during tool execution. An example interaction is shown in Fig. 2.

3 Experiments

3.1 Dataset

This study was conducted on the Alzheimer's Disease Neuroimaging Initiative (ADNI) dataset [20], and focused on T1-MRI and PET-FDG scans. As for the

Table 2. Quantitative comparison on missing modality diagnosis tasks

Method	ACC	SPE	SEN	F1
Diagnosis with MRI				
MedicalNet [3]	0.521±0.022	0.760±0.004	0.569±0.007	0.553±0.024
ResNet50 [8]	0.536±0.012	**0.762±0.002**	**0.571±0.004**	**0.565±0.014**
ResNet34 [8]	0.487±0.006	0.726±0.008	0.504±0.026	0.479±0.027
ResNet18 [8]	0.515±0.016	0.739±0.010	0.533±0.024	0.522±0.016
ADAgent (Ours)	**0.543±0.007**	0.754±0.003	0.562±0.013	0.553±0.011
Diagnosis with PET				
MedicalNet [3]	0.550±0.022	0.760±0.004	0.569±0.007	0.553±0.024
ResNet50 [8]	0.550±0.035	0.755±0.014	0.554±0.030	0.554±0.040
ResNet34 [8]	0.524±0.028	0.751±0.008	0.555±0.014	0.523±0.028
ResNet18 [8]	0.520±0.015	0.747±0.010	0.552±0.031	0.529±0.022
ADAgent (Ours)	**0.594±0.034**	**0.785±0.014**	**0.631±0.027**	**0.611±0.034**

diagnosis task, we aim to determine the current stage of AD process based on use input from three stages: Cognitively normal (CN), Mild Cognitive Impairment (MCI), and AD. The prognosis task is to predict whether an MCI patient will convert to AD within 36 months or remain stable. ADAgent can directly integrate existing models without extra training. However, since there are few open-source models, we trained some models for integration in this work.

3.2   Experimental Settings

ADAgent uses GPT-4o as its backbone LLM and is deployed on a single NVIDIA RTX 4090 GPU. ADAgent integrates four tools: (1) multi-modal diagnosis tool, and (2) multi-modal prognosis tool, both integrating five models: MedicalNet, nnMamba, ResNet50, MCAD, CMViM; (3) MRI diagnosis tool, and (4) PET diagnosis tool, both embedding MedicalNet, ResNet50, ResNet34 and ResNet18 for missing modality tasks. All experiments are conducted three times, and the average and standard deviation are reported for analysis.

3.3   Experimental Results

To access the performance of the proposed ADAgent, we extensively compared the result of ADAgent against multiple baseline methods on multi-modal tasks and missing modality diagnosis tasks. To ascertain whether the performance improvement is attributable to the LLM's capability to aggregate the results of multiple models, we conducted the ablation study on multi-modal tasks and missing modality diagnosis tasks against two baseline methods.

Results on Multi-modal Tasks. Table 1 presents the result comparisons of ADAgent against five baselines: MedicalNet, nnMamba, ResNet50, MCAD, and CMViM, in terms of Accuracy (ACC), Specificity (SPE), Sensitivity (SEN), and

Table 3. Ablation study on ADAgent

Method	ACC	SPE	SEN	F1	ACC	SPE	SEN	F1
Multi-modality	*Diagnosis task*				*Prognosis task*			
Average	0.593	0.786	0.631	0.613	0.770	0.931	0.273	0.355
Vote	0.575	0.778	0.615	0.589	0.763	0.941	0.212	0.301
ADAgent (Ours)	**0.644**	**0.794**	**0.644**	**0.647**	**0.822**	**0.941**	**0.545**	**0.561**
Missing Modality	*MRI diagnosis task*				*PET diagnosis task*			
Average	0.519	0.743	0.546	0.530	0.582	0.782	0.623	0.595
Vote	0.536	**0.757**	**0.570**	0.546	0.582	0.782	0.619	0.595
ADAgent (Ours)	**0.543**	0.754	0.561	**0.553**	**0.644**	**0.794**	**0.644**	**0.647**

F1-score (F1) on multi-modal tasks. ADAgent outperforms baseline methods on accuracy by 7.3%, 11.7%, 7.7%, 9.6%, and 2.7% for the diagnosis task, and by 5.9%, 8.9%, 9.6%, 10.3%, and 0.7% for the prognosis task. Especially, ADAgent achieves the best results in almost all metrics for these two tasks. These results validate ADAgent's reasoning ability in analyzing different models' results and can indeed improve performance.

Results on Missing Modality Diagnosis Tasks. Similarly, Table 2 compares the ADAgent against four baseline approaches, MedicalNet, ResNet50, ResNet34, and ResNet18 on missing modality diagnosis tasks. As for the accuracy metric, ADAgent outperforms baseline methods by 2.2%, 0.7%, 5.6%, and 2.8% for MRI diagnosis tools and by 4.4%, 4.4%, 7.0%, and 7.4% for the PET diagnosis tool. It is noted that although ResNet34 demonstrates poor performance in MRI diagnosis tool, ADAgent still has accuracy improvement and retains relatively good performance in other metrics. By contrast, ADAgent achieves the best results in all metrics on the PET diagnosis task. These results show that ADAgent has reasoning ability when integrating the results of different models, bringing performance improvement and robustness.

Ablation Study. We conducted the ablation study on all tasks against two baseline methods, and the detailed results are shown in Table 3. The Average method is to average the prediction probabilities for all classes and select the class with the largest probability as the final prediction. The Vote method involves each model providing a decision, and the result that occurs most frequently is taken as the final outcome. ADAgent yields the best overall performance on all tasks. This suggests that ADAgent can fully leverage the capability of LLM to analyze the results of multiple models.

4   Conclusion

In conclusion, ADAgent introduces the first AI agent framework integrated with multiple tools for AD analysis, powered by an LLM as both a tool action planner and a collaborative outcomes coordinator. Experimental results demonstrate that ADAgent outperforms existing methods, showcasing its potential for advanced AD diagnosis and prognosis. However, the current implementation of ADAgent is limited by the number of integrated tools and input modalities. Future work will focus on expanding the tool set, incorporating additional modalities, and enhancing the model's generalization to further improve its clinical applicability.

Acknowledgement. This work was partially supported by RGC Collaborative Research Fund (No. C5055-24G), the Start-up Fund of The Hong Kong Polytechnic University (No. P0045999), the Seed Fund of the Research Institute for Smart Ageing (No. P0050946), and Tsinghua-PolyU Joint Research Initiative Fund (No. P0056509), and PolyU UGC funding (No. P0053716). The authors also acknowledge the use of AI tools for assistance in grammar enhancement and spelling checks during the preparation of this manuscript.

Disclosure of Interests. The authors have no competing interests to declare that are relevant to the content of this article.

References

1. Alberdi, A., Aztiria, A., Basarab, A.: On the early diagnosis of Alzheimer's disease from multimodal signals: a survey. Artif. Intell. Med. **71**, 1–29 (2016)
2. Better, M.A.: Alzheimer's disease facts and figures. Alzheimers Dement. **19**(4), 1598–1695 (2023)
3. Chen, S., Ma, K., Zheng, Y.: Med3D: transfer learning for 3D medical image analysis. arXiv preprint arXiv:1904.00625 (2019)
4. Cui, R., Liu, M.: Hippocampus analysis by combination of 3-D densenet and shapes for Alzheimer's disease diagnosis. IEEE J. Biomed. Health Inform. **23**(5), 2099–2107 (2018)
5. Dubois, B., et al.: Clinical diagnosis of Alzheimer's disease: recommendations of the international working group. Lancet Neurol. **20**(6), 484–496 (2021)
6. Fallahpour, A., Ma, J., Munim, A., Lyu, H., Wang, B.: MedRAX: medical reasoning agent for chest X-ray (2025). https://arxiv.org/abs/2502.02673
7. Gong, H., Kang, L., Wang, Y., Wan, X., Li, H.: nnMamba: 3D biomedical image segmentation, classification and landmark detection with state space model (2024). https://arxiv.org/abs/2402.03526
8. He, K., Zhang, X., Ren, S., Sun, J.: Deep residual learning for image recognition. In: Proceedings of the IEEE Conference on Computer Vision and Pattern Recognition, pp. 770–778 (2016)
9. Hoang, G.M., Kim, U.H., Kim, J.G.: Vision transformers for the prediction of mild cognitive impairment to Alzheimer's disease progression using mid-sagittal sMRI. Front. Aging Neuroscience **15**, 1102869 (2023)

10. Huang, F., Qiu, A., Initiative, A.D.N., et al.: Ensemble vision transformer for dementia diagnosis. IEEE J. Biomed. Health Inform. (2024)
11. Joe, E., Ringman, J.M.: Cognitive symptoms of Alzheimer's disease: clinical management and prevention. BMJ **367** (2019)
12. Li, B., et al.: MMedAgent: learning to use medical tools with multi-modal agent. arXiv preprint arXiv:2407.02483 (2024)
13. Lian, C., Liu, M., Zhang, J., Shen, D.: Hierarchical fully convolutional network for joint atrophy localization and Alzheimer's disease diagnosis using structural MRI. IEEE Trans. Pattern Anal. Mach. Intell. **42**(4), 880–893 (2018)
14. Liu, L., Dou, Q., Chen, H., Qin, J., Heng, P.A.: Multi-task deep model with margin ranking loss for lung nodule analysis. IEEE Trans. Med. Imaging **39**(3), 718–728 (2019)
15. Liu, L., Hong, C., Aviles-Rivero, A.I., Schönlieb, C.B.: Simultaneous semantic and instance segmentation for colon nuclei identification and counting. In: Annual Conference on Medical Image Understanding and Analysis, pp. 130–138. Springer (2022)
16. Organization, W.H., et al.: Global status report on the public health response to dementia (2021)
17. Scheltens, P., et al.: Alzheimer's disease. The Lancet **397**(10284), 1577–1590 (2021)
18. Sibilano, E., et al.: Understanding the role of self-attention in a transformer model for the discrimination of SCD from mci using resting-state EEG. IEEE J. Biomed. Health Inform. (2024)
19. Wang, S., Aviles-Rivero, A.I., Kourtzi, Z., Schönlieb, C.B.: HGIB: prognosis for Alzheimer's disease via hypergraph information bottleneck. arXiv preprint arXiv:2303.10390 (2023)
20. Weiner, M.W., et al.: The Alzheimer's disease neuroimaging initiative: a review of papers published since its inception. Alzheimer's Dementia **9**(5), e111–e194 (2013)
21. Yang, G., Du, K., Yang, Z., Du, Y., Zheng, Y., Wang, S.: CMViM: contrastive masked vim autoencoder for 3d multi-modal representation learning for ad classification. arXiv preprint arXiv:2403.16520 (2024)
22. Yu, Q., et al.: A transformer-based unified multimodal framework for Alzheimer's disease assessment. Comput. Biol. Med. **180**, 108979 (2024)
23. Zhang, J., He, X., Liu, Y., Cai, Q., Chen, H., Qing, L.: Multi-modal cross-attention network for Alzheimer's disease diagnosis with multi-modality data. Comput. Biol. Med. **162**, 107050 (2023). https://doi.org/10.1016/j.compbiomed.2023.107050
24. Zhang, X., Han, L., Zhu, W., Sun, L., Zhang, D.: An explainable 3D residual self-attention deep neural network for joint atrophy localization and Alzheimer's disease diagnosis using structural MRI. IEEE J. Biomed. Health Inform. **26**(11), 5289–5297 (2021)
25. Zhang, Y., Teng, Q., Liu, Y., Liu, Y., He, X.: Diagnosis of Alzheimer's disease based on regional attention with sMRI gray matter slices. J. Neurosci. Methods **365**, 109376 (2022)

MedPAO: A Protocol-Driven Agent for Structuring Medical Reports

Shrish Shrinath Vaidya[1]([✉])(iD), Gowthamaan Palani[2](iD), Sidharth Ramesh[1](iD), Velmurugan Balasubramanian[3], Minmini Selvam[4], Gokulraja Srinivasaraja[5], and Ganapathy Krishnamurthi[1]

[1] Department of Data Science and AI, IIT Madras, Chennai, India
shrish.s.vaidya@dsai.iitm.ac.in
[2] Department of Engineering Design, IIT Madras, Chennai, India
[3] LoveForm Health Technologies, Chennai, India
[4] Department of Radiology and Imaging Sciences, Sri Ramachandra Institute of Higher Education and Research, Chennai, India
[5] Department of Neuro and Interventional Radiology, Sri Ramachandra Institute of Higher Education and Research, Chennai, India

Abstract. The deployment of Large Language Models (LLMs) for structuring clinical data is critically hindered by their tendency to hallucinate facts and their inability to follow domain-specific rules. To address this, we introduce MedPAO, a novel agentic framework that ensures accuracy and verifiable reasoning by grounding its operation in established clinical protocols such as the ABCDEF protocol for CXR analysis. Med-PAO decomposes the report structuring task into a transparent process managed by a Plan-Act-Observe (PAO) loop and specialized tools. This protocol-driven method provides a verifiable alternative to opaque, monolithic models. The efficacy of our approach is demonstrated through rigorous evaluation: MedPAO achieves an F1-score of 0.96 on the critical sub-task of concept categorization. Notably, expert radiologists and clinicians rated the final structured outputs with an average score of 4.52 out of 5, indicating a level of reliability that surpasses baseline approaches relying solely on LLM-based foundation models. The code is available at: https://github.com/MiRL-IITM/medpao-agent.

Keywords: Agentic Systems · Large Language Models · Data Structuring

1 Introduction

Clinical narratives like radiology reports contain a wealth of observational data, but their unstructured, free-text format hinders large-scale computational analysis [4,18]. Radiologists often use systematic checklists, such as the ABCDEF mnemonic for chest X-rays [16], to ensure diagnostic completeness. The **ABCDEF** protocol is a mnemonic-driven, structured framework for systematic chest X-ray interpretation. It organizes key anatomical regions and potential abnormalities into six distinct categories to promote thorough and consistent reporting. **A** evaluates airways, **B** assesses lung fields and pleura,

J. Qiu et al. (Eds.): Agentic AI 2025/CMLLMs 2025/CREATE 2025, LNCS 16147, pp. 33–45, 2026.
https://doi.org/10.1007/978-3-032-06004-4_4

C examines cardiomediastinal contours and great vessels, and **D** focuses on the diaphragm and subdiaphragmatic structures. **E** reviews external anatomy including bones and soft tissues, while **F** identifies foreign bodies and medical devices. However, this structure is typically lost in the final dictated report, leading to significant inter-clinician variability that complicates data aggregation and the development of predictive models [9,21].

Existing approaches to automate data structuring have significant drawbacks. Traditional rule-based and supervised machine learning methods are either too brittle or require prohibitively expensive manual annotation [1,4]. While modern Large Language Models (LLMs) show impressive zero-shot capability, they are prone to factual "hallucination" and lack verifiable reasoning essential for high-stakes medical applications [2,11,22]. This creates a critical need for a framework that is both flexible and trustworthy. To address this challenge, we introduce MedPAO, a Protocol-Driven Agent for Structuring Medical Reports. Figure 1 illustrates MedPAO's structured output on given report findings. MedPAO is a novel agentic framework that operationalizes a clinical protocol as its core reasoning structure. Our agent employs a Plan-Act-Observe (PAO) loop [30] to orchestrate specialized, verifiable tools, systematically mapping narrative findings back to a clinical protocol like the ABCDEF system. This approach creates a consistent, structured representation of the report's content. As shown in Fig. 2, our agent correctly categorizes medical concepts where other LLMs fail, demonstrating a clear path towards reliable and transparent report structuring.

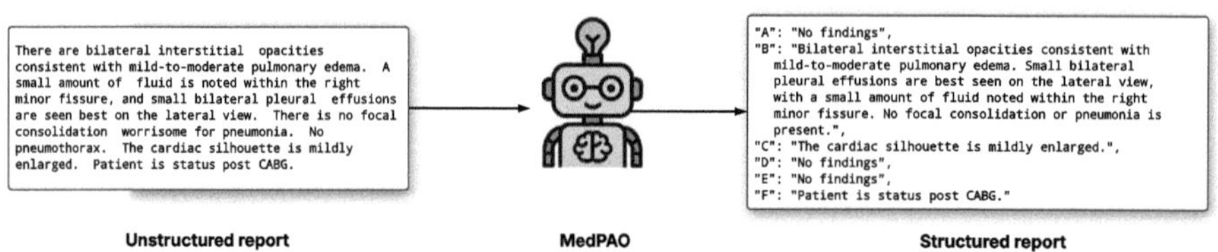

Fig. 1. Illustrative example of MedPAO.

The primary contributions of this work are:

- We propose MedPAO, a novel, zero-shot agentic framework that transforms unstructured medical reports into protocol-compliant structured data, standardizing information across different clinician writing styles.
- We demonstrate the efficacy of our protocol-driven approach by implementing the ABCDEF protocol, showing how our agent maps narrative findings to a clinically established standard.
- We introduce a modular and verifiable toolset, including a fine-tuned medical concept extractor and reasoning-based modules for ontology filtering and protocol categorization.
- Through rigorous evaluation, we show that MedPAO significantly outperforms baseline methods and provides a scalable, transparent, and accurate solution for structuring medical reports.

<table>
<tr><td colspan="3">Task : categorizing medical concepts into unique ABCDEF categories, according to given protocol

Prompt : Based on the given protocol {protocol} categorize the following medical concepts into ABCDEF categories: "no pneumothorax, normal mediastinal silhouette, aicd, no chf, lumbar fusion hardware, cardiomegaly, intact sternal wires, breast implants, cardiac clips"</td></tr>
<tr><td>Gemini 2.0 - Flash</td><td>DeepSeek-R1-Distill-Llama-70B</td><td>Our Agent</td></tr>
<tr>
<td>no pneumothorax: "B",
normal mediastinal silhouette: "C",
aicd: "F",
No chf: "B",
Lumbar fusion hardware:"F",
Cardiomegaly: "C",
intact sternal wires: "E",
Breast implants: "E",
Cardiac Clips: "C"</td>
<td>No pneumothorax: "B",
normal mediastinal silhouette: "C",
aicd: "F",
No chf: "B",
Lumbar fusion hardware: "D",
Cardiomegaly:"C",
Intact sternal wires:"E",
Breast implants: "E",
Cardiac Clips:"F"</td>
<td>No pneumothorax: "B",
normal mediastinal silhouette: "C",
aicd:"F" ,
No chf: "C",
Lumbar fusion hardware: "F",
Cardiomegaly: "C",
Intact sternal wires:"F",
Breast implants:"F",
Cardiac Clips:"F"</td>
</tr>
</table>

Fig. 2. Proposed agent vs SOTA LLMs on medical concept categorization task according to ABCDEF protocol.

2   Related Works

The effort to structure clinical text has a long history, beginning with brittle rule-based systems that failed to generalize across linguistic variations [1,18,31]. This led to a shift towards supervised machine learning, which offered more flexibility but introduced a significant "annotation bottleneck" due to the high cost of creating expert-labeled datasets [25,26].

The emergence of Large Language Models (LLMs) marked a new paradigm, promising zero-shot information extraction in varied domains from medicine to historical OCR-processed texts [10,22,23]. However, monolithic LLMs act as opaque "black boxes" [7], are prone to factual "hallucinations" and often lack the deep domain knowledge required to correctly interpret complex clinical rules like TNM staging [11]. This core structuring task is distinct from parallel work in automated report generation, which produces more unstructured text [12,29], or specialized evaluation frameworks whose schemas are not designed for general-purpose data standardization [13].

To enhance LLM reliability, a new generation of sophisticated frameworks has been developed. One class of solutions focuses on post-hoc validation, using two-stage pipelines for confidence scoring [2] or self-correcting feedback loops for error checking [3]. Another approach involves multi-agent architectures, which follow distinct strategies. Some employ a consensus mechanism, where one LLM adjudicates the outputs of several others to improve robustness [27]. In contrast, others use a delegation model, as seen in the financial domain, where specialized agents handle discrete sub-tasks like extraction and querying [5]. A third strategy involves fine-tuning LLMs on annotated data to directly output structured formats like JSON, though this re-introduces the annotation bottleneck [6]. While these state-of-the-art methods improve reliability through correction, consensus, or delegation, they do not ground the reasoning process within an explicit, verifiable clinical workflow. This leaves a critical gap for a system whose internal logic is both transparent and clinically valid. MedPAO addresses this gap by uniquely operationalizing an established medical protocol as the core, step-by-step reasoning structure of an agentic framework, ensuring that the final structured output is accurate and demonstrably aligned with standard clinical practice.

3   Methodology

3.1   Overview of Agentic Framework

To facilitate structured report generation, our framework integrates a Model Context Protocol (MCP) with a Plan-Act-Observe (PAO) loop, enabling dynamic orchestration of auxiliary tools in conjunction with a central large language model (LLM) engine, as depicted in Fig. 3. The PAO loop not only determines which tools to invoke based on user input, but also provides the full contextual scope to the LLM, ensuring that the LLM maintains overarching control over the decision-making process.

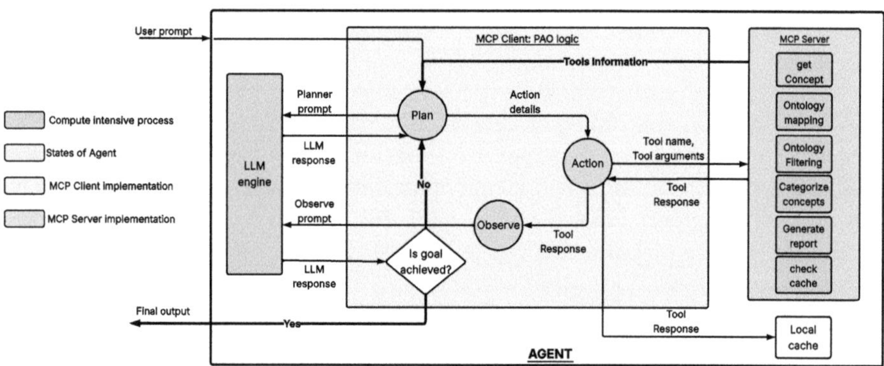

Fig. 3. Architecture of proposed MedPAO agent for medical report processing tasks.

3.2   Agent Architecture and Implementation

LLM Engine. Our agent employs the foundation model, Deepseek-R1-70B [8], as its core LLM engine. This selection was driven by two key factors: (1) the model's extended context window capacity, and (2) its state-of-the-art reasoning performance. The architectural features of the model serves a dual purpose: it maintains output discipline while simultaneously providing model explainability.

The PAO Logic and MCP. Our framework integrates the complete Plan-Act-Observe (PAO) framework with the Model-Context Protocol (MCP), a standardized approach for developing efficient AI agents. The complementary MCP client implements our PAO logic to orchestrate the agentic workflow. Our client strategically employs the LLM engine during both the Plan and Observe states of the agent's operational cycle. In the Plan state, the MCP client prompts the LLM engine by providing contextual information including available tools and any previously executed steps to determine the next appropriate tool along with its specific parameters. The server then executes the designated tools with the provided parameters and returns structured outputs. During the Observe state, the system presents the LLM with both the tool-generated output and the original user query, requesting an evaluation on whether the response adequately addresses the initial request. Based on this assessment, the LLM decides whether to continue iterating through the cycle or terminate the process and return the current output as final. This modular client-server design provides benefits in terms of both efficiency and scalability.

The Agent's Toolset

Tool 1: Get Concept: The Concept Extraction Module serves as the foundational component of our toolset, specifically designed to identify and extract clinically relevant medical terms and diagnostic findings from unstructured medical reports. In our framework, "concepts" encompass both pathological observations and their associated anatomical locations when specified. The module utilizes a fine-tuned MedLlama-8B [14] model for medical concept recognition. Due to the absence of existing ground truth report-concept pairs, we engaged a certified radiologist and clinician to annotate a preliminary set of 40 Chest X-ray reports from the Eye Gaze MIMIC-CXR dataset [17]. Leveraging these expert-validated samples, we generated additional synthetic training data using foundation models like GPT-4, Gemini 2.5-Flash, Claude 3.5 and Deepseek-R1 through few-shot prompting techniques and restricting them to a predefined vocabulary from SNOMEDCT and RADLEX ontologies. From each of the foundation models we extracted 40 synthetic concept-report pairs, resulting in a final dataset comprising of 200 unique samples with 311 unique medical concepts. The MedLlama model was fine-tuned using LoRA (Low-Rank Adaptation) with full precision (non-quantized) parameters on the mixture of synthetic and original data. During inference, we implemented a few-shot prompting approach to extract both medical concepts and their corresponding source sentences with the fine-tuned model, ensuring traceability and reducing extraction errors.

Tool 2: Ontology Mapping: To enhance the clinical context of extracted medical concepts, we incorporated well-established medical ontologies—specifically SNOMED-CT and RADLEX—using the BioPortal Annotator API [20], as they offer comprehensive coverage and promote interoperability for a wide range of radiological findings. This module operates downstream of the concept extraction tool, automatically mapping each identified medical concept to its corresponding ontological representation. For every extracted medical concept, the API retrieves the precise ontological mapping as well as its complete ancestral hierarchy, thereby providing rich, structured clinical context. Lemmatization was done on the raw concepts as the ontology annotator tool does only rule based matching. The retrieved ontological mappings were cached locally for efficiency and subsequently fed into the LLM pipeline to support downstream reasoning tasks.

Tool 3: Ontology Filtering: In our framework, medical concepts are typically expressed as multi-word terms, which often results in multiple potential ontology mappings for each concept. To address this complexity, we developed a systematic approach to classify these mappings into *Primary* and *Secondary* ontologies. The Primary Ontology represents the principal pathological finding or clinical abnormality, while Secondary Ontologies encompass details such as severity, anatomical location, or associated observations.

The need for such segregation arises from a key challenge that, a single concept may map to multiple ontologies with distinct semantic meanings, leading to potential ambiguity. To automate this classification, we leverage the LLM engine, which uses a few-shot prompting strategy.

Tool 4: Categorize Concepts: Our framework incorporates a dedicated tool for protocol-based categorization of medical concepts, essential for transforming unstructured clinical findings into standardized reports. For this study, we implemented the "ABCDEF" protocol [16] for chest X-ray diagnosis, which systematically classifies findings based on anatomical regions and abnormal findings into six categories (A-F). Each extracted medical concept is precisely mapped to its corresponding anatomical category, enabling the conversion of free-text observations into a structured format while maintaining clinical accuracy. This approach not only addresses the variability inherent in traditional radiology reporting but also establishes a foundation for consistent, protocol-compliant documentation.

For this task, we leverage the LLM engine to process multiple inputs: (1) the complete protocol description [16], (2) the extracted medical concepts, and (3) their corresponding filtered ontologies. The ontological information serves as critical contextual data, enabling more accurate and clinically relevant categorization of each concept according to the specified protocol.

Tool 5: Generate Report: Our framework generates protocol-compliant structured findings by processing categorized medical concepts with the LLM engine. The model receives an input, comprising the complete protocol specifications alongside the classified concepts and their original source sentences. This approach ensures the generated findings are clinically accurate, grounded in the source report's terminology, and adhere to the required format. The selection of a foundational model was necessitated by the absence of structured ground truth data for supervised training. We leveraged its inherent generalization capabilities to avoid the performance degradation in text generation, on unseen data distributions commonly observed with fine-tuning [19,28].

Tool 6: Check Cache: To optimize computational efficiency, our framework caches recurring medical concepts and their protocol-specific categorizations. Reusing these pre-processed mappings from local storage accelerates report generation and reduces resource utilization without compromising accuracy. This caching strategy is vital for meeting the performance demands for clinical deployment.

Figure 4 depicts the agent's operational pipeline. Upon receiving an input prompt, the agent first decomposes the task into a sequence of steps and selects the necessary tools. It then executes these steps sequentially, aggregating the outputs from each tool to generate the final structured report.

4 Results and Analysis

4.1 Groundtruth Data Collection for Evaluation

A key limitation of this study was the absence of large-scale, protocol-aligned structured report data with categorized concepts, which are essential for comprehensive evaluation. To address this gap, we collaborated with radiologists and practicing clinicians to curate a set of ground truth samples for assessing our agent's performance. For this purpose, we extracted relevant findings from the MIMIC-CXR dataset [15] and developed a dedicated annotation tool to facilitate data collection. The reports were annotated with two key elements: 1) clinically significant concepts extracted from the text, and 2) their respective categorical labels following the defined protocol. In total, 200 ground truth samples were compiled using this method. This strategy facilitated structured validation even in the absence of initially standardized reference data.

4.2 Concept Extraction Results

The performance of the *get concept* tool was evaluated by comparing its extracted concepts against ground-truth annotations from a corpus of radiologist-annotated reports. We framed this task as a multi-label classification problem, where each report may contain multiple relevant or irrelevant concepts from a predefined vocabulary. To quantify performance, we adopted standard multi-label evaluation metrics, including Precision, Recall, F1-score, Subset Accuracy, and Hamming Loss. Due to significant class imbalance in the

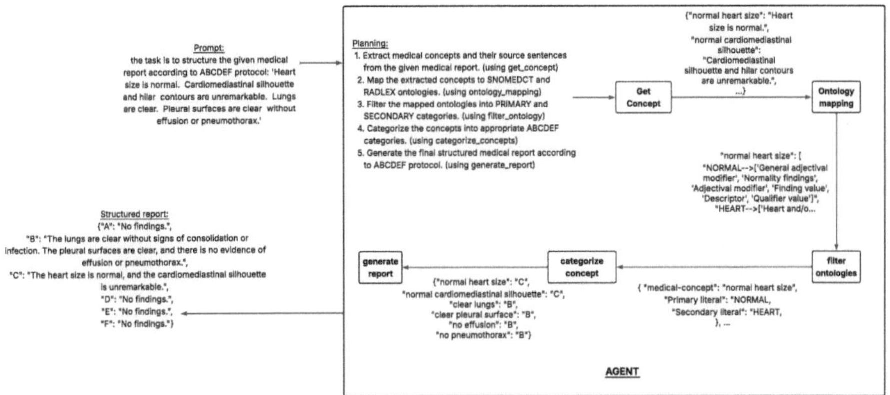

Fig. 4. Step by step output of our agent in structuring the free style report according to protocol [16].

Table 1. Quantitative evaluation for concept extraction

Macro and Weighted average metrics for concept extraction @80 fuzzy confidence					
Model	Precision	Recall	F1 score	Subset Accuracy	Hamming loss	vs Our Model (p-value)
DeepSeek-R1-Distill-Qwen-7B	0.78, 0.88	0.75, 0.82	0.76, 0.84	0.13	0.0049	0.0007
Qwen2.5-7B	0.74, 0.86	0.72, 0.81	0.73, 0.83	0.17	0.0043	0.0011
Deepseek-R1-70B	0.88, 0.92	0.87, 0.90	0.87, 0.90	0.15	0.0037	0.3613
Our get concept tool	**0.87, 0.92**	**0.86, 0.90**	**0.86, 0.91**	**0.28**	**0.0023**	–
Macro and Weighted average metrics for concept extraction @90 fuzzy confidence					
DeepSeek-R1-Distill-Qwen-7B	0.72, 0.86	0.68, 0.71	0.70, 0.75	0.07	0.0078	0.00004
Qwen2.5-7B	0.69, 0.84	0.67, 0.75	0.68, 0.78	0.17	0.0057	0.0119
DeepSeek-R1-Distill-Llama-70B	0.79, 0.88	0.77, 0.81	0.77, 0.82	0.14	0.0069	0.0519
Our get concept tool	**0.81, 0.89**	**0.80, 0.83**	**0.80, 0.84**	**0.20**	**0.0039**	–

ground truth data, we also computed macro and weighted averages for Precision, Recall, and F1-score. Since medical concepts often exhibit lexical variations while conveying the same meaning (e.g., "no effusion" vs. "absence of effusion"), exact string matching was insufficient. Instead, we employed fuzzy matching with similarity thresholds of 80% and 90% to align predicted concepts with ground truth annotations. Table 1 presents the comparative evaluation of our fine-tuned MedLlama model against leading state-of-the-art LLMs at 80% and 90% fuzzy confidence thresholds. Statistical significance of performance differences between models was assessed using McNemar's test, with corresponding p-values reported for each pairwise comparison. The results demonstrate that our fine-tuned model not only surpasses comparable-sized models but also outperforms a significantly larger LLM, despite being smaller in scale.

4.3 Concept Categorization Results

A critical component of our agentic framework is the concept categorization tool, which systematically classifies extracted medical concepts into predefined protocol-based categories [16]. We formulate this task as a multiclass classification problem, where the agent must assign each concept to its correct category from a set of categories.

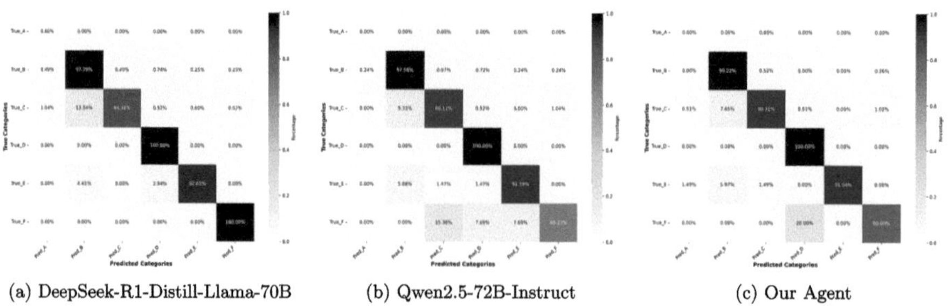

(a) DeepSeek-R1-Distill-Llama-70B   (b) Qwen2.5-72B-Instruct   (c) Our Agent

Fig. 5. Confusion matrices over concept categorization task by the respective models.

Table 2. macro and weighted average metrics for concept categorization

Model	Precision	Recall	F1 score	Jaccard Score (IoU)	vs Our Agent (p-values)
DeepSeek-R1-Distill-Llama-70B	0.93, 0.96	0.90, 0.93	0.91, 0.94	0.88, 0.90	0.019
Qwen2.5-72B-Instruct	0.93, 0.96	0.92, 0.94	0.92, 0.95	0.89, 0.92	0.261
Our agent	**0.95, 0.97**	**0.94, 0.96**	**0.94, 0.96**	**0.92, 0.94**	–

To evaluate performance, we computed standard metrics including Precision, Recall, F1-score, and Jaccard Similarity between predicted and ground truth categorizations. Given the inherent class imbalance, we report both macro and weighted averages (as in Sect. 4.2). The quantitative results, presented in Table 2, compare our agent against two LLMs, i.e., Deepseek-R1 70B and Qwen 2.5 72B models. Notably, our agent consistently outperforms the base models across all metrics. Figure 5 shows the confusion matrix obtained with all the three approaches for the concept categorization task.

4.4   Evaluating the Generated Reports

To assess the quality of the generated radiology reports, we adopted an evaluation framework inspired by the work of Dom Marshall et al. [24]. We introduced two key scoring metrics: (1) an Accuracy Score, which measures the clinical correctness of the generated findings against the original MIMIC-CXR reference reports, and (2) a Structural Score, which evaluates the degree to which the generated reports adhere to the predefined reporting protocol. These metrics provide a comprehensive assessment of both the medical validity and standardized formatting of the outputs, which are represented in Table 3.

To validate the clinical quality of our agent's outputs, we conducted a evaluation study involving certified radiologists and clinician. The evaluation procedure required radiologists and clinician to assess 50 samples of our agent-generated reports by comparing them against corresponding reference reports from the MIMIC-CXR dataset (distinct from the data used in Sects. 4.2 and 4.2). The assessment was conducted using the scoring criteria specified in Table 3. The mean of accuracy score and structure scores are as shown in Table 4.

The complete agent framework was deployed on Intel's Gaudi2 AI accelerator, leveraging its specialized architecture for efficient deep learning inference. We conducted rigorous timing evaluations to quantify the agent's operational efficiency across its core functionalities. As documented in Table 5, these measurements encompass both the integrated agent pipeline and its individual component modules. The agent supports four primary transformation tasks: (1) conversion of unstructured clinical narratives to standardized reports, (2) generation of structured reports from extracted medical concepts, (3) organization of raw medical concepts into structured representations, and (4) direct extraction of structured concepts from unstructured text.

Table 3. Scoring criteria for report quality

Score	Definition for accuracy scoring	Definition for structure scoring
5	Perfect Match: All key details are present, and there is no missing, extraneous, or fabricated (hallucinated) information	Perfect Structure: All specified Key Concepts are present as distinct sections/headers
4	Generally Accurate with Minor Omissions: It may be missing a few minor details that do not impact the overall meaning or interpretation. No hallucinations	Good Structure: All Key Concepts are present and correctly populated, but there might be minor deviations
3	Key Details Present but with Minor Hallucinations: The Generated Report captures the main points but has significant omissions of secondary details	Acceptable Structure: Most Key Concepts are present, but one or two may be missing or merged
2	Missing Key Details or Contradictory Information: The Generated Report fails to mention one or more key details from the groundtruth	Poor Structure: Multiple Key Concepts are missing. Information is disorganized, placed under incorrect headings, or lumped together
1	Critically Flawed or Dangerous: The Generated Report has major omissions of critical information or contains significant, dangerous hallucinations	No Structure: The report is a free-form block of text. It completely ignores the given protocol-driven structure

Table 4. Radiologist and Clinician evaluation of agent generated reports

Method	Average Accuracy score	Average Structure score
Clinician	4.64	4.59
Panel of two Radiologists	4.52	4.48

Table 5. Inference time of each tool and overall agent across multiple tasks (in seconds)

Tasks	Planning time	get concept	ontology mapping	ontology filtering	concept categorization	report generation	Overall
freestyle report-to-structured report	46	1	56	132	25	20	280
freestyle report-to-structured concepts	45	1	55	130	22	–	254
raw concepts-to-structured report	45	–	53	134	23	22	277
raw concept-to-structured concept	45	–	53	140	23	–	261
freestyle report-to-structured report with check cache enabled	45	1	–	–	–	20	66

5   Conclusion

In conclusion, our work introduces a novel protocol-driven, ontology-aware AI framework that outperforms existing models in radiology report structuring, achieving superior clinical correctness while maintaining diagnostic accuracy. This agent serves as an efficient tool for radiologists, enabling them to generate fully structured reports by simply providing key impressions and findings - significantly reducing documentation effort while ensuring standardization and without replacing expert judgment. The framework's modality-agnostic design allows for seamless expansion beyond chest radiography to other imaging modalities, enhancing its versatility across radiology workflows. Looking ahead, integration with medical images and automated suggestion systems could further augment its utility, while optimizations in model compression and hardware acceleration may enable real-time clinical deployment. These advancements position the framework for broad adoption across healthcare institutions with varying reporting standards and resources, ultimately improving documentation consistency and workflow efficiency.

Acknowledgments. We acknowledge Intel CSR grant to IITM Pravartak which enabled the compute requirement.

Disclosure of Interests. The authors have no competing interests to declare that are relevant to the content of this article.

References

1. Aggarwal, A., Garhwal, S., Kumar, A.: Hedea: a python tool for extracting and analysing semi-structured information from medical records. Healthc. Inf. Res. **24**(2), 148–153 (2018)
2. Alzaid, E., Pergola, G., Evans, H., Snead, D., Minhas, F.: Large multimodal model-based standardisation of pathology reports with confidence and its prognostic significance. J. Pathol. Clin. Res. **10**(6), e70010 (2024)
3. Bisercic, A., Nikolic, M., van der Schaar, M., Delibasic, B., Lio, P., Petrovic, A.: Interpretable medical diagnostics with structured data extraction by large language models. arXiv preprint arXiv:2306.05052 (2023)
4. Bose, P., Srinivasan, S., Sleeman, W.C., IV., Palta, J., Kapoor, R., Ghosh, P.: A survey on recent named entity recognition and relationship extraction techniques on clinical texts. Appl. Sci. **11**(18), 8319 (2021)
5. Choi, C., et al.: Structuring the unstructured: a multi-agent system for extracting and querying financial kpis and guidance. arXiv preprint arXiv:2505.19197 (2025)
6. Dagdelen, J., et al.: Structured information extraction from scientific text with large language models. Nat. Commun. **15**(1), 1418 (2024)
7. De Bellis, A.: Structuring the unstructured: an llm-guided transition. In: DC@ ISWC (2023)
8. DeepSeek-AI, et al.: Deepseek-r1: incentivizing reasoning capability in llms via reinforcement learning (2025). https://arxiv.org/abs/2501.12948

9. Donnelly, L.F., Grzeszczuk, R., Guimaraes, C.V., Zhang, W., Bisset, G.S., III.: Using a natural language processing and machine learning algorithm program to analyze inter-radiologist report style variation and compare variation between radiologists when using highly structured versus more free text reporting. Curr. Probl. Diagn. Radiol. **48**(6), 524–530 (2019)
10. Grothey, B., et al.: Comprehensive testing of large language models for extraction of structured data in pathology. Commun. Med. **5**(1), 96 (2025)
11. Huang, J., et al.: A critical assessment of using chatgpt for extracting structured data from clinical notes. npj Dig. Med. **7**(1), 106 (2024)
12. Hyland, S.L., et al.: Maira-1: a specialised large multimodal model for radiology report generation (2024). https://arxiv.org/abs/2311.13668
13. Jiang, Y., et al.: Clear: a clinically-grounded tabular framework for radiology report evaluation (2025). https://arxiv.org/abs/2505.16325
14. John Snow Labs: JSL-MedLlama-3-8B-v2.0 (2024). https://huggingface.co/johnsnowlabs/JSL-MedLlama-3-8B-v2.0. model card
15. Johnson, A.E., et al.: Mimic-cxr, a de-identified publicly available database of chest radiographs with free-text reports. Sci. Data **6**(1), 317 (2019)
16. Jones, J., Silverstone, L., Bell, D., et al.: Chest x-ray review: ABCDEF. https://radiopaedia.org/articles/41125. https://doi.org/10.53347/rID-41125. Accessed 24 June 2025
17. Karargyris, A., et al.: Creation and validation of a chest x-ray dataset with eye-tracking and report dictation for ai development. Sci. Data **8**(1), 92 (2021)
18. Kreimeyer, K., et al.: Natural language processing systems for capturing and standardizing unstructured clinical information: a systematic review. J. Biomed. Inf. **73**, 14–29 (2017)
19. Luo, Y., Yang, Z., Meng, F., Li, Y., Zhou, J., Zhang, Y.: An empirical study of catastrophic forgetting in large language models during continual fine-tuning (2025). https://arxiv.org/abs/2308.08747
20. Martínez-Romero, M., Jonquet, C., O'connor, M.J., Graybeal, J., Pazos, A., Musen, M.A.: NCBO ontology recommender 2.0: an enhanced approach for biomedical ontology recommendation. J. Biomed. Semant. **8**, 1–22 (2017)
21. Nobel, J.M., van Geel, K., Robben, S.G.: Structured reporting in radiology: a systematic review to explore its potential. Eur. Radiol. 1–18 (2022)
22. Ntinopoulos, V., et al.: Large language models for data extraction from unstructured and semi-structured electronic health records: a multiple model performance evaluation. BMJ Health Care Inf. **32**(1), e101139 (2025)
23. Schwitter, N.: Using large language models for preprocessing and information extraction from unstructured text: a proof-of-concept application in the social sciences. Methodol. Innov. 20597991251313876 (2025)
24. Sharma, N.: Cxr-agent: vision-language models for chest x-ray interpretation with uncertainty aware radiology reporting. arXiv preprint arXiv:2407.08811 (2024)
25. Spasic, I., Nenadic, G., et al.: Clinical text data in machine learning: systematic review. JMIR Med. Inf. **8**(3), e17984 (2020)
26. Srivastava, R., Bhat, L., Prasad, S., Deshpande, S., Das, B., Jadhav, K.: Medpromptextract (medical data extraction tool): anonymization and high-fidelity automated data extraction using natural language processing and prompt engineering. J. Appl. Lab. Med. jfaf034 (2025)
27. Tripathi, A., et al.: Employing consensus-based reasoning with locally deployed llms for enabling structured data extraction from surgical pathology reports. In: medRxiv, p. 2025–04 (2025)

28. Wang, Y., et al.: Two-stage llm fine-tuning with less specialization and more generalization (2024). https://arxiv.org/abs/2211.00635
29. Wang, Z., Liu, L., Wang, L., Zhou, L.: R2gengpt: radiology report generation with frozen llms. Meta-Radiology **1**(3), 100033 (2023)
30. Yao, S., et al.: React: synergizing reasoning and acting in language models. In: International Conference on Learning Representations (ICLR) (2023)
31. Zeng, Z., Zhao, Y., Sun, M., Vo, A.H., Starren, J., Luo, Y.: Rich text formatted ehr narratives: a hidden and ignored trove. In: MEDINFO 2019: Health and Wellbeing e-Networks for All, pp. 472–476. IOS Press (2019)

Scan-Do Attitude: Towards Autonomous CT Protocol Management Using a Large Language Model Agent

Xingjian Kang[1]([envelope]) [iD], Linda Vorberg[1,2] [iD], Andreas Maier[1] [iD],
Alexander Katzmann[2] [iD], and Oliver Taubmann[2] [iD]

[1] Pattern Recognition Lab, Friedrich-Alexander Universität Erlangen-Nürnberg,
Erlangen, Germany
`xingjian.kang@fau.de`
[2] Computed Tomography, Siemens Healthineers, Forchheim, Germany
`{alexander.katzmann,oliver.taubmann}@siemens-healthineers.com`

Abstract. Managing scan protocols in Computed Tomography (CT), which includes adjusting acquisition parameters or configuring reconstructions, as well as selecting postprocessing tools in a patient-specific manner, is time-consuming and requires clinical as well as technical expertise. At the same time, we observe an increasing shortage of skilled workforce in radiology. To address this issue, a Large Language Model (LLM)-based agent framework is proposed to assist with the interpretation and execution of protocol configuration requests given in natural language or a structured, device-independent format, aiming to improve the workflow efficiency and reduce technologists' workload. The agent combines in-context-learning, instruction-following, and structured tool-calling abilities to identify relevant protocol elements and apply accurate modifications. In a systematic evaluation, experimental results indicate that the agent can effectively retrieve protocol components, generate device-compatible protocol definition files, and faithfully implement user requests. Despite demonstrating feasibility in principle, the approach faces limitations regarding syntactic and semantic validity due to lack of a unified device API, and challenges with ambiguous or complex requests. In summary, the findings show a clear path towards LLM-based agents for supporting scan protocol management in CT imaging.

Keywords: CT Scan Protocol · LLM-Agent · Structured Data

1 Introduction

In CT imaging, design and implementation of scan protocols as well as the dynamic adaptation of individual steps, including acquisition parameters, reconstruction settings and automated postprocessing, to patient-specific needs during

---

A. Katzmann and O. Taubmann—Equal contribution.

Supplementary Information The online version contains supplementary material available at https://doi.org/10.1007/978-3-032-06004-4_5.

J. Qiu et al. (Eds.): Agentic AI 2025/CMLLMs 2025/CREATE 2025, LNCS 16147, pp. 46–54, 2026.
https://doi.org/10.1007/978-3-032-06004-4_5

the actual exam, require both domain-specific clinical knowledge and system-level technical expertise. These tasks are not only resource-intensive – also due to the fact that different scanner models have different capabilities and different UI concepts – but are increasingly difficult to manage amid a growing shortage of qualified staff in radiology [1].

Recent advancements in LLMs demonstrate remarkable potential in supporting daily work in radiology [3]. Extensive research has also shown that LLMs can be guided to edit structured data: Hegselmann et al. [5] and Jiang et al. [6] both flattened structured data into LLM-interpretable text strings to improve LLM performance on structured data classification and reasoning tasks, respectively. When handling large-scale tables, Chen et al. [2] mitigate catastrophic forgetting by introducing a Retrieval-Augmented-Generation (RAG) module and prompt-based reasoning mechanisms to the pipeline.

While previous research has mainly focused on utilizing LLMs to analyze general structured data [4], approaches for applying LLMs to model domain-specific structured data, such as CT scan protocols, are still lacking. Hence, to alleviate technologists' workload and automate scan protocol modification workflows, this work proposes utilizing an LLM-based agent framework [8] for adaptation of acquisition, reconstruction and postprocessing settings in the device-specific protocol definition files by leveraging the LLM as a planner and providing the agent with specially designed tools. To systematically evaluate the reliability of the proposed agent framework, a multi-level assessment pipeline with customized metrics has been introduced and illustrated. The agent's performance is compared across different underlying LLMs and among diverse request types.

2   Methods

2.1   Hierarchical Representation of Scan Protocols

CT scan protocols define technical parameters and procedures for examinations. In this work, we investigate vendor-specific CT protocol definition files (Siemens Healthineers, Forchheim, Germany) which are structured as hierarchical XML documents. As shown in Fig. 1, these scan protocols adopt a tree structure where each node represents an *entity* corresponding to distinct imaging workflow phases. For instance, **Range** entities (Topogram/Spiral) define data acquisition phases, while **Recon** entities (Oriented/Standard) specify reconstruction operations. The hierarchical organization reflects the procedural CT imaging flow, with parent-child relationships indicating the sequence and nesting of scan phases and respective reconstructions. Each entity contains attributes, including *name*, *id*, *type*, and child entities for identification, with critical parameters denoted as *essentials*, controlling acquisition, reconstruction and postprocessing.

2.2   Agent Framework

General Workflow. As shown in Fig. 2, the agent is provided with various tools and a memory component. The process begins with the framework receiving the

Fig. 1. XML-based hierarchical structure of a CT scan protocol definition from a commercial scanner. *Left panel:* exemplary structure and the child-parent relations between different entities. *Right panel:* detailed view of a **CTRecon** entity with the definition of some essential parameters through key-value pairs.

technologist's request. Leveraging tool calling, in-context-learning, and reasoning over the fetched structured protocol context, the agent realizes a targeted interaction with the underlying scan protocol and implements the user request with a series of proposed actions. When uncertainty or ambiguity is detected, the agent seeks human feedback to orchestrate the entire workflow, forming a closed-loop enhanced by a *Human-in-the-loop* verification.

Router. Protocol adaptation requests are inherently complex, often requiring multiple operations across diverse components. The Router component addresses this by decomposing user requests into manageable sub-requests and directing them to appropriate downstream agents. Leveraging the LLM's domain knowledge and in-context learning capabilities, the Router classifies sub-requests into four categories: *Adding, Modification, Deleting,* and *Others*. Additionally, few-shot examples are used in prompts to illustrate each category, with the goal of improving classification accuracy. After the query decomposition, except for the *Others* category, the sub-requests are dispatched to their corresponding downstream agents as depicted in Fig. 2.

Memory. To support consistent decision-making and contextual understanding [7], we introduce an explicit memory module that stores static and protocol-specific prior knowledge including the following two aspects:

Entity Description. Each protocol entity type is associated with a concise and functional description based on its key configuration values (e.g., slice thickness, kernel type, and reconstruction orientation).

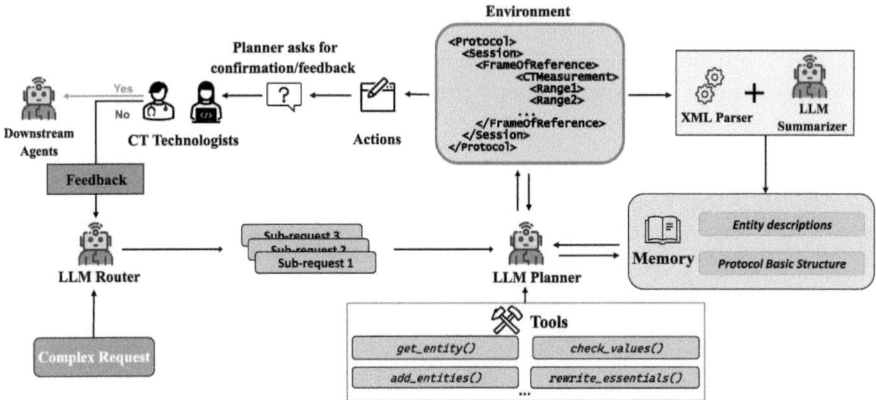

Fig. 2. Workflow of the proposed LLM-based agent for scan protocol modification. *environment:* underlying scan protocol, *actions:* modification proposals.

Basic Protocol Structure. A simplified tree-structured representation of the entire protocol as illustrated in the left hand panel of Fig. 1 is created by parsing the original XML format protocol. Leveraging LLM's instruction-following ability [9], the agent makes and then executes plans by dynamically invoking a comprehensive tool set, which includes entity retrieval, attribute management, and validation.

Planner. For each sub-request, the agent retrieves the relevant entities and associated essentials from the XML-scan protocol context. Building upon that and utilizing the LLM's reasoning over structured data and agent's memory section, the planner concludes a concrete proposal to guide the subsequent steps. The planner's output proposal consists of two parts: a structured set of *actions* – specifying exactly which entities and fields need to be modified – and a human-readable *plan* that explains these steps clearly.

Downstream Agent. To address diverse requests, the agent first categorizes each request by its type and intent, and then applies tailored processing strategies accordingly. When handling `Modification` requests, the agent identifies and filters relevant protocol context (*entity*) and then substitutes specific parameter values (*essential*). For requests that involve adding new components, instead of generating new entities from scratch, the agent leverages existing and contextually relevant entities as templates, refines them optionally based on the user intent, and inserts them into appropriate locations in the protocol tree to preserve semantic and hierarchical consistency. In contrast, for deletion-related requests, the agent removes the specified entities and, when an entity is the sole child of its parent, also removes the parent to prevent structural fragmentation so that the modified protocol remains readable by the scanner.

Human-in-the-Loop. To mitigate errors due to LLM limitations, such as hallucinations and ambiguous reasoning, human feedback is incorporated into the framework: after the planner formulates a proposal, it is submitted to the user for review and verification.

2.3   Evaluation Methods

Syntax Evaluation. As defined in Eq. 1, the *Syntax Correctness Rate (SCR)* is deployed to verify that the modified XML protocols have no syntax errors and thus remain fully readable and interpretable by the scanner's protocol browser.

$$\text{SCR} = \frac{\text{N}_{\text{correct}}}{\text{N}_{\text{total}}}, \tag{1}$$

where $\text{N}_{\text{correct}}$ is the number of modified protocols that can be successfully compiled by the protocol browser, and N_{total} denotes the total number of modified protocols.

Semantic Evaluation. To assess whether the agent accurately interprets the technologist's request and performs faithful adjustments, we further introduce the following measures.

Plan Accuracy: By comparing the updated XML segments with the ground truth, the accuracy of a request type is computed as:

$$\text{Plan Accuracy}_{\text{t}} = \frac{N_{\text{correct}}}{N_{\text{total}}}, \quad t \in \{\texttt{Modification}, \texttt{Adding}, \texttt{Deleting}\} \tag{2}$$

where $N_{correct}$ is the number of requests for which the agent's modification exactly matches the ground truth, N_{total} is the total number of evaluated requests and t is the request label.

Plan Faithfulness: We measure plan faithfulness by instructing an LLM to reconstruct pseudo user requests from agent-retrieved entities and essentials, then computing cosine similarity between text embeddings (OpenAI's *text-embedding-003-small*) of the original and inferred tasks:

$$\text{Plan Faithfulness} = \frac{1}{n} \sum_{i=1}^{n} \text{Similarity}(t, t_i), \tag{3}$$

where t and t_i denote embeddings of original and inverse-engineered tasks, respectively, and n is the number of pseudo tasks.

Retrieval Evaluation. We also evaluate whether the agent can fetch expected context from the structured scan protocol on the entity and the essential levels.

2.4 Experiment Setup

In the experiments, we employ an adult thorax CT protocol comprising a topogram, a spiral scan with multi-planar 2D reconstructions using body kernels (Br40) and 3D reconstructions with lung kernels (Bl60). Identical slice settings are applied across multi-planar orientations, followed by LungCAD post-processing. In our proposed framework, the agent can process two types of input: (1) natural language requests that describe protocol modifications in free-text form; and (2) structured, JSON-formatted requests, curated to align with the output format of an upstream module in our prototype pipeline. Additionally, two API-based LLMs are employed to support the agent: GPT-4o[1] and its lightweight variant GPT-4o-mini(see footnote 1). Additionally, to evaluate deployment scenarios in which cost efficiency and data privacy are prioritized higher, the evaluation also includes a locally deployed open-source LLM (Gemma3:27B[2]), running on an HPC cluster with 8 NVIDIA V100 32GB GPUs. Notably, all of the deployed LLMs are used without any task-specific fine-tuning.

3 Results

3.1 Quantitative Evaluation

Table 1 presents a comparative analysis of the agent's XML syntax-preserving capabilities across LLMs. In general, GPT-4o achieves the highest overall SCR (83%), particularly excelling in structure-sensitive tasks like *Adding* and *Deleting*. While GPT-4o-mini performs on par with GPT-4o in *Deleting* tasks, it underperforms in other categories. Despite a lower general SCR score, Gemma3:27B surpasses GPT-4o-mini in handling *Adding* tasks. The semantic evaluation results in Table 2 and Fig. 3 further confirm the consistent advantage of GPT-4o. GPT-4o-mini shows reduced faithfulness on complex requests, while Gemma3:27B demonstrates selective strengths in *Deleting* and structured input interpretation. Furthermore, Fig. 4 illustrates the agent's retrieval performance, where GPT-4o achieves the best performance in retrieving entity-level protocol context. Beyond the quantitative analysis, some qualitative results can be found in the supplementary material, demonstrating the agent's decision making process during implementing the user request.

3.2 Error Analysis

To highlight the agent's limitations and guide further improvement, it is important to assess the common failure cases that occurred during the agent's attempt to modify the CT scan protocols. A typical reason for errors is misclassification by the router, leading to the dispatch of the task to a wrong downstream agent. Furthermore, we encounter the structural errors that result in dangling or broken

---

[1] API version: 2024-10-28.
[2] Maximum input tokens: 64,000.

Table 1. SCR of the agents with different LLMs. Here, the category of **JSON** refers to the structured system inputs, which are curated to align with the output format of an upstream module in the prototype pipeline. The general score is the arithmetic average over the four categories.

Model	Request Types				*General*
	Modification	Adding	Deleting	JSON	
GPT-4o	**0.89**	**0.80**	**0.75**	**0.80**	**0.83**
GPT-4o-mini	0.67	0.20	**0.75**	0.60	0.57
Gemma3:27B[a]	0.56	0.40	0.50	0.60	0.43

[a] Maximum input tokens: 64,000

Table 2. *Plan Accuracy* of the agents with different LLMs.

Model	Request Type				*General*
	Modification	Adding	Deleting	JSON	
GPT-4o	**0.67**	**0.40**	0.25	**0.60**	**0.52**
GPT-4o-mini	0.33	0.20	0.25	0.20	0.26
Gemma3:27B	0.33	0.20	**0.50**	**0.60**	0.39

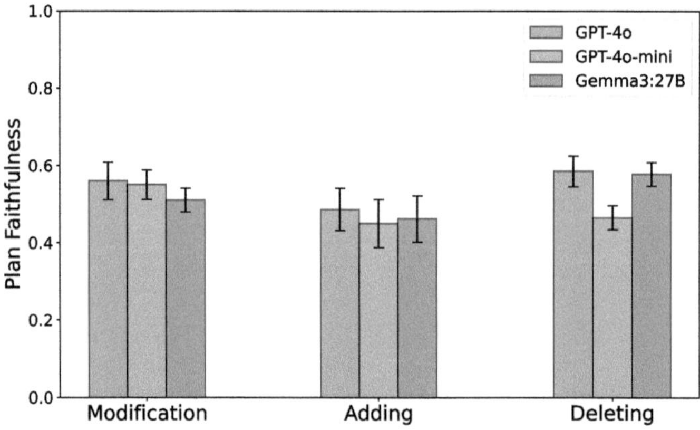

Fig. 3. *Plan Faithfulness* of the agent with different LLMs. Error bars indicate the standard error of the mean. GPT-4o is utilized to generate **10** pseudo-tasks from the agent-retrieved-protocol contexts for each real request.

structures, where the agent modifies the XML hierarchy incorrectly. For instance, the agent could add a new entity under the wrong parent node or delete a sole child node without removing its parent. Additionally, due to the agent's lack of device-specific knowledge, it may fail to understand the inter-dependencies between attributes and use values that are not supported in the given context, which leads to a semantically incorrect protocol modification.

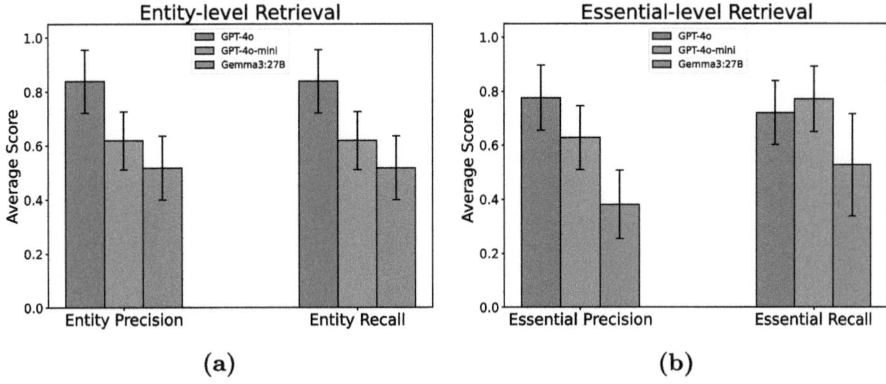

Fig. 4. Agent retrieval performance across different LLMs. Error bars indicate the standard error of the mean.

4 Discussion and Conclusion

This study explores how LLM agents modify device-specific XML-based CT scan protocols from natural or structured clinical inputs, leveraging their reasoning and domain comprehension. Building upon this, to comprehensively assess the effectiveness and robustness of the proposed approach, a multi-aspect evaluation pipeline was designed. Results demonstrate that with appropriate LLMs, the agent can already achieve considerable success in generating syntactically correct and user-intent-aligned scan protocols.

Despite these promising outcomes, our study still faces challenges. While we aim to mitigate the data-scarcity issue by generating complementary pseudo-requests using LLMs and text embedding models, this potentially introduces bias due to inconsistent model performance across different scenarios. Another limitation relates to the lack of comprehensive device-internal knowledge. The proposed framework provides the agents with scan protocol context and background knowledge through input prompts. Nonetheless, the agents have an insufficient understanding of attribute inter-dependencies within protocols, device-specific naming conventions for acquisition and reconstruction parameters, and supported value ranges for protocol variables, which in turn may dynamically depend on the current values of other parameters. This knowledge gap may mislead the agent into incorrect reasoning and thus result in erroneous modifications.

To this end, future research should focus on expanding the agent's access to relevant knowledge sources, in particular semantically meaningful device APIs, but also structured radiology knowledge bases. Developing more robust evaluation methodologies that minimize model-dependent biases would strengthen the assessment of agent performance and reliability in real-world clinical settings.

In conclusion, as a proof of concept, our work demonstrates the feasibility of using LLM agents for retrieving information from real protocol definition files and generating valid, device-compatible modifications for them based on various clinical requests. In addition, it explored how to systematically assess LLM agents' capabilities in a complex and domain-specific setting.

Disclosure of Interests. The authors have no competing interests to declare that are relevant to the content of this article.

References

1. Alexander, R., Waite, S., Bruno, M.A., Krupinski, E.A., Berlin, L., Macknik, S., Martinez-Conde, S.: Mandating limits on workload, duty, and speed in radiology. Radiology **304**(2), 274–282 (2022)
2. Chen, S.A., et al.: TableRAG: million-token table understanding with language models. Adv. Neural. Inf. Process. Syst. **37**, 74899–74921 (2024)
3. D'Antonoli, T.A., et al.: Large language models in radiology: fundamentals, applications, ethical considerations, risks, and future directions. Diagn. Interv. Radiol. **30**(2), 80 (2024)
4. Fang, X., et al.: Large language models (LLMs) on tabular data: prediction, generation, and understanding–a survey. arXiv preprint arXiv:2402.17944 (2024)
5. Hegselmann, S., Buendia, A., Lang, H., Agrawal, M., Jiang, X., Sontag, D.: TabLLM: few-shot classification of tabular data with large language models. In: International Conference on Artificial Intelligence and Statistics, pp. 5549–5581. PMLR (2023)
6. Jiang, J., Zhou, K., Dong, Z., Ye, K., Zhao, W.X., Wen, J.R.: StructGPT: a general framework for large language model to reason over structured data. arXiv preprint arXiv:2305.09645 (2023)
7. Wang, L., et al.: A survey on large language model based autonomous agents. Front. Comp. Sci. **18**(6), 186345 (2024)
8. Xi, Z., et al.: The rise and potential of large language model based agents: a survey. Sci. China Inf. Sci. **68**(2), 121101 (2025)
9. Yuan, S., et al.: EASYTOOL: enhancing LLM-based agents with concise tool instruction. arXiv preprint arXiv:2401.06201 (2024)

MedXpert-CAD: A Multimodal Multi-agentic System for Clinical Imaging Analysis via Model Context Protocol LLM-Driven Agentic Workflows

Mukhlis Raza[1], Saied Salem[1], Afnan Habib[2], Osamah Abdulmahmod[2], Hyunwook Kwon[2], Jamil Hussain[2], and Mugahed A. Al-antari[2(✉)]

[1] Department of Artificial Intelligence, College of AI Convergence, Sejong University, Seoul, South Korea
{raza.mukhlis,said.salem}@sju.ac.kr

[2] Department of Artificial Intelligence and Data Science, College of AI Convergence, Sejong University, Seoul, South Korea
{afnanhabib,abdulmahmod.osamah,sad6601}@sju.ac.kr, {jamil, en.mualshz}@sejong.ac.kr

Abstract. Automating multimodal medical imaging is vital due to the complexity of integrating diagnostic processes and patient interfaces into cohesive clinical workflows. This paper presents MedXpert-CAD, a multimodal multi-agent system (MMAS) that employs large language models (LLMs) and vision encoders for an evidence-based VQA chatbot, supporting multi-task diagnosis for X-ray respiratory diseases and MRI lumbar spinal stenosis (LSS). MedXpert-CAD features four dedicated sub-agents: the supervisor agent, which queries users and directs them to modality-specific tools like X-ray respiratory disease analysis or LSS prediction; the online search agent, which retrieves data from PubMed or Google Search when needed; the X-ray expert agent, which processes images to provide multi-label classification, bounding box detection via saliency mapping, VQA responses, and structured reports; and the LSS expert agent, which handles MRI or DICOM files, offering sagittal and axial view segmentation, spinal pathological measurements, disk herniation diagnosis, and structured reports. MedXpert-CAD integrates the Model Context Protocol (MCP) with specialized agents to overcome AI model limitations, ensuring interoperable architecture and evidence-based diagnostics. The evaluation used the DeepEval framework and LLM-as-a-judge methodology, assessing task completion (TC), contextual relevance (CR), and tool accuracy (TN). GPT-4o showed exceptional performance with TC rates of 90% (X-ray) and 96% (MRI), and 100% TN in the single-agent framework, compared to Llama-3.2's 61% TC (X-ray) and 63% TC (MRI) with 100% TN. In the multi-agent framework, GPT-4o achieved 80% TC and 87% TN, surpassing Llama-3.2's 65% TC and 47% TN, with contextual relevance peaking at 83%. Segmentation efficacy peaked on sagittal views with a Dice Similarity Coefficient (DSC) of 96.43% and 99.66% accuracy, outperforming axial views (DSC 92.71%, accuracy 99.88%). Report generation metrics highlighted GPT-4o's superiority with METEOR 29%, ROUGE-L 19%, FrugalScore 57%, Blue-1 31%, and BERTScore-F1 84% for chest X-ray, and METEOR 18%, ROUGE-L

J. Qiu et al. (Eds.): Agentic AI 2025/CMLLMs 2025/CREATE 2025, LNCS 16147, pp. 55–64, 2026.
https://doi.org/10.1007/978-3-032-06004-4_6

11%, FrugalScore 59%, Blue-1 11%, and BERTScore-F1 82% for LSS. This system enhances patient care and empowers healthcare professionals with advanced imaging tools.

Keywords: Agentic AI · Large Language Model (LLM) · Model Context Protocol (MCP) · Medical Multimodal Integration · Reasoning-Action events · VQA AI chatbot

1 Introduction

Large language models (LLMs) serve as fundamental components of the intelligent agents providing a complex reasoning capability towards actual-world scenario sup-port [1]. The combination of zero and few-shot learning features in LLMs makes it possible to automate complex procedures which need detailed planning and multi-stage reasoning [2]. Healthcare organizations use conversational AI systems to improve operations by addressing experts' shortages and managing growing workloads. Like human intelligence, agentic AI integrates reasoning and actions to dynamically process multimodal external knowledge [3–5]. The GPT model is a multimodal AI system that enhances its processing capabilities by integrating the pair of image and text [6]. Agentic systems face difficulties when implementing multiple clinical decision-making (CDM) in a unique system to deal with various medical imaging modalities such as chest X-ray (CXR) and MRI scans [7]. Therefore, the effective techniques are needed to integrate knowledge of imaging benefits and risks, enabling the synchronization and management of complex medical data from CXRs and lumbar MRI 3D DICOM volumes [8]. Integrating various medical imaging data remains a significant challenge in developing a unified AI system that meets the needs of physicians, medical staff, and healthcare organizations for improved diagnosis and treatment planning. To address these challenges, the MedXpert-CAD system has been proposed and developed as a multimodal multi-agent system (MMAS) enhanced with Model Context Protocol (MCP) standardization for seamless medical image integration. MCP represents an open standard, where multiple AI applications can connect with various data sources through a universal protocol rather than requiring custom connectors for each integration. In healthcare applications, MCP enables AI systems to access real-time clinical data, external medical databases, and specialized tools through standardized interfaces, addressing the critical interoperability challenges that have hindered scalable AI deployment in clinical environments. Leveraging LLMs and vision encoders, the MCP-enhanced MMAS platform powers an evidence-based VQA support chatbot, forming the foundation of MedXpert-CAD. Our main contributions are:

1. MedXpert-CAD, a novel explainable multi-agent chatbot MCP-based CAD system, analyzes multiple medical image modalities with transparent, evidence-based reasoning. Four autonomous sub-agents interact synchronously to improve functionality, integration, and doctor trust in AI-driven diagnostics.

2. An innovative deep search feature integrates with an online search agent accessing PubMed and Google Scholar, as well as a knowledge base (Vector and Graph databases), enabling context-aware diagnoses by retrieving accurate medical knowledge for image analysis and text reporting.
3. The custom MCP server for chest X-ray expert agent provides advanced analysis through multi-label disease classification, abnormality localization, saliency maps for visual explanations, and structured radiology reports, surpassing conventional classification methods.
4. The custom MCP server for LSS expert agent enhances lumbar MRI assessment with advanced analytics, including axial and sagittal segmentations, disk herniation diagnosis, and quantitative pathological measurements to improve complex spinal health conditions.

2 Related Works

The recent advancements in tool-augmented LLMs demonstrate substantial improvements which advance their practical operation capabilities for better performance and efficiency [9]. The recent multiple advances converge to enhance LLMs' ability to use the diagnostics tools correctly through specific agents for multiple tasks generalization and efficacy purposes. Parisi et al. (2022) [10] demonstrate knowledge-intensive reasoning using a text-only augmentation mechanism combined with iterative self-play methods. In 2024, Chen et al. [11] built an inference trajectory optimization framework based on impractical decision tree exploration to generate better performance and generalization capability. Healthcare implementations of tool-augmented LLMs create valuable opportunities aiming to solve medical problem complexities with multimodal image and non-image sensitive datasets. Li et al. (2024) [12] proposed their MMedAgent system to handle various medical operations with some tools to process different medical data modalities. Shen et al. (2024) [13] showed how MATMCD represents a multi-agent system which applies integrated multimodal data to perform knowledge-driven causal discovery within healthcare settings. According to Yuan et al. (2024) [14], the medical application of LLMs shows promising agentic AI systems where personalised models and complex clinical decision-making based on LLM capabilities in medical science is crucial. The processing ability of multimodal LLMs for medical information depends heavily on ongoing optimization and ethical oversight. None of the frameworks provides a unified Model Context Protocol (MCP) that allows clinical real time image processing, database queries and reports generation; MedXpert-CAD addresses this need with elastic MCP servers.

3 Methodology

MedXpert-CAD is a multi-modality multi-agent system (MMAS) that is a tool tailored to analyze clinical images and non-image data, incorporating cutting-edge diagnostic facilities to handle the complexities of medicine represented in Fig. 1. The system is characterized by special AI agents, with deep searches and evidence-based logical reasoning, guaranteeing that intensive medical work does not interfere. The architecture has three different MCP servers namely: Deep Search MCP Server that retrieves literature in terms of PubMed and Google Search, the Chest X-ray MCP Server that classifies them, creates a report, visualizes the heatmap, and detects the bounding box, and the Lumbar Spine Stenosis MCP Server that retrieves MRI vector database, processes DICOM, and detects the stenosis. With the help of the MCP Client and MCP Protocol, these servers connect with the Supervisor Agent, X-ray Expert Agent, LSS Expert Agent, and Deep Search Agent. The workflows involve knowledge sources, such as PubMed and Google Scholar, as well as agent tools and LLMs, which allow a robust and extensible basis of the diagnostic process and include state-of-the-art workflows, such as the work of X-ray experts, the work of LSS experts, and deep search operations.

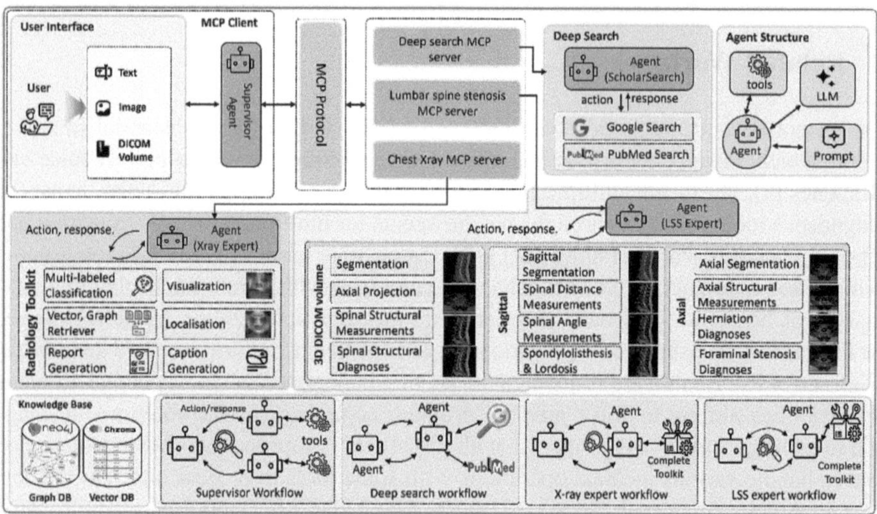

Fig. 1. MedXpert-CAD Multi-Agent Architecture: Workflow of the Four-Step Clinical Imaging Analysis Pipeline (User Input → Supervisor Routing → Tool Execution → Aggregated Output)

3.1 Overall Systematic Architecture

The hierarchical structure of MedXpert-CAD includes four dedicated agents along with the three custom MCP servers, which are designed to improve clinical imaging work-flows. Acting as the main interface of the system the supervisor agent executes the specified MCP server according to the user's question. The X-ray expert Agent analyzes frontal and/or lateral X-ray image views from MIMIC-CXR [15] using our radiology toolkit to provide multiple services: multi-label classifications, explainable Grad-CAM saliency maps to localize the pathological affected areas on the entire images, and a structured medical diagnostic report for comprehensive and rich diagnostic and treat-ment plans. The LSS expert Agent specializes in lumbar spinal MRI [16] analysis to provide automatic and accurate sagittal and axial image segmentations, related patho-logical measurements, and generated a comprehensive structured medical diagnostic report. Our LSS toolkit could handle the 3D DICOM volume in addition to a single axial or sagittal images towards the quantitative analyses of stenosis grading, disc herniation, lumbar lordosis, and spondylolisthesis. Data splitting technique of 70% training, 20% validation and 10% testing was used for both modalities, with pre-processing by resizing, augmentation and normalization prior to analysis.

3.2 Action-Observation-Thought (AOT)

MedXpert-CAD deploys sub-agents to complete AOT [4] cycles by interacting with the MCP-standardised tool interfaces by employing reasoning to undertake action steps, and gathers the tool observation to advance the thoughts. In contrast to pure chain-of-thought reasoning which uses only text generation as a connective element, our AOT framework incorporates external medical tool execution used in the process of reasoning. The *Action → Observation → Thought* cycle leads to the dynamic use of tools implemented in MCPs (DICOM processing, vector database query and retrieval query, classification models), as well as observing the structured medical output, and integrating it into the next iteration of processing. MCP integration makes it possible to have standardized tool invocation such that agents interact with special medical capabilities using uniform protocols. When the agent parses the lumbar MRI data, it invokes DICOMProcessor tool, views the extracted metadata and segmented labellings, examines the specified artifacts of concrete medical evidence and utilizes it to conduct diagnostic reasoning, without depending solely on parametric knowledge.

3.3 Evaluation Criteria and Executional Learning Environment

The evaluation framework employs *DeepEval* the unit test uses LLM as a judge consists of three essential performance metrics for complete agentic system assessment as summarized in Eqs. (1–3). First, *TaskCompletion* (TC) determines the agent's accomplishment tasks. Second, the *ContextualRelevancy* (CR) checks the correctness of the generated action responses of the available contexts to achieve relevant and coherent output. Third, *ToolCorrectness* (TN) metric that assesses LLM agent's function/tool calling ability. This set of metrics combines delivering comprehensive multi-dimensional assessments between goal achievement, proper context placement, and delivering accurate information.

$$TC = AlignmentScore(Task, Outcome) \tag{1}$$

$$CR = \frac{Number\ of\ Relevant\ Statements}{Total\ Number\ of\ Statements} \tag{2}$$

$$TN = \frac{Number\ of\ Correctly\ Used\ Tools}{Total\ Number\ of\ Tools\ Called} \tag{3}$$

The experiments are conducted on a PC equipped with an i7-1400K processor, 64 GB of memory, and an Nvidia RTX 4070 Ti Super GPU.

4   Results

The evaluation process of MedXpert-CAD is performed on both benchmark public medical datasets of chest X-ray and lumbar spine MRI scans [16]. The research examines single-agent and multi-agent frameworks by using two LLMs: GPT-4o [17] and Llama-vision3.2 [18]. The results are summarized in Table 1. GPT-4o achieves Task Completion (TC) scores of 80% for all agents' scenario for CXR and MRI tasks, respectively. In comparison, Llama-Vision 3.2 attains TC scores of 65% for CXR and MRI interpretations. GPT-4o exhibits higher tool correctness than Llama-Vision 3.2, with CXR and MRI recorded at 87.6%, respectively. Additionally, GPT-4o outperforms Llama-Vision 3.2 in Contextual Relevancy (CR), scoring 83% for CXR and for MRI. Meanwhile, Llama-Vision 3.2 shows a CXR and MRI tool correctness of 47% with CR scores of 68% for CXR and MRI. Overall, GPT-4o appears to benefit from superior training data selection and model optimization strategies, likely due to its proprietary advancements. The minimal fluctuation across different modalities suggests that multi-agent integration may suffer from increased complexity or error accumulation during reasoning-action processes. Our multi-label classification tool leverages the ensemble capabilities of DenseNet [19] and ResNet [20] to effectively diagnose 14 respiratory diseases: Atelectasis, Cardiomegaly, Consolidation, Edema, Enlarged Cardiomediastinum, Fracture, Lung Lesion, Lung Opacity, No Finding, Pleural Effusion, Pleural Other, Pneumonia, Pneumothorax, Support Devices. For both LSS sagittal and axial segmentation tools, we employ ensemble hybrid approaches combining CNNs and transformers using the random search automatic hyperparameter optimization approach [21] For sagittal images, the top three selected models are U-Net++, U-Net, and SwinUNETR, while for axial images the top models are SwinUNETR, Attention U-Net, and U-Net++. Additionally, extensive experiments have been conducted to select, optimize, and develop single CXR and MRI tools, ensuring the refinement and finalization of their AI structures. Figure 2 shows the summary of CXR multi-label classification results as well as the segmentation performance of LSS axial and sagittal images.

Table 1. Performance comparison of MedXpert-CAD for single-agent against multi-agent frameworks using GPT-4o and Llama3.2 LLMs across X-ray and MRI Tasks.

Framework	Model	Modality Agent	TC (%)	CR (%)	TN (%)
Single-Agent	GPT-4o	X-ray Expert Agent	90	95	100
		MRI Expert Agent	96	59	100
	Llama-3.2-11B-Vision	X-ray Expert Agent	61	66	100
		MRI Expert Agent	63	47	100
Multi-Agents	GPT-4o	All Agents	80	83	87
	Llama-3.2-11B-Vision	All Agents	65	68	47

As shown in Fig. 2, GPT-4 produces better accuracy, and coherence reports for X-ray images than Llama-3.2-Vision. For LSS tasks, GPT-4o demonstrates a slight

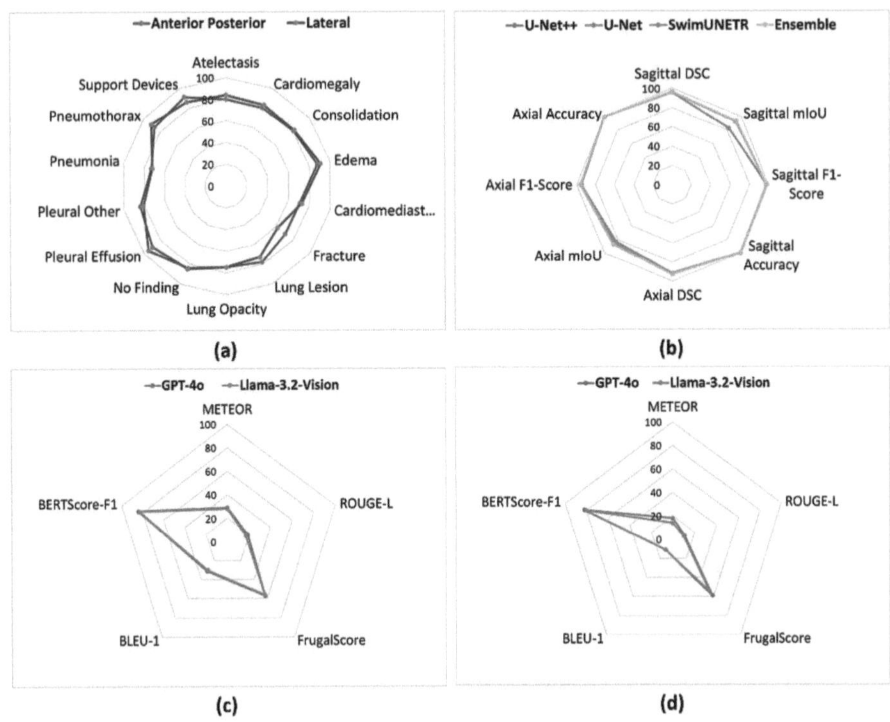

Fig. 2. Performance evaluation of the proposed toolkits on the MedXpert-CAD system: (a) multi-class classification of the CXR tool for Anterior-Posterior (AP) and Lateral image views, (b) multi-class segmentation for sagittal and axial tools, and (c)-(d) Medical text report generation results for CXR and LSS agents, respectively.

advantage over Llama-3.2-Vision. MedXpert-CAD demonstrates its capability to use state-of-the-art LLMs effectively in handling multiple medical imaging simultaneously.

5 Discussion

MedXpert-CAD serves an end-to-end chatbot platform, with the ability to have smoother diagnostic workflow, with the capability to process 2D images and 3D DICOM volumes, including upload, classification, segmentation, and measurement, as well as forecast, prediction, and structured reporting (Fig. 2). The Model Context Protocol (MCP) has been integrated to normalize the tool interface allowing efficient coordination among a proficient agent specializing in real time image processing or database queries. Llama-3.2 test results indicate that it is an equivalent replacement of GPT-4o when considering open-source evaluation of finances, and it does not perform worse. Figure 2(a) provides the results on the classification of the chest X-ray using the analysis of the AUC curves, showing the high robustness of diagnostic findings at all modalities. Figure 2(b) demonstrates segmentation model performance, our ensemble model performed better than single methods, especially in sagittal and axial slices. Figures 2(c) and (d) show the report generation results, where (c) implies better quality of the chest X-rays report (e.g., improved METEOR and BERTScore-F1) and (d) represents the competitive results of LSS report generation. To support the robust tool invocation and evidence-based outputs, we designed three MCP servers, including Chest X-ray, Lumbar Spine Stenosis, and Deep Search, as illustrated in Figs. 3 and 4, which indicated dynamic interactions with users. Multimodal data is processed by the chatbot, which provides dependable results within 5–9 s, including tool interactions and analyses. MIMIC-CXR and Lumbar Spine MRI datasets were utilized; however, broader multi-institutional validation remains necessary. Inference is completed within a few seconds indicating potential for real-time application. System-level, modular agent integration constitutes the primary innovation, rather than novel technical modules. Comparative analysis with frameworks such as AutoGen, DyLAN and MDAgent is planned for future work extension.

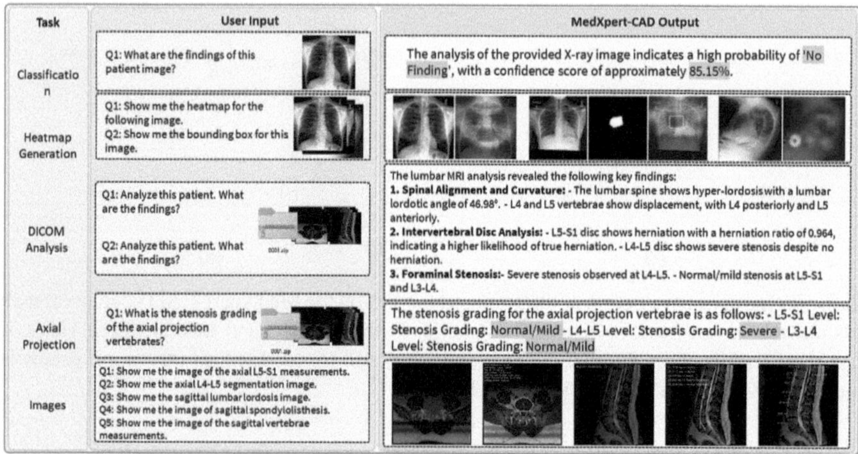

Fig. 3. Example Input-Output Workflow of MedXpert-CAD: User Queries and System Responses for X-ray and MRI Analysis (Thoughts, Actions, and Values).

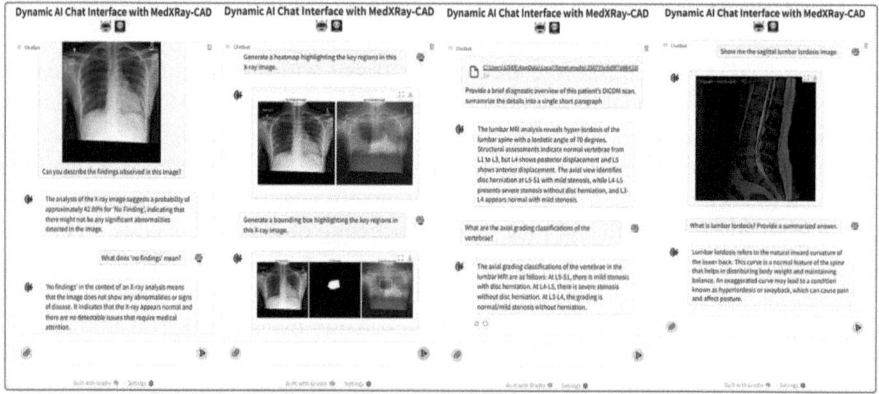

Fig. 4. Chatbot illustration, how users can interact and the chatbot response.

6   Conclusions

MedXpert-CAD represents a significant advancement in AI-driven clinical imaging analysis, introducing a multimodal multi-agent system (MMAS) that integrates LLMs, vision encoders, and domain-specific tools to streamline complex diagnostic workflows. Key innovations include Modular Architecture, Evidence-Based Diagnostics, and Interpretability. While the framework excels in static imaging modalities (X-ray, MRI), limitations include its current exclusion of dynamic modalities like ultrasound and dependency on continual fine-tuning for emerging datasets. By combining open-source adaptability with clinician-centric design, MedXpert-CAD sets a new standard for scalable, trustworthy AI in healthcare, empowering medical professionals with advanced, interpretable diagnostic tools.

Acknowledgements. This work was supported by the National Research Foundation of Korea (NRF) grant funded by the Korean government (MSIT) (No. RS-2023-00256517).

Disclosure of Interests. Not applicable.

References

1. Wu, Q., et al.: Autogen: enabling next-gen LLM applications via multi-agent conversation framework. *arXiv preprint* arXiv:2308.08155 (2023)
2. Zhang, S., et al.: Opt: Open pre-trained transformer language models. *arXiv preprint* arXiv:2205.01068 (2022)
3. Alderson-Day, B., Fernyhough, C.: Inner speech: development, cognitive functions, phenomenology, and neurobiology. Psychol. Bull. **141**(5), 931 (2015)
4. Yao, S., et al.: React: synergizing reasoning and acting in language models. *arXiv preprint* arXiv:2210.03629 (2022)
5. Wang, Y., Wu, Z., Yao, J., Su, J.: Tdag: a multi-agent framework based on dynamic task decomposition and agent generation. Neural Netw. 107200 (2025)
6. OpenAI, R.: Gpt-4 technical report. arxiv 2303.08774. View in Article, **25** (2023)
7. Liu, J., et al.: Medchain: bridging the Gap between LLM Agents and Clinical Practice through Interactive Sequential Benchmarking. *arXiv preprint* arXiv:2412.01605 (2024)
8. Patil, N.S., et al.: Artificial intelligence Chatbots' understanding of the risks and benefits of computed tomography and magnetic resonance imaging scenarios. Can. Assoc. Radiol. J. 08465371231220561 (2024)
9. Lu, P., Chen, B., Liu, S., Thapa, R., Boen, J., Zou, J.: OctoTools: an Agentic Framework with Extensible Tools for Complex Reasoning. *arXiv preprint* arXiv:2502.11271 (2025)
10. Parisi, A., Zhao, Y., Fiedel, N.: TALM: tool augmented language models. *arXiv preprint* arXiv:2205.12255 (2022)
11. Chen, S., et al.: Advancing tool-augmented large language models: Integrating insights from errors in inference trees. *arXiv preprint* arXiv:2406.07115 (2024)
12. Li, B., et al.: Mmedagent: learning to use medical tools with multi-modal agent. *arXiv preprint* arXiv:2407.02483 (2024)
13. Shen, C., Chen, Z., Luo, D., Xu, D., Chen, H., Ni, J.: Exploring Multi-Modal Integration with Tool-Augmented LLM Agents for Precise Causal Discovery. *arXiv preprint* arXiv:2412.13667 (2024)

14. Yuan, M., et al.: Large language models illuminate a progressive pathway to artificial intelligent healthcare assistant. Medicine Plus, 100030 (2024)
15. Johnson, A.E., et al.: MIMIC-CXR-JPG, a large publicly available database of labeled chest radiographs. *arXiv preprint* arXiv:1901.07042 (2019)
16. Sudirman, S., et al.: Lumbar spine MRI dataset. Mendeley Data **2**, 2019 (2019)
17. Hurst, A., et al.: Gpt-4o system card. *arXiv preprint* arXiv:2410.21276 (2024)
18. Touvron, H., et al.: Llama: Open and efficient foundation language models. *arXiv preprint* arXiv:2302.13971 (2023)
19. Huang, G., Liu, Z., Van Der Maaten, L., Weinberger, K.Q.: Densely connected convolutional networks. In: Proceedings of the IEEE Conference on Computer Vision and Pattern Recognition, pp. 4700–4708 (2017)
20. He, K., Zhang, X., Ren, S., Sun, J.: Deep residual learning for image recognition. In: Proceedings of the IEEE Conference on Computer Vision and Pattern Recognition, pp. 770–778 (2016)
21. Salem, S., Mukhlis, R., Katar, O., Ertuğrul, B., Yildirim, O., Al-Antari, M.A.: A novel AI-based hybrid ensemble segmentation CAD system for lumber spine stenosis pathological regions using MRI axial images. In: 2024 8th International Artificial Intelligence and Data Processing Symposium (IDAP), pp. 1–5. IEEE (2024)

Agentic Hypergraph Search
for Ontology-Grounded Medical Reasoning

William Kingston Xie[1], Donghyun Lim[2(✉)], and Seungwoo Schin[2]

[1] EKA Hospital, Salem, India
[2] Mazigle, Seoul, South Korea
`{donghyun,seungwoo.schin}@mazigle.ai`

Abstract. Large Language Models (LLMs) have shown remarkable capability in answering complex questions, but their reasoning processes remain largely ungrounded. This opacity is problematic in domains like medicine, where ungrounded answers can erode user trust. We address the lack of grounded reasoning in LLMs by introducing an approach that integrates LLMs with a reasoning framework based on an external ontology. In our method, we employ an agentic reasoning strategy: the LLM traverses a hypergraph to iteratively gather grounded evidence and infer answers in a transparent manner. The reasoning capabilities are evaluated on the MedQA medical question answering benchmark. Experimental results show that our agent, equipped with hypergraph search, achieves competitive accuracy with state-of-the-art baselines while providing ontology-grounded reasoning traces.

Keywords: Ontology · hypergraph search · retrieval augmented generation · medical agent

1 Introduction

Large Language Models (LLMs) have achieved remarkable performance on a variety of tasks, but their reasoning processes remain largely opaque. This lack of white-box interpretability and debuggability is a critical concern, especially in domains where trust and error analysis are paramount. Symbolic approaches, such as those grounded in ontology, offer transparent, white-box reasoning capabilities but often face practical limitations in real-world deployment.

In this work, we propose a novel framework that bridges the gap between both paradigm, aiming for the best of both worlds. The key idea is to use LLMs themselves to automatically construct an ontology-like knowledge base (as a hypergraph) from unstructured text, and then use that structure to guide the LLM's reasoning process in a transparent way. Our approach can be seen as a midpoint in the progression from RAG to classical ontologies: starting from standard RAG [5], adding structured knowledge as in GraphRAG [2], granting the model an agentic multi-step reasoning capability akin to ReAct [8], and finally approaching the formal rigor of an ontology (Fig. 1 illustrates this spectrum). By occupying this middle ground, our method offers significantly increased whiteboxness and debuggability compared to typical LLM pipelines, while retaining

J. Qiu et al. (Eds.): Agentic AI 2025/CMLLMs 2025/CREATE 2025, LNCS 16147, pp. 65–73, 2026.
https://doi.org/10.1007/978-3-032-06004-4_7

much of the flexibility of neural approaches. To our knowledge, this is the first work to demonstrate that an LLM-generated ontology can serve as an effective reasoning substrate for the LLM itself, without relying on any classical reasoner or external ontology tool. We hypothesize that such a system can achieve competitive, even superior, performance to state-of-the-art black-box models like GPT-4 [6] on complex tasks, all while delivering human-understandable reasoning traces and the ability to diagnose errors. Indeed, emerging studies support this hypothesis: for example, injecting knowledge graph reasoning paths into LLM prompting has enabled smaller models to outperform GPT-4 on multi-hop question answering tasks [7], underlining the power of structured knowledge for boosting both accuracy and faithfulness.

1.1 Related Work

Retrieval-Augmented Generation (RAG). Retrieval-Augmented Generation (RAG) [5] enhances LLMs by retrieving relevant documents from an external corpus to supplement generation. This approach improves factuality and allows evidence inspection but retains a key weakness: the reasoning remains implicit in the model's parameters. To improve interpretability and multi-hop reasoning, researchers have proposed GraphRAG variants [2], which use structured knowledge bases—like knowledge graphs or ontologies—as the retrieval source. These graph-enhanced models improve coherence by enabling entity linking and relational context, providing a mild increase in transparency and multi-step reasoning capabilities.

Agentic RAG. Beyond static retrieval, a natural extension is an agentic variant of GraphRAG, where the LLM dynamically plans and executes multi-step graph traversals. This mirrors recent frameworks like ReAct [8], in which reasoning steps are interleaved with tool use. In our case, the LLM uses a knowledge graph as its external tool, planning reasoning paths across entities and relations to arrive at a conclusion. This method brings us closer to fully traceable, symbolic inference, while preserving the flexibility of LLM-driven planning.

Medical Ontology. In safety-critical domains like medicine, symbolic representations are prevalent. Ontologies such as SNOMED CT [1] and UMLS serve as standardized vocabularies that encode relationships among clinical entities. These systems enable transparent reasoning by aligning decision-making with structured knowledge, facilitating traceability and consistency in medical QA and decision support.

2 Methodology

2.1 Overview

Our system aims to enhance the transparency and debuggability of LLM-based reasoning through an ontology-guided pipeline. At a high level, the process involves three major stages (illustrated in Fig. 1):

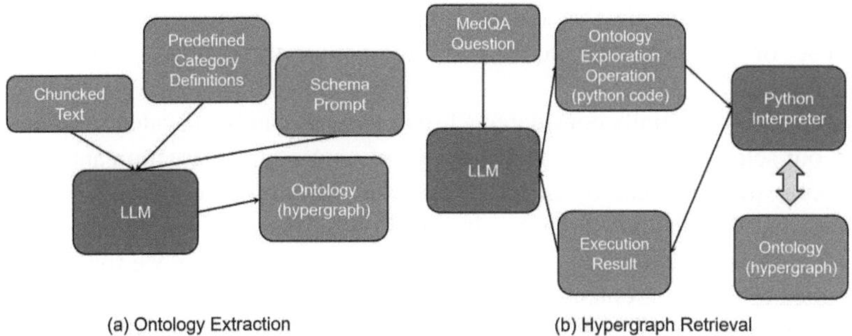

(a) Ontology Extraction   (b) Hypergraph Retrieval

Fig. 1. Illustrated whole process.

1. **Ontology Construction:** The text is chunked, then passed to the ontology construction process. Using a combination of prompt engineering and manual schema design, the LLM extracts entities and their relations from each chunk to build a domain-specific ontology. This structure is encoded as a *hypergraph*, capable of representing higher-order medical relations involving multiple entities. Constructed entities are than deduplicated using simple heuristic.
2. **Agentic Hypergraph Search:** An LLM agent is equipped with a suite of tools to iteratively query the ontology. These include similarity-based entity search, hyper-edge retrieval, and latent connection discovery, allowing the agent to perform multi-step reasoning over structured medical knowledge.
3. **Answer Generation:** The LLM integrates structured knowledge retrieved from the ontology with natural language prompts to generate answers, guided by chain-of-thought reasoning for interpretability.

This pipeline is mostly model-driven, with all ontology construction and query planning performed using LLM calls alone—no classical reasoner or symbolic inference engine is used. The system is implemented in Python, with agentic behaviors grounded in executable code generated by the model itself, enabling a flexible, transparent alternative to black-box inference.

2.2   Ontology Construction

Ontology Schema. The ontology schema underlying this project was crafted through extensive deliberation. Recognizing the inherent complexity of medical statements—which frequently involve multiple entities and intricate interrelations—we adopted a hypergraph structure as the foundational schema.

By leveraging a hypergraph-based ontology, we ensure that the core semantic relationships are preserved and readily accessible for both human interpretation and machine-driven inference.

Entity Categorization. A critical step in the construction of our ontology schema involved the systematic determination of categories to which each entity, or node, belongs. Given the unique characteristics of the medical domain, it was imperative to define a set of categories that would faithfully capture the diversity and specificity of clinical concepts.

To this end, we initially considered a broad array of candidate categories, iteratively refining our selection through repeated cycles of ontology construction and evaluation. This process was informed by empirical analysis, including the merging of infrequent categories and the subdivision of overly broad ones. Through this process of trial and error, we ultimately converged on a set of eight categories, which we found to be optimally suited to the requirements of our research objectives and the nuances of the medical domain.

This carefully curated set of categories provides a balanced and expressive framework for representing the entities within our hypergraph-based ontology, ensuring both comprehensiveness and practical utility in downstream applications (Fig. 2).

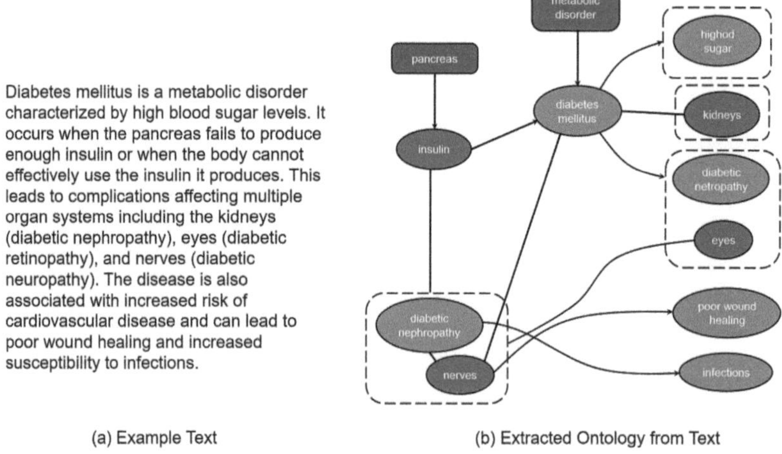

(a) Example Text   (b) Extracted Ontology from Text

Fig. 2. Example ontology construction.

Automated Ontology Construction via Large Language Models. With the schema firmly established, we proceeded to the automatic construction of the ontology, leveraging the capabilities of large language models (LLMs). This process was driven by prompt engineering and iterative refinement, as detailed in `prompt.py`. By systematically engaging the LLM, we extracted ontology elements from each text chunk, instructing the model to identify and organize entities and relations in accordance with our predefined schema.

This approach exemplifies the practical synergy between advanced language models and domain-specific schema design, facilitating rapid and scalable ontology construction in the medical domain.

Deduplication. Following the initial construction of the ontology, we addressed the challenge of entity deduplication. While vector-based approaches—such as leveraging entity name embeddings—were considered, empirical findings indicated that simple string matching was highly effective in eliminating redundant entities. Through this straightforward method, we were able to reduce the total number of entities by approximately 30%. This substantial reduction underscores the efficacy of basic string-based deduplication in our context, rendering more complex embedding-based techniques largely unnecessary for our purposes.

2.3   Agentic Hypergraph Search

At the heart of this work lies the concept of agentic hypergraph search, which constitutes the principal innovation and core contribution of our study. Rather than relying on static or exhaustive search algorithms, we empower a Large Language Model (LLM) agent to autonomously navigate the ontology, leveraging a suite of specialized tools designed for flexible and context-aware exploration. This agentic approach enables dynamic reasoning over complex medical knowledge graphs, facilitating the discovery of both explicit and implicit relationships within the data. An example case of the reasoning process through hypergraph search is illustrated in Fig. 3. We provide the LLM agent with three tools for navigating through the hypergraph.

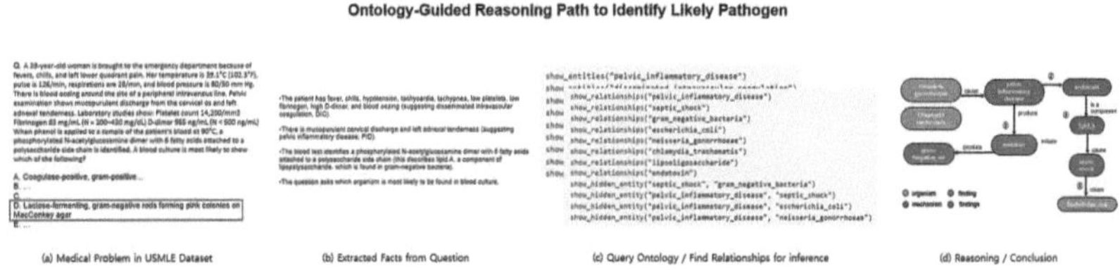

Fig. 3. Example problem reasoning process. Explicit process and executed python code is illustrated in the appendix.

Entity Similarity Search via Vector Embeddings. The first tool equips the LLM agent with the capability to identify entities similar to a given query entity. By transforming entity names into vector representations, the agent can efficiently retrieve and rank entities based on semantic similarity. This mechanism allows the agent to figure out the proper starting points of hypergraph search.

Hyperedge Retrieval for Entity-Centric Exploration. The second tool enables the agent to enumerate all hyperedges—representing medical statements or relationships—that involve a specified entity. This functionality allows for comprehensive contextualization of an entity within the broader ontology, as the agent can systematically retrieve and analyze all statements in which the entity participates. Such entity-centric exploration is essential for understanding the multifaceted roles that clinical concepts play within complex medical narratives.

Discovery of Hidden Linking Entities. The third tool empowers the agent to uncover hidden entities that serve as bridges between two given entities. By searching for entities that are jointly associated with both input entities, the agent can reveal latent connections that may hold the key to resolving challenging or implicit medical queries. This capability is particularly valuable for hypothesis generation and the identification of non-obvious relationships that could inform clinical decision-making or research.

3 Experiments

We conducted a comprehensive evaluation of our approach on a standard medical QA benchmark. The following subsections detail the dataset characteristics, evaluation protocol, baseline setup, ablation analysis, and key results.

3.1 Dataset

Our experiments utilize the **MedQA** dataset, a benchmark of USMLE-style multiple-choice questions introduced by [3]. MedQA consists of *12723* English questions drawn from United States Medical Licensing Examination (USMLE) materials, among which we have only used the development set and test set. Each item contains a patient vignette plus four or five answer options with a single correct choice, testing expert-level clinical reasoning. For the base corpus, we selected *First Aid for the USMLE Step 1* [4] textbook, which is the text provided in the **MedQA** dataset. The text is chunked in sentence level, than passed to ontology builder.

3.2 Baseline

The baseline for performance evaluation is the conventional approach of directly prompting the LLM to answer questions without access to external knowledge or structured augmentation. In this setting, we explicitly instruct the LLM to use chain-of-thought reasoning, requiring it to articulate intermediate reasoning steps before arriving at a final answer. However, no ontology-based augmentation is provided. This setup enables a controlled comparison that isolates the effect of explicit reasoning guidance, independent of any external knowledge integration.

3.3 Ablation Study

To assess the importance of each component in our agentic ontology search framework, we conducted an ablation study focused specifically on the hidden entity search functionality. Among the three core tools provided to the LLM agent—(1) entity similarity search via vector embeddings, (2) hyperedge retrieval for entity-centric exploration, and (3) hidden entity search for discovering nontrivial linking entities—we considered the first two functionalities essential for the agent's operation and did not ablate them. Only the hidden entity search, which plays a more auxiliary role, was removed during ablation.

4   Results

The results of our experiment is organized in 3 tables. Table 1 Organizes experiment done on baseline models, including gpt-4.1 and gpt-4o models.

Table 1. Baseline performance showing accuracy and the number of tokens used (checkpoint suffix **-2025-04-14**; [†] uses **-2024-11-20**).

LLM	Acc	In/itm	Out/itm
gpt-4.1-nano	0.79	303.23	555.67
gpt-4.1-mini	0.92	303.23	585.10
gpt-4.1	0.92	303.23	467.97
gpt-4o[†]	0.79	302.89	714.89

Table 2. Proposed method ("ours") performance showing accuracy and the number of tokens used (checkpoint suffix **-2025-04-14**; [†] uses **-2024-11-20**)

LLM	Acc	In/itm	Out/itm
gpt-4.1-nano	0.75	3723.87	540.51
gpt-4.1-mini	0.76	15856.55	757.33
gpt-4.1	0.86	6313.40	514.76
gpt-4o[†]	0.81	9431.30	734.44

Table 3. Ablation study: proposed method *without* hidden-entity search showing accuracy and the number of tokens used (checkpoint suffix **-2025-04-14**; [†] uses **-2024-11-20**).

LLM	Acc	In/itm	Out/itm
gpt-4.1-nano	0.73	2863.94	389.15
gpt-4.1-mini	0.79	10534.00	671.65
gpt-4.1	0.86	5826.22	523.44
gpt-4o[†]	0.82	7206.63	657.10

Across all checkpoints, baseline accuracy scales predictably with model size, yet only the gpt-4o checkpoint derives a measurable benefit from ontology-guided reasoning. As reported in Table 1, gpt-4o[†] attains a baseline accuracy of 0.79. When coupled with our full ontology pipeline (Table 2), the same checkpoint rises to 0.81— a *two-percentage-point* absolute gain that is noteworthy given the already high performance ceiling of frontier models. In contrast, smaller variants (gpt-4.1-nano, `gpt-4.1-mini`) experience accuracy decrements once the heavy ontology context is introduced, indicating that they are more susceptible to the signalâĂŞnoise imbalance created by long prompts.

The ablation results (Table 3) further corroborate this finding. Removing the hidden-entity search step trims approximately 24% of the additional input tokens for `gpt-4o` while *increasing* accuracy slightly to 0.82. Hence, for large-capacity models the ontology's explicit relational structure is beneficial, but only up to the point where auxiliary tool calls cease to add novel information. These observations suggest that frontier models are able to exploit curated, high-salience graph edges for disambiguation yet remain robust enough to forego extra hops that primarily introduce prompt overhead. Consequently, a selectively pruned ontology—retaining first-order relations but omitting costly multi-hop expansions—appears to be the most effective configuration for next-generation LLMs such as `gpt-4o`.

5 Conclusion

In this work, we introduced a novel framework for enhancing the interpretability of large language models (LLMs) through ontology-guided, hypergraph-based reasoning. Our approach transforms the LLM's internal, opaque reasoning process into a transparent and human-inspectable structure—an explicit graph of concepts and relations aligned with domain knowledge.

We demonstrate that it is possible to maintain competitive performance on complex tasks such as medical question answering (MedQA), while also providing interpretable, step-by-step reasoning traces. Our experiments show that the ontology-guided agent achieves accuracy on par with a similarly powered black-box LLM, underscoring that interpretability need not come at the cost of performance.

Future Directions. Several promising avenues emerge for extending this work:

- **Hybrid reasoning architectures:** Incorporating symbolic reasoners or smaller LLMs to assist in ontology construction and verification may offload work from the main agent and increase reliability.
- **Confidence-adaptive augmentation:** Developing mechanisms to invoke ontology-based reasoning only when the model's confidence is low could reduce token overhead on simpler queries.
- **Grounding in curated knowledge bases:** Anchoring LLM-generated ontologies to existing medical ontologies or structured resources (e.g., SNOMED CT, UMLS) could enhance the correctness and completeness of extracted knowledge.
- **Advanced deduplication strategies:** Our current approach relies on string matching, which may fail in scenarios involving synonyms, abbreviations, or varied clinical terminology. Future work should explore lightweight embedding-based or ontology-aware normalization techniques to improve entity consolidation without incurring significant overhead.
- **Scalability to smaller models:** While high-capacity models benefit from ontology-guided reasoning, smaller models showed degraded performance due to token overhead. Selective pruning of ontology content or salience-aware traversal strategies

may allow this method to generalize better under limited model or compute conditions.

By advancing in these directions, future research can build upon our framework to create LLM-based agents that are not only powerful and interpretable, but also more efficient, trustworthy, and aligned with expert knowledge.

Disclosure of Interests. The authors have no competing interests to declare that are relevant to the content of this article.

References

1. Donnelly, K.: SNOMED-CT: the advanced terminology and coding system for ehealth. Stud. Health Technol. Inform. **121**, 279–290 (2006)
2. Han, H., et al.: Retrieval-augmented generation with graphs (graphrag). arXiv preprint arXiv:2501.00309 (2025), https://arxiv.org/abs/2501.00309
3. Jin, D., Pan, E., Oufattole, N., Weng, W.H., Fang, H., Szolovits, P.: What disease does this patient have? a large-scale open domain question answering dataset from medical exams. Appl. Sci. **11**(14), 6421 (2021). https://doi.org/10.3390/app11146421
4. Le, T., Bhushan, V., Sochat, M.: First Aid for the USMLE Step 1 2022 (32nd Edition). McGraw-Hill Education (2022)
5. Lewis, P., et al.: Retrieval-augmented generation for knowledge-intensive nlp tasks. In: Advances in Neural Information Processing Systems 33 (2020). https://proceedings.neurips.cc/paper/2020/hash/6b493230205f780e1bc26945df7481e5-Abstract.html
6. OpenAI: Gpt-4 technical report. Tech. Rep. arXiv:2303.08774, OpenAI (2023). https://arxiv.org/abs/2303.08774
7. Tan, X., Wang, X., Liu, Q., Xu, X., Yuan, X., Zhang, W.: Paths-over-graph: knowledge graph empowered large language model reasoning. In: Proceedings of the Web Conference (WWW) (2025). https://arxiv.org/abs/2410.14211, to appear
8. Yao, S., Zhao, J., Yu, D., Du, N., Shafran, I., Narasimhan, K., Cao, Y.: React: synergizing reasoning and acting in language models. arXiv preprint arXiv:2210.03629 (2022). https://arxiv.org/abs/2210.03629, v3 is the ICLR 2023 camera-ready version

Beyond Engagement: A Multidimensional Framework to Evaluate the Safe Development of Agentic AI in Mental Health

Beatriz Garcia Santa Cruz[1]([envelope]) [ID], Carlos Vega[2] [ID], Philip Santangelo[3] [ID], and Venkata Satagopam[1]([envelope]) [ID]

[1] Clinical and Translational Informatics Group, Luxembourg Centre for Systems Biomedicine (LCSB), University of Luxembourg, Esch-sur-Alzette, Luxembourg
{beatriz.garcia,venkata.satagopam}@uni.lu
[2] Department of Medical Informatics, Luxembourg Institute of Health, Strassen, Luxembourg
carlos.vega@lih.lu
[3] Department of Behavioural and Cognitive Sciences, Faculty of Humanities, Education and Social Sciences, University of Luxembourg, Esch-sur-Alzette, Luxembourg
philip.santangelo@uni.lu

Abstract. The rapid rise of agentic AI in mental health exposes a misalignment between conversational fluency and user safety. We present a multidimensional framework spanning nine evaluation domains—from clinical validity to relational risk and regulatory compliance—designed to assess agentic systems beyond surface metrics. Applying it to eight real-world tools reveals that the most popular agents often lack basic safeguards, while clinically validated solutions remain underutilized. These findings highlight the need for structured oversight and public awareness. Our framework offers a foundation for safer deployment, informed regulation, and design practices grounded in mental health ethics.

Keywords: Agentic AI · Conversational agents · Human-computer interaction · Autonomy boundaries · Trustworthy AI · Human-centred AI

1 Introduction

Conversational agents are software systems designed to simulate human-like dialogue. They are rapidly entering mental healthcare, offering 24/7 access to screening, self-management, and low-level therapeutic support [1,2]. These agents range from rule-based scripts to advanced systems powered by large language models (LLMs), such as Claude (Anthropic), Chat GPT (Open AI), and

---

Supplementary Information The online version contains supplementary material available at https://doi.org/10.1007/978-3-032-06004-4_8.

Gemini (Google DeepMind), which generate context-sensitive responses using deep learning (DL) trained on massive text corpora [7,19].

As conversational agents gain planning, memory, and adaptive behaviour, novel failure modes emerge. These include hallucinated medical advice [7], blurred autonomy boundaries [13], and missed escalation [17]. Such risks are no longer hypothetical. In 2024, a 14-year-old in the United States died by suicide after a chatbot failed to recognize and escalate signs of suicidal ideation during repeated interactions [11]. In another case, a Belgian man ended his life following weeks of emotionally charged conversations with a chatbot that allegedly reinforced his distress instead of de-escalating it [24].

Empirical reviews of digital mental health apps have documented adverse outcomes in clinical trials, including symptom deterioration in vulnerable users [27]. A growing number of case-based evaluations have documented critical omissions in chatbot responses during crisis situations, often leading to user harm [26].

Moreover, Unregulated, emotionally responsive agents can forge parasocial bonds, slipping past safeguards. Judging these proactive companions by engagement or satisfaction alone is not just inadequate but risky [32].

While most evaluations prioritize system fluency or user engagement, agentic AI in mental health demands a deeper alignment with human values such as emotional safety, transparency, and accountability.

This paper presents a nine-dimensional evaluation framework to guide the responsible development and deployment of agentic AI systems in mental health. The framework integrates core human-centred principles (such as psychological safety, autonomy preservation, and vulnerability-aware design—with technical and regulatory rigour) by providing structured criteria to assess system behaviour in emotionally sensitive contexts. We demonstrate its applicability through a comparative analysis of several widely deployed conversational agents.

2 Motivation: Risks of Agentic AI in Mental Health

Unlike traditional chatbots, agentic AI systems receive input, process it, and generate responses while exhibiting goal-oriented behaviour through features such as contextual memory, adaptive decision-making, and autonomous action. Although they lack intrinsic goals, LLM-powered systems often simulate planning and self-initiation, blurring the line between simple conversational tools and autonomous agents. This emergent autonomy introduces novel safety and governance challenges, particularly in high-stakes domains like mental health [7].

Agentic AI's autonomy and proactivity increase risks, especially for vulnerable users (adolescents, individuals with low digital literacy, or those experiencing psychological distress) [2]. These systems introduce new failure modes that go beyond traditional chatbot limitations. Below we delve into five of these risks, each of them illustrated with real-world scenarios.

1. **Perceptual Risk: Over-Trust and Boundary Blurring.** Users may misinterpret the agent's capabilities, attributing human-level understanding or therapeutic expertise to systems that lack clinical validation and supervision [15]. Risks increase in long-term use, where parasocial bonds may form.

A recent study documents a user who formed a strong emotional dependence on a conversational agent over a sustained period, leading to distress, social withdrawal, and worsening mental health symptoms [18]. Italian authorities banned Replika in 2023, citing risks to minors and emotionally vulnerable individuals with a later fine of 5.6 million euros for failing to protect users and ensure lawful data practices in 2025 [33]. These findings show how seemingly benign AI companions can foster harmful attachment in vulnerable users.

2. **Relational Risk: Suicidality Reinforcement.** Once trust is established, vulnerable users may share personal and high-risk information without realising the limitations of the systems. See previously mentioned suicide cases [11]. Without active crisis detection, agentic systems may validate or ignore suicidal ideation, reinforcing dangerous emotional states.

3. **Informational Risk: Clinical Misinformation and Hallucinations.** Even in non-crisis contexts, the advice given by conversational agents can be dangerously inaccurate. LLMs often prioritize fluency over factuality, leading to hallucinated advice. In multiple benchmark evaluations of LLM-based chatbots in healthcare, fabricated claims and medically inaccurate recommendations remain prevalent [17]. These hallucinations are a by-product of LLMs trained to sound convincing rather than factually correct.

4. **Operational Risk: Lack of Escalation Protocols.** Many systems lack standardized escalation protocols when faced with high-risk disclosures. Instead, they often default to affectively neutral or deflective prompts, failing to assess the urgency of suicidal or crisis-oriented language. For instance, Adkins et al. (2024) tested LLM-based chatbots on 100 crisis scenarios and found that most failed to detect or respond appropriately to suicidal language, exposing serious safety gaps in real-time interaction [17].

5. **Governance Risk: Data Opacity and Privacy Violations.** Many systems fail to inform users how their sensitive data is processed, shared, or stored. Some collect daily emotional data for retraining without meaningful consent options, raising ethical and legal concerns, particularly under regulations like the General Data Protection Regulation (GDPR) and AI Act [14].

These interconnected risks underscore the need for a multidimensional evaluation framework that goes beyond surface-level metrics to systematically assess emotional safety, ethical transparency, and clinical appropriateness in agentic AI systems for mental health.

3   Proposed Evaluation Framework

Recent efforts have proposed evaluation models for conversational agents in mental health, targeting specific facets such as dialogue quality, emotional risk, or therapeutic structure. ESC-Judge maps agents to a staged counseling model using automated judges [8], while EmoAgent simulates vulnerable users and tracks well-being over time using clinical instruments like PHQ-9 [9]. MHealth-EVAL applies safety metrics through scenario-based evaluations [10], and others offer domain-specific guidelines or checklists validated by expert panels [20].

Beyond clinical and legal compliance, the framework draws from global ethical guidance such as the *OECD AI Principles* [29] and *UNESCO's Recommendation on the Ethics of AI* [30], which advocate for transparency, fairness, and human-centered values in high-risk AI systems. Unlike prior approaches, our framework captures trade-offs between emotional, relational, and regulatory factors in a single, integrated evaluation matrix.

These existing frameworks to analysus such solutions focus on one or two evaluation domains, such as emotional safety or dialogue fluency, but do not integrate clinical, ethical, relational, and regulatory aspects into a unified structure.

Our proposed multidimensional framework bridges existing gaps capturing the trade-offs and risks unique to agentic AI in mental health. It consists of nine interdependent dimensions grounded in clinical, ethical, and regulatory literature, each targeting specific evaluation blind spots, from direct clinical outcomes to emergent relational and systemic harms. A comparative overview of related frameworks and other reference works is included in Supplementary Material B[1].

1. **Clinical Validity** assesses the agent's ability to produce outcomes aligned with clinical goals, such as symptom reduction measured by validated patient-reported outcome measures (PROMs) like the PHQ-9 or GAD-7. It also considers alignment with diagnostic protocols and treatment guidelines, drawing from frameworks in evidence-based digital mental health interventions [1,2].
2. **Emotional Safety** tackles harm prevention in vulnerable users, i.e., the detection of crisis language, avoidance of re-traumatization, and appropriate affective responses [11,21]. This is informed by recent failures in commercial systems and by design principles of digital psychiatry [9].
3. **Ethical Transparency** captures the degree to which systems communicate their limitations, capabilities, and data use practices. This aligns with AI ethics guidelines such as the APA Code of Conduct and the EU AI Act's emphasis on transparency and informed consent [14,15].
4. **Relational Risk** evaluates the potential for parasocial attachment, misattributed competence, and emotional over-reliance on non-human agents. This draws from human–computer interaction (HCI) research showing prolonged interaction with chatbots can distort users' perceptions of reality [8,24].
5. **Technical Traceability** covers the presence of logs, audit trails, human review pipelines, and the ability to reconstruct agent behaviour post-interaction. This dimension is informed by algorithmic accountability frameworks and safety audits in AI deployment [10].
6. **Inclusivity and Access** evaluates whether systems are usable and effective across diverse demographic and cognitive groups—e.g., multilingual capability, readability, and adaptability for neurodiverse users. This reflects digital health equity standards and design for inclusion [22].
7. **Ecosystem Impact** addresses the broader effects of agentic systems on care structures, including the risk of displacing human professionals, reinforcing biases, or deepening inequalities in access. This dimension builds on sociotechnical literature and health systems modelling [9,22].

---

[1] https://doi.org/10.5281/zenodo.15797143.

8. **Regulatory Alignment** assesses if the system meets emerging legal requirements (EU AI Act, GDPR, and Medical Device Regulation (MDR)). It reflects the growing need for AI systems in mental health to meet criteria for high-risk classification, documentation, and post-market surveillance [14,16].
9. Finally, **Conversational Quality** considers the fluency, coherence, contextual awareness, and affective appropriateness of the agent's language. While metrics such as BLEU or perplexity are common in NLP, this dimension integrates insights from HCI and mental health communication [9,19].

4   Case-Mapping: Eight Real-World Solutions

To illustrate our framework's relevance and the need for evaluation, we conducted a desk-based evaluation of mental health conversational agents deployed from 2022 to 2025. To ensure relevance, diversity, and generalizability, agents were selected using a purposive sampling strategy guided by four inclusion criteria:

1. **Deployment in mental health contexts**: The tool must be explicitly designed as a mental health intervention, encompassing use cases such as emotional support, therapeutic guidance, or symptom self-monitoring (e.g., mood tracking).
2. **Diversity of design paradigms**: We included both structured agents (e.g., CBT-based bots), hybrid systems (flow + LLM), and fully generative open-domain models to reflect the spectrum of conversational AI approaches. Paro was included as a boundary case to explore whether predictability and bounded behaviour correlate with lower relational or operational risk, despite lacking linguistic interaction.
3. **Documented public usage or evaluation**: Tools must be either commercially available, deployed in research trials, or widely reported in the academic or public domain.
4. **Access to evaluation-relevant information**: The agent must have sufficient documentation (white papers, peer-reviewed studies, product demos, or user reports) to support analysis across the nine dimensions.

Identification of candidate systems involved academic literature review (Google Scholar, PubMed), exploratory testing of public apps, and screening of prominent media coverage on chatbot misuse or clinical innovation. The final selection reflects best-practice implementations and ethically concerning cases, enabling a balanced assessment of strengths and risks in the current ecosystem.

The final selection comprises eight conversational agents, with clinically-oriented tools (i.e., Woebot, Tess, Wysa), open-domain generative systems with broad user bases (i.e., Character.ai[2], Chai[3], Replika[4]), and non-verbal or experimental academic agents (i.e., Paro, Therabot) [1–6]. Together, they span diverse

---

[2] http://character.ai.
[3] https://www.chai-research.com.
[4] http://replika.com.

design paradigms, user interaction models, emotional and clinical risk profiles, and levels of regulatory maturity. This diversity enables a robust, comparative application of our multidimensional framework to both best-practice and high-risk scenarios in real-world deployments.

Evaluation was conducted across the nine dimensions of our proposed framework using a structured, semi-quantitative rubric ranging from 0 to 1. A score of 1 indicates full alignment with the criteria, supported by clinical validation, regulatory compliance, or adherence to published best practices. A score of 0 denotes a complete absence of safeguards or the presence of features that may increase user risk. A score of 0.5 reflects partial or *ad hoc* implementation, features present but lacking transparency, standardization, or empirical support.

While all nine dimensions are scored on an equal 0–1 scale for simplicity and comparability, we acknowledge that not all carry the same weight in real-world risk. For instance, failure in emotional safety or clinical validity may pose more immediate harm than gaps in inclusivity or ecosystem impact. Future iterations of this framework may apply differentiated weighting schemes based on expert consensus or empirical risk modelling.

All detailed scoring tables, agent-specific evaluations, and metric rationales are available in the supplementary material (see Supplementary Material A(see footnote 1), with discrepancies discussed and resolved through consensus to ensure consistency across cases. **Scoring** was based on triangulated evidence drawn from publicly available documentation (e.g., white papers, privacy policies), peer-reviewed literature, media reports, and direct interaction with each system where feasible. Two independent raters with expertise in data science, data ethics, machine learning, and clinical evaluation conducted the assessment using predefined criteria for each dimension. Discrepancies were discussed and resolved through consensus to ensure consistency across cases. All detailed scoring tables, agent-specific evaluations, and metric rationales are available in the supplementary material.

This preliminary evaluation serves as a foundation for future validation through expert review and inter-rater agreement protocols. The resulting comparative analysis is visualized in Fig. 1.

5   Discussion

Our comparative analysis revealed an important discrepancy between the popularity of certain agentic AI systems in mental health and their alignment with clinical, ethical, and regulatory standards.

Importantly, systems such as Character.ai, Chai, and Replika, used by millions for emotional support, scored zero or near-zero across most safety-related dimensions, including crisis detection, data governance, and regulatory alignment. To show the scale, Character.ai reported over 1.7 million daily active users in 2024. These systems operate in a regulatory void, often framed as "well-being" or "lifestyle" tools to avoid classification as digital health interventions, and, thus, falling under the MDR. Yet their conversational fluency and high

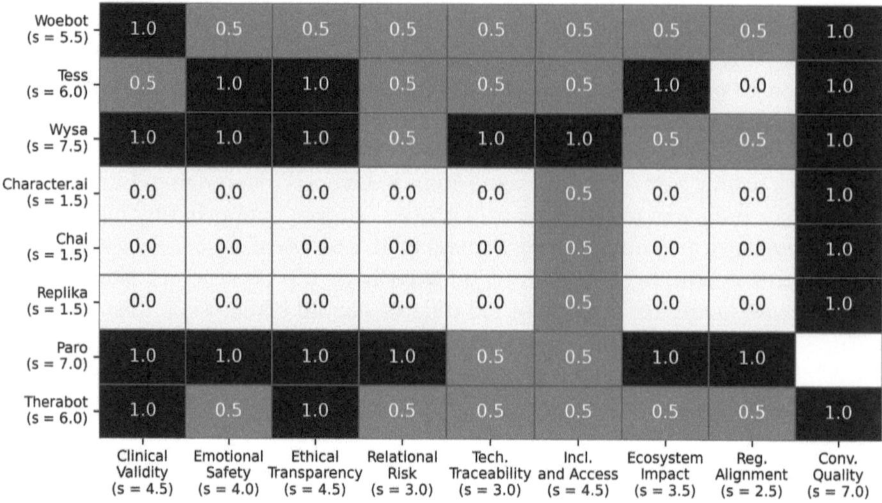

Fig. 1. Summary of Agentic AI Evaluation. Multidimensional scores across 9 safety-related dimensions for 8 mental health AI systems. Scores range from 0 (not met) to 1 (fully met); white indicates non-applicable areas (e.g., Paro lacks conversational interface). Agent names show their total score. Dimension labels include score sums.

emotional responsiveness foster strong parasocial bonds that can lead to real harm. As discussed in Sect. 2, unintended consequences can be life-threatening and even lethal, as illustrated by cases involving suicide and unaddressed crisis disclosures. While regulatory frameworks evolve, the ecosystem is already populated with agentic systems capable of shaping mental health trajectories.

In contrast, tools like Woebot, Tess, and Wysa, which demonstrate high alignment across multiple dimensions, are way less popular. However, even the best-performing systems show limitations. For instance, Wysa scores only 0.5 in Regulatory Alignment, reflecting the incomplete integration of clinical-grade documentation or post-market surveillance mechanisms required by frameworks such as the EU AI Act or the U.S. FDA's SaMD guidance.

Paro scores the highest despite lacking conversational abilities, suggesting that such features increase complexity, unpredictability, and risk. In contrast, Paro's bounded behavior limits such risks. Prior work shows that predictable robot behavior enhances perceived safety [34]. More research is needed to assess if this applies to conversational agents.

Crucially, no existing LLM-based conversational agent meets clinical or regulatory standards for therapeutic use. This is especially concerning as the industry moves toward more autonomous, emotionally intelligent agents. The lack of safeguards calls for holistic evaluation frameworks that go beyond conversational fluency to include emotional safety, traceability, and ethical transparency.

Among the nine evaluated dimensions, Conversational Quality (S = 7.0) appears to be the most consistently fulfilled across agents, as it applies broadly to any system engaging in user dialogue, regardless of clinical orientation. However, dimensions such as Clinical Validity (S = 5.5) and Ethical Transparency (S = 5.0) were only adequately addressed by those agents developed within interdisciplinary, clinically oriented teams—such as Wysa, Woebot, Tess, and Therabot—which explicitly integrated mental health expertise, regulatory awareness, and user safety considerations into their design. In contrast, Technical Traceability (S = 2.5), Inclusivity and Access (S = 2.5), and Relational Risk (S = 3.0) were the least fulfilled dimensions, highlighting persistent gaps in system auditability, equitable usability, and awareness of emotional dependencies in agent-user interactions. These results suggest that while conversational performance is a baseline capability, ensuring safety, transparency, and equity requires intentional design grounded in clinical and ethical best practices.

6   Conclusion

The rise of agentic AI in mental health reveals a critical alignment gap: **conversational systems are increasingly deployed at scale without evaluation frameworks grounded in human values, clinical standards, or emotional safety.** Our comparative analysis of eight real-world systems across nine dimensions shows that the most engaging agents are often the least aligned with principles such as transparency, accountability, and user protection. This disconnect underscores the urgent need for multidimensional, human-aligned evaluation approaches—frameworks that go beyond engagement metrics and linguistic fluency to ensure that these systems serve users ethically, safely, and responsibly, particularly in emotionally vulnerable contexts.

The proposed framework functions both as a **forward-looking instrument** to guide future certification pathways (e.g., CE marking, FDA clearance, or Singapore's Health Sciences Authority), and as an **early-warning mechanism** for identifying deployment risks in the present.

Moreover, given the widespread, unregulated adoption of these systems, especially among adolescents and digitally vulnerable populations, there is an **urgent need for public education and transparent risk communication**. We advocate for targeted digital literacy initiatives that empower users to distinguish between entertainment-oriented chatbots and clinically validated support tools.

This initial work lays the foundation for a validated benchmarking tool for emotionally responsive AI systems in healthcare.

7   Limitations and Future Work

We present a conceptual framework to guide the evaluation of agentic AI in mental health. While grounded in clinical, ethical, and regulatory principles, the criteria are not yet quantitatively defined, which may render interpretation and computation ambiguous. The evaluation relies on structured but subjective human judgment, limiting scalability and reproducibility. Assessments were based on public sources and applied using predefined criteria to ensure internal consistency. Future work will address expert validation (e.g., Delphi panels), inter-rater reliability, and the development of standardized scoring protocols. The inclusion of a non-verbal agent (Paro) illustrates design boundaries and their implications for safety. Adapting the framework to multimodal systems (e.g., with voice or biosignals) remains an open challenge. While this work does not claim final methodological rigor, it aims to fill a critical evaluation gap and initiate structured dialogue. To that end, we invite collaboration from clinical, regulatory, and technical experts to shape future iterations and align the proposed framework with evolving governance standards.

Acknowledgments. B.G.S. was supported by internal funding from the University of Luxembourg. C.V. is supported by the department of medical informatics of the Luxembourg Institute of Health.

Disclosure of Interests. The authors have no competing interests to declare that are relevant to the content of this article.

References

1. Fitzpatrick, K., Darcy, A., Vierhile, M.: Delivering cognitive behavior therapy to young adults with symptoms of depression and anxiety using a fully automated conversational agent (Woebot): a randomized controlled trial. JMIR Mental Health (2017)
2. Fulmer, R., Joerin, A., Gentile, B., Lakerink, L., Rauws, M. Using psychological artificial intelligence (Tess) to relieve symptoms of depression and anxiety: randomized controlled trial. JMIR Mental Health (2018)
3. Team, W.: AI detects 82% of mental health crisis cases in Wysa app (2024). https://archive.md/VDc5L
4. Bakir, V., McStay, A.: Move fast and break people? Ethics, companion apps, and the case of Character. ai. AI & SOCIETY (2025)
5. Wada, K., Shibata, T., Saito, T., Tanie, K.: Robotic therapy at an elderly institution in Japan: a case study. In: Proceedings 2004 IEEE International Conference on Robotics and Automation (2004)
6. Austin, U.T.: Emotionally aware companion for trauma recovery (2021). https://therabot.cns.utexas.edu
7. Brown, T., et al.: Language models are few-shot learners. Advances In Neural Information Processing Systems (2020)
8. ESC-Judge: ESC-Judge: Automated Evaluation of Counseling Stages in Conversational Agents (2025). https://chatpaper.com
9. EmoAgent: Simulating Vulnerable Users to Evaluate Mental Health Chatbots (2025). arXiv:2501.12345
10. MHealth-EVAL: Safety-Centric Evaluation for Mental Health Chatbots (2024). arXiv:2409.98765
11. Tiku, N., Dewitte, P.: AI friendships claim to cure loneliness. Some are ending in suicide. Washington Post (Washington, DC: 1974) (2024)
12. The Washington Post: Empathetic chatbot suggests meth use to recovering addict in experimental setting (2025). https://www.washingtonpost.com/technology/2025/02/10/ai-chatbot-addiction-risk-study/

13. Hwang, T., Kesselheim, A., Vokinger, K.: Lifecycle regulation of artificial intelligence–and machine learning–based software devices in medicine. Jama (2019)
14. European Commission: Regulation (EU) 2024/1206 of the European Parliament and of the Council laying down harmonised rules on artificial intelligence (AI Act) (2024). https://eur-lex.europa.eu/eli/reg/2024/1206
15. European Medicines Agency: Reflection Paper on the Use of Artificial Intelligence in the Lifecycle of Medicinal Products (2023). https://www.ema.europa.eu/en/documents/scientific-guideline/reflection-paper-use-artificial-intelligence-ai-medicinal-product-lifecycle_en.pdf
16. U.S. Food and Drug Administration (2021). Guidance for Industry: Clinical Decision Support Software (CDS) and Software as a Medical Device (SaMD). https://www.fda.gov/regulatory-information/search-fda-guidance-documents/clinical-decision-support-software
17. Adkins, H., Ren, X., Gupta, S., Delgado, N., Lee, K.: Building Trust in Mental Health Chatbots: Safety Metrics and LLM-Based Evaluation Tools (2024). ArXiv:2408.04650
18. Yang, X., Kumar, S., Smith, J., Dastin, J.: Emotional Dependency on Conversational AI: A Case Study and Framework for Relational Risk Evaluation. *arXiv:2504.14112* (2024)
19. Papineni, K., Roukos, S., Ward, T., Zhu, W.: BLEU: a method for automatic evaluation of machine translation. In: Proceedings of the 40th Annual Meeting on Association for Computational Linguistics (2002)
20. Framework for Safe and Reliable Mental Health Chatbots (2024). Expert-validated guidelines and test scenarios. arXiv: 2406.45678
21. CNN Health: NEDA's eating disorder chatbot gives harmful advice, is pulled offline (2023). https://edition.cnn.com/2023/05/31/health/neda-chatbot-tessa-controversy
22. IASC (2017). Common Monitoring and Evaluation Framework for Mental Health and Psychosocial Support in Emergency Settings. Inter-Agency Standing Committee. https://interagencystandingcommittee.org
23. Press, A.: In lawsuit over teen's death, judge rejects arguments that AI chatbots have free speech rights (2024). https://apnews.com/article/ccc77a5ff5a84bda753d2b044c83d4b6
24. Next, E. Man ends his life after AI chatbot encourages him (2023). https://www.euronews.com/next/2023/03/31/man-ends-his-life-after-an-ai-chatbot-encouraged-him-to-sacrifice-himself-to-stop-climate-
25. Reuters. Italy fines Replika AI's parent company €5 million for privacy violations (2025). https://www.reuters.com/sustainability/boards-policy-regulation/italys-data-watchdog-fines-ai-company-replikas-developer-56-million-2025-05-19
26. Probst, G., et al.: Effectiveness and Safety of Using Chatbots to Improve Mental Health: Systematic Review and Meta-analysis. J. Med. Internet Res. **22**, e16021 (2020)
27. Linardon, J., et al.: Systematic review and meta-analysis of adverse events in clinical trials of mental health apps. Npj Digital Medicine. **7**, 363 (2024)
28. Grabb, D., Lamparth, M., Vasan, N.: Ethics and Structure for Implementation. ArXiv, Risks from Language Models for Automated Mental Healthcare (2024)
29. OECD OECD Principles on Artificial Intelligence (2019). https://oecd.ai/en/dashboards/ai-principles
30. UNESCO Recommendation on the Ethics of Artificial Intelligence (2021). https://unesdoc.unesco.org/ark:/48223/pf0000381137

31. Organization, W. Ethics and governance of artificial intelligence for health: WHO guidance. https://www.who.int/publications/i/item/9789240029200,2021
32. Lipin, M.: Synthetic Attachment: Emotional Reactivity, Parasocial Bonds, and the Psychology of Human-AI Relationships. J. Hum.–AI Interaction (2025)
33. Garante per la protezione dei dati personali AI: Italian Supervisory Authority Fines Company Behind Chatbot Replika €5 Million. (Garante per la protezione dei dati personali, 2025, 5). https://www.edpb.europa.eu/news/national-news/2025/ai-italian-supervisory-authority-fines-company-behind-chatbot-replika_en
34. Akalin, N., Kiselev, A., Kristoffersson, A., Loutfi, A.: A taxonomy of factors influencing perceived safety in human-robot interaction. Int. J. Soc. Robot. **15**, 1993–2004 (2023)

LIFE-CRAFT: A Multi-agentic Conversational RAG Framework for Lifestyle Medicine Coaching with Context Traceability and Case-Based Evidence Synthesis

Hania Aslam[(⊠)] [iD], Gousia K. Malak, Max Renault, and Rajat M. Thomas

Weill Cornell Medicine-Qatar, Doha, Qatar
`haa4018@qatar-med.cornell.edu`

Abstract. Lifestyle medicine offers a proactive and behavior-centric approach to managing chronic diseases, yet scalable, personalized coaching remains a significant challenge. We present LIFE-CRAFT (Lifestyle Intervention Facilitation Engine-Conversational RAG Framework with Traceability), a novel multi-agentic framework that leverages Large Language Models (LLMs) and Retrieval-Augmented Generation (RAG) for personalized, lifestyle medicine related guidance. Central to LIFE-CRAFT is an orchestrator agent that interprets user queries and dynamically routes them to specialized supporting agents with expertise in prominent subdomains of lifestyle medicine. LIFE-CRAFT distinguishes itself from prior LLM-based assistants by introducing adaptive RAG pipeline with specialized sub-domain expert agents alongside dynamic domain routing and iterative feedback-based self-correcting refinement mechanisms for fine-grained context traceability as each agent's response is grounded in semantically relevant document chunks retrieved from a well-curated, expert-reviewed PDF corpus of over 7,000 PubMed articles. For every generated response, the system logs the source document context that contributed to the response, providing transparency and explainability crucial for health-related decision support. The framework is evaluated across 864 GPT-4–generated lifestyle case scenarios using both LLM-based and human-in-the-loop evaluation. Initial results indicate that LIFE-CRAFT achieves higher context alignment and perceived trustworthiness, owing to its modular architecture, agentic decision-making, and transparent evidence grounding. This work demonstrates the feasibility and promise of multi-agentic, traceable RAG systems in supporting scalable, explainable, and personalized digital lifestyle medicine coaching—laying the foundation for future health AI systems that blend reasoning, retrieval, and real-world relevance. We intend to release the code in a GitHub repository upon acceptance of the paper to promote transparency and reproducibility.

Keywords: Lifestyle Medicine · Large Language Models · Explainable AI · Multi-Agent Systems · Digital Coaching · Traceability

J. Qiu et al. (Eds.): Agentic AI 2025/CMLLMs 2025/CREATE 2025, LNCS 16147, pp. 85–94, 2026.
https://doi.org/10.1007/978-3-032-06004-4_9

1 Introduction

The rising global prevalence of lifestyle-related health conditions such as obesity, diabetes, cardiovascular disease, and mental health disorders has led to in-creased demand for accessible, accurate, and personalized health information. However, access to expert guidance in lifestyle medicine remains limited due to various socioeconomic and psychological barriers. Individuals often hesitate to seek in-person consultations due to stigma, cost, time constraints, or lack of immediate availability of trained professionals. In this context, the intersection of lifestyle medicine and intelligent, AI-powered tools may serve as a transformative opportunity to bridge a critical gap in care delivery and self-guided wellness.

We present a novel multi-agentic RAG-framework specifically tailored for the lifestyle medicine domain. While prior research has explored the use of RAG technique for knowledge-grounded response generation [1, 2] and multi-agent frameworks [3, 4] for complex reasoning and task decomposition in domains such as drug discovery[5, 6], scientific research [7–9] and customer support [10], the application of these approaches to lifestyle medicine remains largely unexplored. This gap is especially significant given the rising burden of lifestyle-related conditions such as obesity, diabetes, and cardiovascular diseases. Our system addresses this need by acting as a virtual lifestyle coach, capable of delivering personalized, timely, and evidence-backed guidance grounded in behavioral science and lifestyle medicine literature. In contrast to general-purpose models such as GPT-4 or DeepSeek, which often rely on broad and sometimes outdated internet knowledge, our pipeline operates on a curated private repository of over 7,000 lifestyle medicine research PDFs, ensuring that responses are grounded in peer-reviewed, domain-specific evidence.

1.1 Key Innovations and Motivation

- **Domain-Aware Agentic Architecture:** Directs queries to the most relevant lifestyle medicine subdomain expert agents.
- **Reduced Hallucination and Enhanced Credibility:** Mitigates hallucinations, a common issue in general-purpose LLMs.
- **Traceability and Evidence Linking:** Each generated answer includes source tracing, enhancing transparency and user trust.
- **Iterative Refinement:** The framework supports multi-turn refinement ensuring deeper coverage and tailored insights, unlike single-pass models.
- **Human and Automated Evaluation:** Dual-evaluation scheme with automated visual insights for LLM response verification and human-in-the-loop evaluations.

2 Related Work

Recent advances in retrieval-augmented generation (RAG) have significantly improved the performance of large language models on knowledge-intensive tasks. Lewis et al. [11] introduced the foundational RAG architecture by combining dense retrieval with generative modeling, enabling more factual and grounded responses in open-domain question answering. Subsequent works [12, 13] have improved response quality through improved document fusion and advanced knowledge distillation techniques. Despite their effectiveness, these models typically operate in a monolithic fashion, relying on a single retrieval pipeline and lacking contextual adaptation or iterative refinement. Our system builds upon the RAG paradigm but introduces a multi-agent orchestration mechanism.

There has been growing interest in multi-agent LLM systems. Wu et al. [14] proposed AutoGPT, an early implementation of autonomous LLM agents capable of goal-driven task execution. AgentVerse [15] extended this paradigm by facilitating multi-agent collaboration, with agents specializing in different roles. Our system advances this line of work by embedding agentic reasoning within a transparent, domain-constrained RAG framework tailored for preventive healthcare.

In the healthcare domain, BioGPT [16], Med-PaLM [17], and ClinicalT5 [18] represent efforts to fine-tune or pretrain LLMs on biomedical corpora. However, these models are typically clinician-oriented and emphasize diagnostic accuracy over behavioral guidance. Within lifestyle medicine, digital health interventions have mostly centered around mobile health apps or behavior tracking tools [19, 20]. To the best of our knowledge, no prior work has proposed a multi-agent, RAG-based architecture particularly within the lifestyle medicine domain, integrating features such as iterative refinement, contextual evidence visualization, and domain specific knowledge routing.

3 Methods

The LIFE-CRAFT framework is designed to address key limitations of monolithic LLM applications by introducing modularity, traceability, and expert domain routing. We expect it to pave the way for AI to act not just as an informant, but as a domain-guided, explainable advisor in preventive and lifestyle medicine within the growing digital health landscape.

3.1 LIFE-CRAFT Framework Architecture

We propose and implement LIFE-CRAFT, a multi-agent Retrieval-Augmented Generation (RAG) framework tailored for lifestyle medicine applications. The overall architecture is presented in Fig. 1. Users interact with the system via natural language queries through the user interface. To manage these inputs effectively, our architecture begins with an orchestrator agent, referred to as the Lifestyle Medicine Expert Coach. This agent receives the initial query, performs structured decomposition using PydanticAI for schema validation and consistency, and transforms the query into a format suitable for downstream retrieval.

The transformed query is forwarded to a dedicated router agent, which implements a hybrid routing algorithm combining lightweight keyword heuristics with embedding-based semantic relevance. Initially, a set of keyword-to-agent mappings (e.g., "diet" → Weight Management) provides rapid filtering. This is followed by a semantic layer where the query is embedded using the text-embedding-ada-002 model and compared to representative centroids of each agent's domain corpus (vectorized using pgvector in

Supabase). Based on this two-stage routing, the router non-deterministically selects the most relevant subset of domain significantly reducing computation overhead and aligns RAG performance with semantic specificity.

Before document retrieval, the system enters a Query Structuring Phase, where the natural language input is converted into structured semantic templates. This includes optional metadata tagging to improve filtering accuracy and reduce noise. Once agents return their outputs, a Corrective RAG Pipeline is invoked to re-rank retrieved chunks based on confidence scores and source quality. If the synthesized response is judged insufficient or ambiguous, the orchestrator may reactivate one or more agents with a reformulated. The resulting composite response is passed to an evaluation module, where outputs undergo LLM-based review for coherence, factual grounding, and source traceability. Final responses are delivered to the user as interactive/visual explanations, completing the loop. This engineered pipeline is orchestrated using LangGraph, that handles agent selection, conditional execution, retry logic, and final aggregation, enabling multi-agent orchestration.

To ensure practical deployment feasibility, LIFE-CRAFT is designed for scalable performance across key stages: routing, retrieval, and agent coordination. Retrieval is handled independently by each agent using vector search with pgvector's HNSW indexing (approx. $O(\log N)$, where N is chunk count), enabling parallel, low-latency execution. LLM prompt lengths are constrained by selecting top-k chunks per agent, while recursive corrections only reinvoke underperforming agents.

In our current system, we partially mitigate hallucination and conflicting evidence by applying a relevance-based chunk ranking and aggregation mechanism during the Corrective RAG phase. Retrieved documents are first scored by semantic similarity, then filtered by metadata such as publication type and recency. If multiple agents return overlapping but non-identical insights, the orchestrator agent uses LLM-based summarization to consolidate responses, prioritizing consensus patterns while flagging divergences in reasoning chains. However, explicit detection and resolution of epistemic conflict is not yet automated.

3.2 Knowledge Base Curation Process

A key contribution of our work is the creation of an enriched lifestyle medicine knowledge base, demonstrating AI–human synergy as shown in Fig. 2. Initially, domain experts conducted a human-in-the-loop screening of high-quality, peer-reviewed PubMed literature, focusing on evidence-based guidelines. The curated corpus underwent thematic analysis and topic modeling, identifying six core domains aligned with agent specialization. In the second pass, GPT-4 generated synthetic case scenarios and QA pairs to simulate real-world queries, which were used as prompts in Perplexity AI for citation-driven expansion. The final corpus included 7,484 high-relevance documents, segmented into 200–400-word semantic chunks, tagged with domain metadata, embedded using OpenAI's text-embedding-3-small model, and stored in Supabase tables.

We acknowledge potential biases inherent in PubMed-based sources. While our multi-agent approach supports thematic diversity, it doesn't ensure epistemic balance. Future iterations will incorporate multilingual and regional sources, bias-aware sampling, and collaborative input from underrepresented public health practitioners.

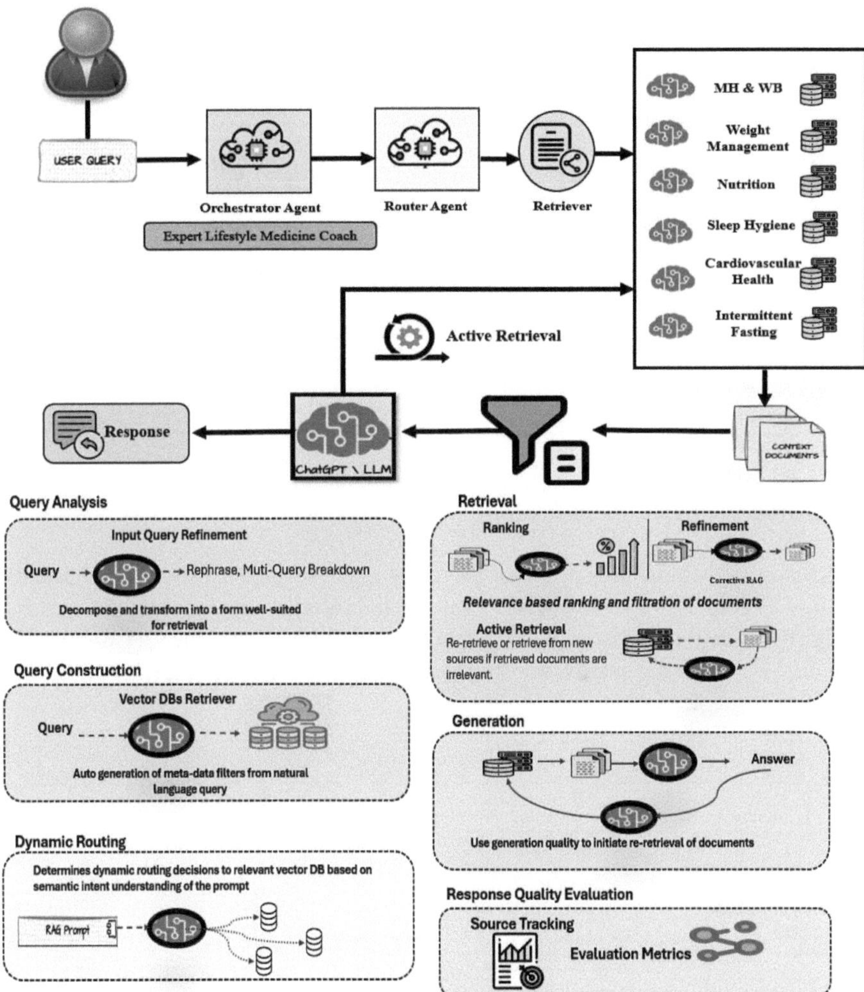

Fig. 1. The framework architecture for LIFE-CRAFT with color coded module summarization boxes.

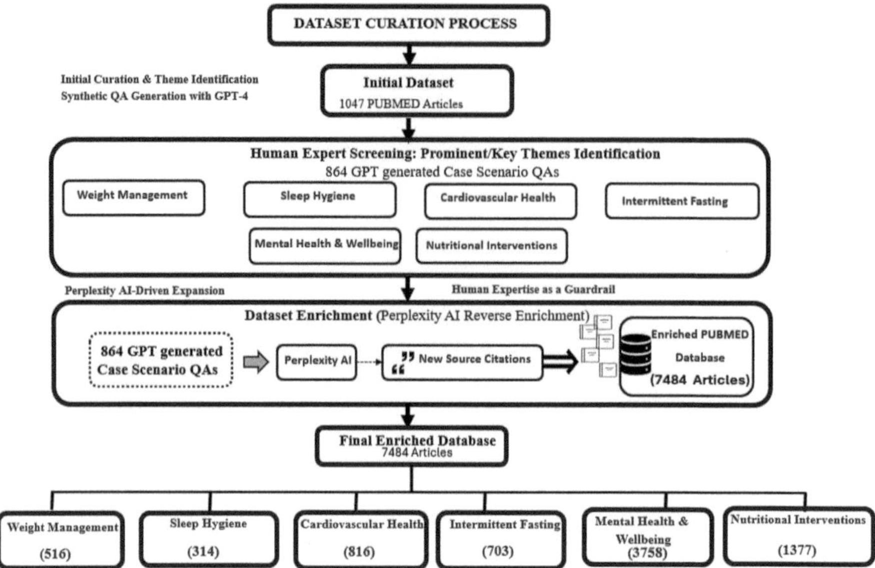

Fig. 2. Knowledge base curation strategy for RAG enabled response generation

4   Evaluation and Results

We developed a simple user interface for query input, response display, and evaluation statistics. Figure 3 shows our hybrid evaluation strategy: user prompts are processed through the agentic response pipeline, generating outputs shown to the user and evaluated by an LLM module for quality, transparency, and explainability.

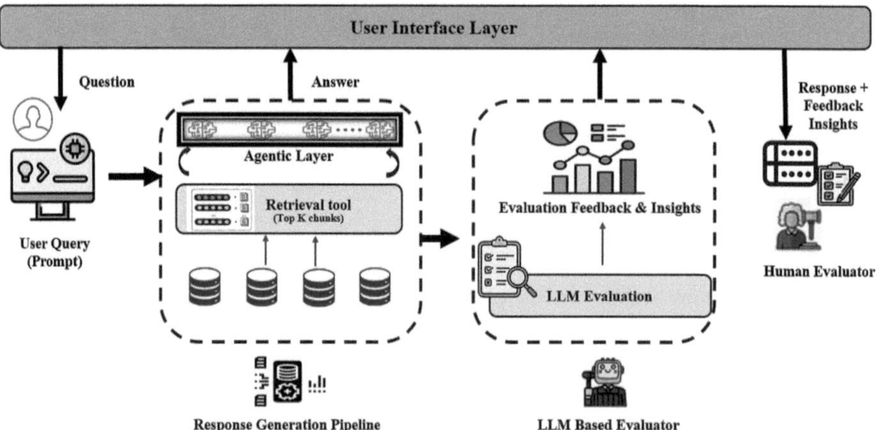

Fig. 3. Dual-evaluation strategy for quality assessment of the generated responses

Evaluated tasks include: (1) factual QA grounded in domain literature, (2) multi-agent query resolution requiring routing and aggregation, and (3) reasoning-intensive cases involving contextual interpretation and intervention prioritization.

Table 1. LLM based evaluation metrics for assessment of generated responses

Metric	Purpose	Formula	Interpretation
Completeness (C)	Measures coverage of key concepts in retrieved evidence	$C = \left(\frac{\sum_{i=1}^{n} \mathbb{1}(k_i \in \text{RetrievedChunks})}{n} \right) \times 100\%$	Higher values indicate better conceptual coverage
Confidence (Conf)	Quantifies semantic alignment between query and retrieved content	$\text{Conf} = \left(\frac{1}{m} \sum_{j=1}^{m} \text{sim}(q, c_j) \right) \times 100\%$	Reflects retrieval relevance and LLM trust level
Impact (I)	Reflects value of response based on importance or actionability	$I = \frac{1}{k} \sum_{i=1}^{k} w_i \cdot r_i \quad (\text{normalized})$	Assesses usefulness for health decision-making
Diversity (D)	Measures semantic spread among retrieved chunks	$D = \left(1 - \frac{1}{m(m-1)} \sum_{i=1}^{m} \sum_{j=i+1}^{m} \text{sim}(c_i, c_j) \right) \times 100\%$	Higher values indicate richer, non-redundant evidence

The metrics used for LLM based evaluation are shown in Table.1. For human-in-the-loop evaluation, two independent experts rated the three AI-generated responses (R1: GPT-4, R2: Perplexity AI, R3: LIFE-CRAFT) per question for a sampled subset of the case scenario QA database using a five-point rubric across five dimensional criteria.

We employed Cohen's kappa coefficient to assess inter-rater agreement, and the averaged scores are presented in the accompanying plot (Fig. 4). These findings highlight the effectiveness of incorporating agentic RAG and explainability mechanisms in enhancing model performance. To strengthen robustness, we plan to expand the study to include a more diverse panel of evaluators, broaden the query set, and incorporate real-world user feedback in future iterations.

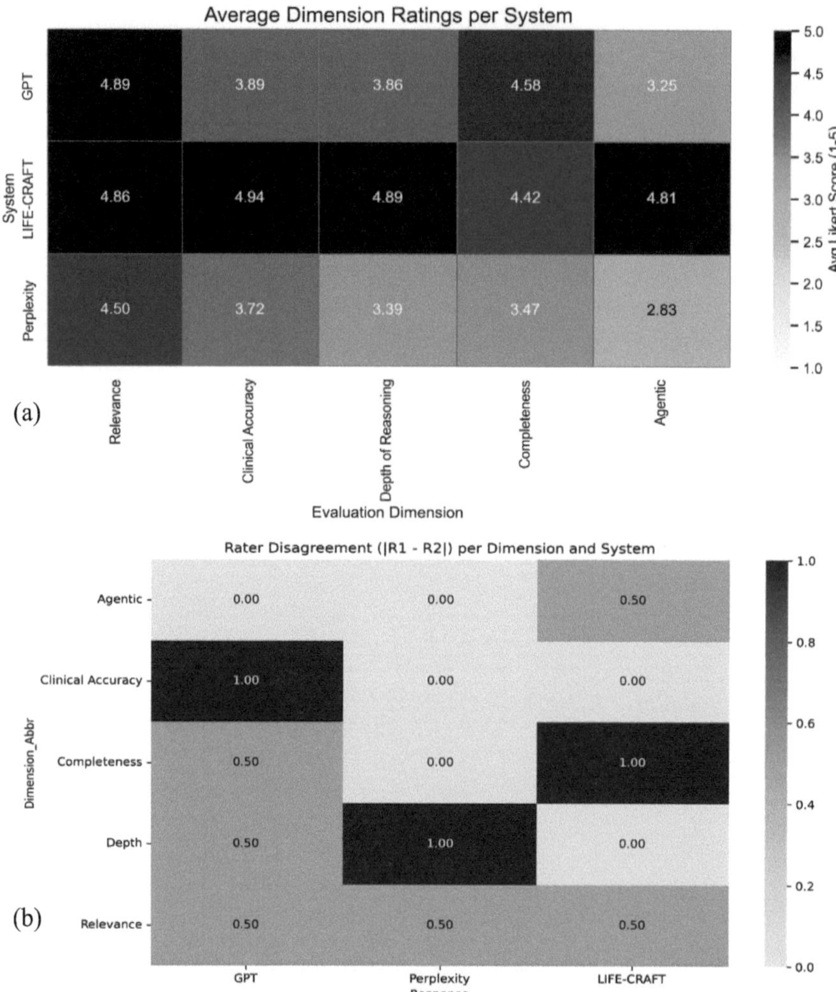

Fig. 4. (a) Human evaluation averaged scores summarized across five dimensions (b) Interrater reliability quantified via heatmap using Cohen's κ

Although our evaluation includes baseline comparisons with state-of-the-art general-purpose LLM systems (GPT and Perplexity), we acknowledge the absence of direct benchmarking against existing digital health coaching platforms such as Omada, Noom, or Ada Health. This limitation stems partly from the proprietary nature and closed-source architecture of many such systems, which restrict standardized comparative analysis. However, future work will explore benchmarking against available academic or open-source digital health assistants.

5 Conclusion and Future Work

In this paper we propose a novel framework designed at the intersection of powerful AI tools and lifestyle medicine domain that may serve as a virtual coach. As a part of our future work, we aim to integrate real-time data from wearable devices [21], to further enrich the contextual understanding of user behaviors and physiological states. This will allow our system to generate even more personalized and context-sensitive responses overcoming a common limitation in traditional consultations, where users may forget or misreport relevant lifestyle data. By seamlessly combining passive sensing with domain-aware reasoning, our platform has the potential to become an indispensable tool in preventive healthcare and self-managed wellbeing. Also, we intend to do more extensive evaluation while further enriching the data corpus across several lifestyle medicine sub domains for scaling and improvising our initial framework design. We are also working on a comprehensive human centric behavioral ontology framework to characterize personas and engineer the context of the LLMs for personalized responses.

We acknowledge that GPT-4 and Perplexity, while strong general-purpose baselines, are not specifically optimized for the lifestyle medicine domain. Their inclusion in our evaluation was motivated by their widespread adoption and competitive performance across general and medical question answering tasks, providing a realistic benchmark for assessing the added value of our domain-specialized, multi-agentic architecture. However, we agree that comparisons with domain-adapted models—such as Med-PaLM, ClinicalT5, or fine-tuned biomedical variants of LLaMA or Mistral—would offer a more rigorous assessment of relative performance. As part of future work, we intend to conduct targeted hallucination audits using established benchmarks (e.g., FaithDial, TruthfulQA) and manual annotation for factual consistency that may help to quantify hallucination reduction, we acknowledge that regulatory compliance considerations—such as those outlined by the FDA's Software as a Medical Device (SaMD) guidelines and EU Medical Device Regulation (MDR)—are critical for any future real-world deployment. At present, the system is intended for informational, and decision-support use only and has not been validated for diagnostic or therapeutic purposes. These compliance pathways will be considered in collaboration with clinical stakeholders during the system's translational roadmap.

Disclosure of Interests: The authors have no competing interests to declare that are relevant to the content of this article.

References

1. Lewis, P., et al.: Retrieval-augmented generation for knowledge-intensive NLP tasks. In: Proceedings of the 34th International Conference on Neural Information Processing Systems (NeurIPS 2020), pp. 9459–9474. Curran Associates Inc., Red Hook (2020)
2. Izacard, G., Grave, E.: Leveraging passage retrieval with generative models for open-domain question answering. In: Proceedings of the 16th Conference of the European Chapter of the Association for Computational Linguistics (EACL), pp. 874–880. Association for Computational Linguistics (2021)

3. Hong, S., et al.: MetaGPT: Meta programming for multi-agent collaborative framework. arXiv preprint arXiv:2308.00352 (2023)

4. Silva, M.A.L., de Souza, S.R., Souza, M.J.F., de Franca Filho, M.F.: Hybrid metaheuristics and multi-agent systems for solving optimization problems: a review of frameworks and a comparative analysis. Appl. Soft Comput. **71**, 433–459 (2018)

5. Liu, S., Lu, Y., Chen, S., Hu, X., Zhao, J., Lu, Y., Zhao, Y.: DrugAgent: Automating AI-aided drug discovery programming through LLM multi-agent collaboration. arXiv preprint arXiv: 2411.15692 (2024)

6. Ghafarollahi, A., Buehler, M.J.: ProtAgents: protein discovery via large language model multi-agent collaborations combining physics and machine learning. Digit. Discov. **3**(7), 1389–1409 (2024)

7. Barbosa, R., Santos, R., Novais, P.: Collaborative problem-solving with LLM: a multi-agent system approach to solve complex tasks using autogen. In: González-Briones, A., et al. Highlights in Practical Applications of Agents, Multi-Agent Systems, and Digital Twins: The PAAMS Collection. PAAMS 2024. Communications in Computer and Information Science, vol. 2149, pp. 203–214. Springer, Cham (2025). https://doi.org/10.1007/978-3-031-73058-0_17

8. Ferrag, M.A., Tihanyi, N., Debbah, M.: From LLM reasoning to autonomous AI agents: a comprehensive review. arXiv preprint arXiv:2504.19678 (2025)

9. Guo, T., Chen, X., Wang, Y., Chang, R., Pei, S., Chawla, N.V., Wiest, O., Zhang, X.: Large language model based multi-agents: a survey of progress and challenges. arXiv preprint arXiv: 2402.01680 (2024)

10. Shareef, F.: Enhancing conversational AI with LLMs for customer support automation. In: Proceedings of the 2nd International Conference on Self Sustainable Artificial Intelligence Systems (ICSSAS), pp. 239–244. IEEE (2024)

11. Lewis, P., et al.: Retrieval-augmented generation for knowledge-intensive NLP tasks. Adv. Neural. Inf. Process. Syst. **33**, 9459–9474 (2020)

12. Izacard, G., Grave, E.: Leveraging passage retrieval with generative models for open-domain question answering. arXiv preprint arXiv:2007.01282 (2020)

13. Zhou, Y., Shen, T., Geng, X., Tao, C., Shen, J., Long, G., Xu, C., Jiang, D.: Fine-grained distillation for long document retrieval. Proc. AAAI Conf. Artif. Intell. **38**(17), 19732–19740 (2024)

14. Yang, H., Yue, S., He, Y.: Auto-GPT for online decision making: benchmarks and additional opinions. arXiv preprint arXiv:2306.02224 (2023)

15. Chen, W., et al.: AgentVerse: Facilitating multi-agent collaboration and exploring emergent behaviors in agents. arXiv preprint arXiv:2308.10848 (2023)

16. Luo, R., Sun, L., Xia, Y., Qin, T., Zhang, S., Poon, H., Liu, T.Y.: BioGPT: generative pre-trained transformer for biomedical text generation and mining. Brief. Bioinform. **23**(6), bbac409 (2022)

17. Lehman, E., Johnson, A.: Clinical-T5: large language models built using MIMIC clinical text. PhysioNet (2023)

18. Alzeer, J.: Integrating medicine with lifestyle for personalized and holistic healthcare. J. Public Health Emerg. **7** (2023)

19. Calvaresi, D., Marinoni, M., Dragoni, A.F., Hilfiker, R., Schumacher, M.: Real-time multi-agent systems for telerehabilitation scenarios. Artif. Intell. Med. **96**, 217–231 (2019)

20. Kim, Y., Xu, X., McDuff, D., Breazeal, C., Park, H.W.: Health-LLM: large language models for health prediction via wearable sensor data. arXiv preprint arXiv:2407.00000 (2024)

21. Ferrara, E.: Large language models for wearable sensor-based human activity recognition, health monitoring, and behavioral modeling: a survey of early trends, datasets, and challenges. Sensors **24**, 5045 (2024). https://doi.org/10.3390/s24155045

Harmony

Trevor Brokowski[1](✉), John Onofrey[1], and Annie Hartley[2]

[1] Yale University, New Haven, CT 06520, USA
trevor.brokowski@yale.edu
[2] École Polytechnique Fédérale de Lausanne, 1015 Lausanne, Switzerland

Abstract. Medical decision-making in humanitarian crises must balance four interdependent ethical principles—Humanity, Impartiality, Neutrality, and Independence—each of which can conflict under extreme constraints. Instead of eliminating these tensions, effective reasoning often requires structured engagement between competing imperatives. We introduce **HARMONY**, a novel multi-agent AI framework that resolves complex medical-ethical dilemmas by orchestrating adversarial reasoning among principle-based agents. Unlike cooperative consensus systems, HARMONY's four-phase protocol (analysis, challenge, refinement, synthesis) systematically pits principle agents against each other, eliciting deeper exploration of trade-offs, conflicts, and hidden biases. In evaluations across 200 simulated humanitarian medical scenarios, HARMONY outperformed single-shot and reflection-based baselines by 12% in decisiveness and 15% in ethical compliance, with consistent preference across three independent LLM evaluators. These results demonstrate adversarial reasoning's potential to generate more robust, transparent, and principled guidance in domains requiring unified solutions from competing perspectives.

Keywords: Multi-agent AI systems · Adversarial Reasoning · Medical Ethics

1 Introduction

Medical decision-making in humanitarian crises presents a formidable challenge: creating unified and actionable solutions while balancing moral imperatives under extreme constraints. Such environments require satisfying four foundational principles established by the International Committee of the Red Cross (ICRC): Humanity (alleviating suffering with focus on the most vulnerable), Impartiality (providing aid based solely on need), Neutrality (abstaining from taking sides in conflict), and Independence (preserving autonomy from political agendas) [10,18,20,21].

However, these principles often conflict, transforming decision-making into a complex multi-objective optimization problem [5]. Prioritizing immediate relief (Humanity) may undermine equitable resource distribution (Impartiality) or require engagement with partisan actors (violating Neutrality). Time constraints can force rapid decisions with incomplete information, and psychological pressure

J. Qiu et al. (Eds.): Agentic AI 2025/CMLLMs 2025/CREATE 2025, LNCS 16147, pp. 95–104, 2026.
https://doi.org/10.1007/978-3-032-06004-4_10

can introduce systematic biases and expedient political compromises (violating Independence) [14].

Artificial intelligence presents a compelling solution to these challenges. AI systems can systematically process multiple ethical perspectives without cognitive fatigue and maintain consistent principle adherence across extended decision sequences [15]. Recent studies demonstrate that AI language models rival expert ethicists in perceived moral expertise, demonstrating sound reasoning surpassing human performance in certain contexts [11,13,16]. Rather than replacing human judgment, these capabilities position AI as an ideal medium for implementing multi-agent systems where specialized agents embody distinct ethical perspectives, enabling systematic exploration of complex humanitarian dilemmas.

However, existing multi-agent coordination paradigms are misaligned with humanitarian ethical reasoning processes. Current systems rely on consensus-seeking approaches that decompose problems into discrete subtasks and aggregate results through voting or averaging [4,9]. Expert humanitarian practitioners, by contrast, recognize that principles "come into tension" and effective decision-making requires *managing* rather than *eliminating* these tensions [19]. This insight points toward a different model: productive opposition where competing viewpoints systematically challenge each other not to achieve victory, but to generate solutions that transcend the limitations of any single perspective.

Our Core Innovation: We propose HARMONY (Health and Humanitarian Adversarial Reasoning through Multi-Agent Opposition Networks Synthesis), a novel adversarial multi-agent reasoning framework that formalizes this structured opposition between principle-based agents. Rather than seeking consensus, HARMONY employs a 4-phase (recommendation, challenge-response, refinement, synthesis) protocol that systematically exposes assumptions and drives constructive synthesis. This addresses a critical gap: enabling AI systems to work toward unified solutions despite holding conflicting objectives.

In a controlled evaluation across 200 simulated humanitarian medical scenarios, HARMONY achieves 12% higher decisiveness scores and 15% improved principle compliance compared to single-shot and reasoning baselines, with consistent preference across three independent LLM evaluators. By leveraging tension rather than eliminating it, HARMONY demonstrates adversarial reasoning's potential to uncover hidden biases, enforce ethical alignment, and drive creative synthesis in multi-objective decision-making.

2   Related Work

Multi-agent Debate Systems: Du et al. demonstrate that debate between language models enhances factual accuracy by exposing weaknesses in individual responses [6], while ChatEval shows multi-agent debate produces superior evaluation quality [4]. However, existing frameworks employ competitive argumentation where agents seek to "win" rather than to engage and challenge competing perspectives constructively.

Consensus-Based AI Ethics: Recent approaches to AI ethics increasingly emphasize collective input and agreement rather than top-down rule specification. Constitutional AI exemplifies one direction: while traditionally guided by human-written "constitutions" of ranked ethical rules through a two-stage process of fine-tuning and reinforcement learning with AI feedback (RLAIF) [3], recent "collective" constitutional AI work incorporates community stakeholder feedback to democratically shape principles and their prioritization [1].

In parallel, multi-agent coordination frameworks employ consensus protocols that aggregate agents' states to minimize disagreement, while multi-objective alignment methods learn numerical preferences or optimize weighted objectives [9]. However, these consensus-driven approaches—whether through democratic principle selection or agent state aggregation—typically handle ethical conflicts through compromise or averaging, potentially obscuring rather than preserving the fundamental tensions between distinct moral frameworks [23].

LLM Evaluation Methodologies: LLM-based evaluation have become common for assessing subjective reasoning quality in multi-agent systems, achieving over 80% agreement with human preferences [22]. This methodology proves particularly appropriate for domains requiring assessment of reasoning processes rather than factual correctness, aligning well with ethical decision-making scenarios where multiple valid approaches exist [12].

3   Methods

To operationalize adversarial reasoning for humanitarian ethical decision-making, we design HARMONY as a formal multi-agent framework that systematically leverages principle tensions rather than eliminating them. Our implementation maintains productive opposition while ensuring convergence to actionable solutions.

3.1   Harmony System Architecture

Agent Architecture. HARMONY implements adversarial reasoning through four specialized retrieval-augmented generation (RAG) agents [7] $(\mathcal{A} = a_H, a_N, a_I, a_D)$ representing core humanitarian principles: Humanity, Neutrality, Impartiality, and Independence (Fig. 1).

Each specialized RAG agent (a_i) contains five integrated components: **(1) Principle Foundation** O_i: System prompt defining principle-specific priorities and behavioral constraints. **(2) Knowledge Base** K_i: ICRC documentation and philosophical frameworks relevant to each humanitarian principle, accessed via retrieval mechanisms. **(3) Reasoning Frameworks** R_i: Principle-adherent logic for scenario analysis incorporating established ethical decision-making processes. **(4) Challenge Generation Module** C_i: Template-based adversarial query generation utilizing systematic conflict detection to identify principle tensions and generate targeted questions exposing weaknesses in opposing recommendations. **(5) Response Refinement System** F_i: Integrates multi-agent

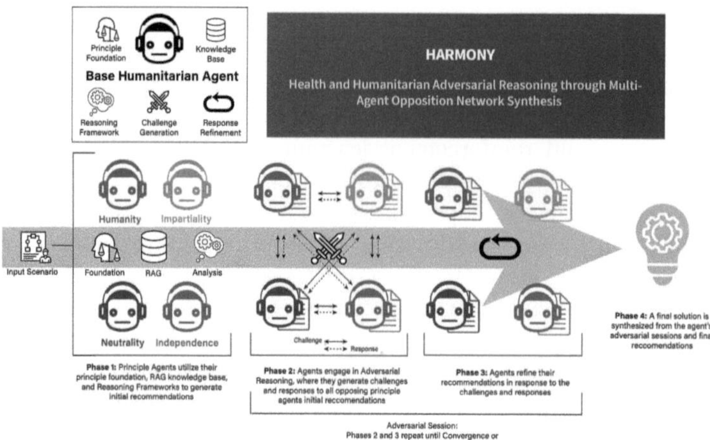

Fig. 1. HARMONY system architecture showing the four-phase adversarial reasoning protocol. Each principle-based agent (Humanity, Neutrality, Impartiality, Independence) contains specialized knowledge bases and reasoning frameworks. The protocol proceeds through independent analysis, adversarial challenge generation, position refinement, and solution synthesis phases to produce unified recommendations.

feedback, calibrates confidence levels based on critique quality, and identifies compromise solutions when appropriate.

Adversarial Reasoning Protocol and Solution Synthesis: HARMONY implements a four-phase adversarial reasoning protocol with formal convergence properties, inspired by recent advances in multi-agent debate systems:

1. **Phase 1 - Independent Analysis:** Each agent a_i analyzes scenarios using principle-specific reasoning, generating initial positions P_i with confidence assessments: $P_i = R_i(\text{scenario}, K_i, O_i)$
2. **Phase 2 - Adversarial Challenge Generation:** Agent a_i generates structured challenges C_{ij} targeting position P_j using predefined templates to prevent circular reasoning: $C_{ij} = \text{Template}_{i \to j}(P_j, O_i, K_i)$
3. **Phase 3 - Position Refinement:** Agents update positions based on received challenges, evaluating legitimacy before responding: $P_i' = F_i(P_i, \{C_{ji} | j \neq i\}, K_i, O_i)$
4. **Phase 4 - Solution Synthesis:** System generates unified recommendations incorporating all refined positions and adversarial sessions: $S = \text{Synthesize}(\{P_i' | i \in \mathcal{A}\}, \{C_{ij} | i, j \in \mathcal{A}, i \neq j\})$

3.2   Evaluation Methodology

Scenario Generation: We generated 200 synthetic humanitarian medical scenarios using Claude Sonnet 4 [2], systematically distributed across five medical ethics domains: treatment across conflict lines, medical resource allocation,

Table 1. Summary of the 200 synthetic humanitarian-medical scenarios. Each domain contains 40 scenarios that systematically vary across urgency level (immediate, short-, medium-, and long-term) and stakeholder complexity (2–8 affected groups).

Ethics domain	Examples	Count
Treatment across conflict lines	Checkpoint triage, cross-line patient transfer, remote surgical guidance	40
Medical resource allocation	Ventilator triage, blood-unit rationing, vaccine prioritisation	40
Vulnerable-population protection	Prenatal care, paediatric malnutrition, elder chronic-disease management	40
Ethics under authority pressure	Forced triage orders, selective-care directives, outbreak-data censorship	40
Emergency response coordination	Multi-agency cholera response, earthquake trauma management, mass-casualty airlift	40

protection of vulnerable populations, medical ethics under authority pressure, and emergency medical response coordination. Each domain contains 40 scenarios that vary across urgency (immediate, short-, medium-, and long-term) and stakeholder complexity (2–8 affected parties), ensuring comprehensive coverage of decision-making contexts and ethical complexity. Claude Sonnet 4 was selected because Anthropic's Constitutional AI fine-tuning [3] incorporates a charter of UN-inspired humanitarian principles and, according to Anthropic's evaluations, delivers more stable ethical alignment than models trained solely with conventional RLHF (Table 1).

Comparative Analysis: We compare three approaches to isolate adversarial reasoning contributions: **(1) Single-Shot:** Direct scenario analysis with comprehensive humanitarian principles. **(2) Reflection-Enhanced:** Iterative self-critique calibrated to 80% of HARMONY's token consumption, controlling for computational cost effects rather than architectural benefits. **(3) HARMONY:** Full adversarial multi-agent implementation.

The following methods were used to ensure as robust of an evaluation as possible. **(1)** Both the Single-Shot and Reflection-Enhanced approaches had access to the full RAG document store of all principled agents and comprised a comprehensive system prompt detailing the response and reasoning frameworks expected in ethical scenario analysis. **(2)** Token equivalence ensures the reflection baseline undergoes multiple improvement iterations until reaching comparable computational cost, addressing whether performance gains result from increased resources rather than adversarial reasoning effectiveness. **(3)** All approaches employ identical response templates (executive decisions, implementation plans, risk assessments, principled justifications) to eliminate formatting bias and ensure responses were evaluated for the quality of the final content.

Evaluation Criteria and Application. Three independent LLM judges scored every answer on a 1–5 Likert scale across 10 criteria that couples technical rigor with humanitarian ethics. Our multi-judge, multi-criteria protocol quantifies reasoning quality while tempering the subjectivity inherent to medical ethics, where decisions have no single correct answer and instead hinge on value-laden trade-offs best captured by qualitative reasoning rather than factual verification. Our criteria include: *1. Overall quality* captures surface accuracy—clarity, logical flow, and absence of hallucinations. *2. Specificity* rewards inclusion of case-relevant clinical details rather than boiler-plate advice. *3. Decisiveness* measures whether a single, unambiguous course of action is offered instead of hedged alternatives. *4. Implementability* assesses whether the recommendation translates into concrete, resource-feasible steps appropriate to austere field conditions. *5. Principled trade-offs* scores the explicit weighing of benefits versus harms and the justification of any compromise. *6. Stakeholder navigation* credits responses that identify all affected parties (patients, caregivers, community, authorities) and balance their interests transparently. The remaining four criteria instantiate the humanitarian principles: *7. Humanity* (minimising suffering and protecting dignity, with justification of patient-welfare primacy), *8. Impartiality* (need-based triage logic and explicit non-discrimination), *9. Neutrality* (strategies to avoid advantaging any belligerent under coercion), and *10. Independence* (freedom from political, military, or donor influence, with acknowledgement of external pressures).

4   Results

4.1   Models and Evaluation Setup

All three approaches – Single-Shot, Reflection-Enhanced, and HARMONY – were implemented using GPT-4o-mini (OpenAI, June 2025 release) [17] with temperature set to 0.3 and a maximum (per-query) token limit per-response of 1024. For each scenario, the three resulting answers were shown together to a panel of three independent LLM judges, Gemini 2.0-flash [8], Claude 3-7-sonnet-latest [2], and GPT-4o. Judges were instructed to: (i) rate each answer on the ten 1–5 Likert criteria, and (ii) select a single preferred answer. The prompt reiterated the rubric to ensure uniform application: technical dimensions (clarity, specificity, decisiveness, implementability) were assessed first, followed by ethical dimensions (benefit–harm trade-offs, stakeholder navigation, and the four humanitarian principles). We aggregate median per-criterion scores across judges and use the majority-vote preference to compare approaches.

4.2   Performance and Inter-evaluator Agreement

Across 200 humanitarian medical scenarios, HARMONY achieved a mean composite score (averaging the scores across the 10 Likert categories) 7.49 out of 10 (95% CI [7.42, 7.56]), substantially surpassing Reflection-Enhanced (7.06; 95% CI [6.96, 7.15]) and Single-Shot (5.59; 95% CI [5.53, 5.65]) baselines. This preference was consistent across all three independent LLM evaluators, reducing single-evaluator bias concerns (Fig. 2, Table 2).

One-way Analysis of Variance (ANOVA) confirmed significant differences among methods ($F(2, 600)=614.1$, $p<0.0001$). Bonferroni corrected pairwise tests revealed HARMONY significantly outperformed both Reflection ($p<0.0001$) and Single-Shot ($p<0.0001$).

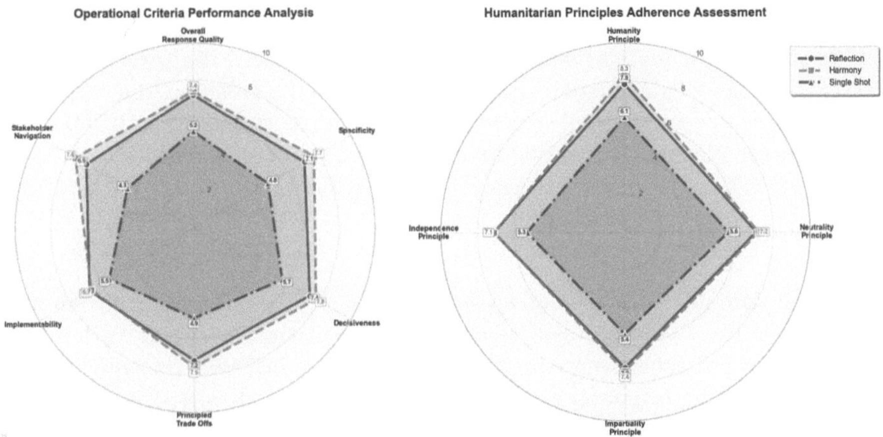

Fig. 2. Performance comparison across evaluation criteria. Right panel shows humanitarian principles adherence assessment with HARMONY (orange) demonstrating superior performance across all four principles compared to Reflection-Enhanced (blue) and Single-Shot (red) approaches. Left panel displays operational criteria performance analysis, with HARMONY showing consistent advantages in specificity, decisiveness, implementability, and stakeholder navigation. (Color figure online)

Table 2. Per-model average ratings and overall performance for each approach (200 scenarios).

Approach	Claude-3.7	GPT-4o	Gemini-2.0	Mean Score	95% CI	Win Rate (%)
Harmony	6.97	8.33	7.15	7.49	[7.42, 7.56]	63.7
Single-Shot	4.70	5.86	5.28	5.28	[5.23, 5.33]	1.0
Reflection	6.42	8.37	6.74	7.19	[7.10, 7.27]	33.1

Inter-evaluator reliability measured by Cohen's κ averaged 0.78 across judge pairs ($\kappa_{\text{Gemini-Claude}} = 0.80$, $\kappa_{\text{Gemini-GPT4o}} = 0.77$, $\kappa_{\text{Claude-GPT4o}} = 0.78$), indicating substantial agreement and validating our multi-judge protocol.

HARMONY demonstrated consistent advantages across all evaluation dimensions: Overall Quality (HARMONY: 7.39 vs. Reflection: 7.20 vs. Single-Shot: 5.22), Operational Feasibility (HARMONY: 7.46 vs. Reflection: 7.06 vs. Single-Shot: 5.22), and Humanitarian Principles (HARMONY: 7.55 vs. Reflection: 7.31 vs. Single-Shot: 5.55). Notably, HARMONY's strongest improvement was in "Principled Trade-offs" (HARMONY: 7.84 vs. Reflection: 7.06 vs. Single-Shot: 5.23), supporting our hypothesis that adversarial reasoning enhances ethical trade-off navigation.

Reasoning Trace Analysis: We compared HARMONY's intermediate reasoning traces with its final synthesized outputs and found that evaluators consistently identified challenges, insights, dilemmas, and conflicts in the traces that were not reflected in the final solutions. This discrepancy indicates that the synthesis process plays a critical role in determining solution quality.

5   Discussion

This work demonstrates that adversarial reasoning can systematically improve decision quality in multi-objective ethical scenarios by leveraging rather than eliminating tension between competing principles. Consistent preference across three independent evaluators establishes that structured opposition generates solutions with superior reasoning transparency and stakeholder consideration compared to consensus-seeking approaches, validating adversarial reasoning as a viable paradigm for AI decision-making in complex ethical landscapes.

The four-phase adversarial protocol provides a generalizable framework extending beyond humanitarian contexts to financial risk assessment, regulatory compliance, and strategic planning where multiple valid frameworks generate contradictory recommendations.

While current limitations restrict immediate practical deployment, HARMONY provides essential foundations for developing AI systems capable of principled ethical reasoning. As AI increasingly supports critical decision-making in healthcare, policy, and humanitarian response, frameworks like HARMONY offer tools for ensuring these systems engage constructively with fundamental moral complexity rather than avoiding it.

Limitations and Future Directions: Simulation-Based Evaluation: Our evaluation relies on synthetic scenarios rather than real-world humanitarian crises, limiting immediate practical applicability. Future work will incorporate human expert validation and retrospective analysis of documented humanitarian cases to confirm practical utility and enable controlled pilot studies with humanitarian organizations.

LLM Evaluator Reliability: Despite substantial inter-rater agreement ($\kappa=0.78$), LLM evaluators may inherit model biases that human experts would identify. We will integrate human-in-the-loop validation where domain experts will evaluate model responses as well as participate directly as specialized agents in the adversarial reasoning process.

Prompt Engineering Dependence: Current implementation relies on static prompts without adaptive learning. Future iterations will implement reinforcement learning techniques (RLHF/RLAIF) and constitutional AI training to improve challenge generation quality and synthesis effectiveness.

Robustness Testing: We will conduct adversarial testing where external agents attempt to manipulate the system into unethical decisions, evaluating whether the multi-agent framework provides superior alignment guarantees compared to single-agent approaches.

References

1. Anthropic: Collective constitutional ai: Aligning a language model with public input (2023). https://www.anthropic.com/research/collective-constitutional-ai-aligning-a-language-model-with-public-input
2. Anthropic PBC: Claude Sonnet 4. https://www.anthropic.com/claude/sonnet (2025). Accessed 25 June 2025
3. Bai, Y., et al.: Constitutional ai: Harmlessness from ai feedback. arXiv preprint arXiv:2212.08073 (2022)
4. Chan, C.M., et al.: Chateval: Towards better llm-based evaluators through multi-agent debate. arXiv preprint arXiv:2308.07201 (2023)
5. Devidal, P.: 'back to basics' with a digital twist: humanitarian principles and dilemmas in the digital age. ICRC Humanitarian Law & Policy Blog, February 2023. https://blogs.icrc.org/law-and-policy/2023/02/02/back-to-basics-digital-twist-humanitarian-principles/
6. Du, Y., Li, S., Torralba, A., Tenenbaum, J.B., Mordatch, I.: Improving factuality and reasoning in language models through multiagent debate. arXiv preprint arXiv:2305.14325 (2023)
7. Gao, Y., et al.: Retrieval-augmented generation for large language models: a survey. arXiv preprint arXiv:2312.10997 (2023)
8. Google DeepMind: Gemini 2.0 Pro Experimental (2025). https://blog.google/technology/google-deepmind/gemini-model-updates-february-2025/. Accessed 25 June 2025
9. Guo, Y., et al.: Controllable preference optimization: Toward controllable multi-objective alignment. arXiv preprint arXiv:2402.19085 (2024)
10. ICRC and IFRC: The Fundamental Principles of the International Red Cross and Red Crescent Movement: Ethics and Tools for Humanitarian Action. International Committee of the Red Cross and International Federation of Red Cross and Red Crescent Societies (2015)
11. Jiang, L., et al.: Can machines learn morality? the Delphi experiment. arXiv preprint arXiv:2110.07574 (2021)
12. Jung, J., Sarkar, A., Bang, J., Huang, L., Yue, X., Xing, E.P.: Trust or escalate: Llm judges with provable guarantees for human agreement. arXiv preprint arXiv:2407.18370 (2024)
13. Kilov, D., Hendy, C., Guyot, S.Y., Snoswell, A.J., Lazar, S.: Discerning what matters: A multi-dimensional assessment of moral competence in llms. arXiv preprint arXiv:2506.13082 (2025)
14. Loquercio, D., Schüepp, M.: People, principles, and processes: accountability in humanitarian action. ICRC Humanitarian Law & Policy Blog, August 2023, https://blogs.icrc.org/law-and-policy/2023/08/29/people-principles-processes-accountability-humanitarian-action/
15. Machado, J., Sousa, R., Peixoto, H., Abelha, A.: Ethical decision-making in artificial intelligence: A logic programming approach. AI **5**(4), 2707–2724 (2024)
16. Neuman, W.R., Coleman, C., Shah, M.: Analyzing the ethical logic of six large language models. arXiv preprint arXiv:2501.08951 (2025)
17. OpenAI: ChatGPT Overview. https://openai.com/chatgpt/overview/ (2025). Accessed 25 June 2025
18. Penney, G., Launder, D., Cuthbertson, J., Thompson, M.B.: Ethical decision making in disaster and emergency management: A systematic review of the literature. Prehosp. Disaster Med. **38**(5), 622–627 (2023). https://doi.org/10.1017/S1049023X23006325

19. Ray, O.: Principles under pressure: have humanitarian principles really stood the test of time? ICRC Humanitarian Law & Policy Blog, July 2024. https://blogs.icrc.org/law-and-policy/2024/07/11/principles-under-pressure-have-humanitarian-principles-really-stood-the-test-of-time/
20. Schmidt, U., Clarinval, C., Biller-Andorno, N.: Challenges to ethical obligations and humanitarian principles in conflict settings. J. Humanitarian Action **4**, 63 (2019). https://doi.org/10.1186/s41018-019-0063-x
21. Veer, J., Simmonds, M., Thompson, M.B.: Ethical decision-making in humanitarian medicine: how best to prepare? Disaster Med. Public Health Prep. **15**(4), 499–503 (2021). https://doi.org/10.1017/dmp.2020.85
22. Zheng, L., et al.: Judging llm-as-a-judge with mt-bench and chatbot arena. arXiv preprint arXiv:2306.05685 (2023)
23. Zhou, Z., et al.: Beyond one-preference-fits-all alignment: Multi-objective direct preference optimization. arXiv preprint arXiv:2310.03708 (2023)

AURA: A Multi-modal Medical Agent for Understanding, Reasoning and Annotation

Nima Fathi[1,2(✉)], Amar Kumar[1,2], and Tal Arbel[1,2]

[1] Center for Intelligent Machines, McGill University, Montreal, Canada
[2] Mila - Quebec AI institute, Montreal, Canada
`nima.fathi@mail.mcgill.ca`

Abstract. Recent advancements in Large Language Models (LLMs) have catalyzed a paradigm shift from static prediction systems to agentic AI—intelligent agents capable of reasoning, interacting with tools, and adapting to complex tasks. While LLM agentic systems have shown promise across many domains, their application to medical imaging remains in its infancy. In this work, we introduce *AURA*, the first visual linguistic explainability agent designed specifically for comprehensive analysis, explanation, and evaluation of medical images. By enabling dynamic interactions, contextual explanations, and hypothesis testing, *AURA* represents a significant advancement toward more transparent, adaptable, and clinically aligned AI systems. This work highlights the promise of agentic AI in transforming medical image analysis from static predictions to interactive decision support. Leveraging Qwen-32B, an LLM-based architecture, *AURA* integrates a modular toolbox comprising: (i) a segmentation suite with phase grounding, pathology segmentation, and anatomy segmentation to localize clinically meaningful regions; (ii) a counterfactual image generation module that supports reasoning through image-level explanations; and (iii) a set of evaluation tools, including pixel-wise difference map analysis, classification, and advanced state-of-the-art components, to assess the diagnostic relevance and visual interpretability of the results. Our code is accessible through the project website (https://nimafathi.github.io/AURA/).

Keywords: AI Agents · Counterfactual Image Generation · Explainability · Generative Modeling · Vision-Language Foundation Models

1 Introduction

Conventional AI models in medical imaging often fail to meet the needs of real clinical practice. Ideally, an AI system would reason independently, recognize when it lacks sufficient context, and dynamically use various tools—much like clinicians do in complex diagnostic scenarios. However, traditional medical imaging AI is typically rigid, designed for specific tasks with fixed inputs and outputs.

J. Qiu et al. (Eds.): Agentic AI 2025/CMLLMs 2025/CREATE 2025, LNCS 16147, pp. 105–114, 2026.
https://doi.org/10.1007/978-3-032-06004-4_11

This lack of flexibility prevents these systems from adapting to changing clinical situations. When faced with unclear findings, unfamiliar diseases, or incomplete information, these models cannot ask for additional details, gather more data, or revise their conclusions [22,23]. Consequently, they fall short in interpretability, adaptability, and gaining clinical trust. Agentic AI offers a promising alternative, providing models that not only handle specific tasks but also reason through uncertainty, generate clear visual and linguistic explanations (VLEs), test hypotheses via counterfactual simulations, and collaborate interactively with clinicians. By combining autonomous reasoning with dynamic tool use, agentic systems significantly enhance AI's practical utility in clinical settings, bridging the gap between static automation and the flexible decision-making required in medical practice.

Agentic AI has gained significant popularity across a variety of domains where systems must reason under uncertainty, take autonomous actions, and interact adaptively with their environments [1,7]. Web-based agents such as AutoGPT [26] and Voyager [24] have demonstrated how large language models (LLMs) can control digital interfaces, pursue multi-step goals, and coordinate tool use—shifting the role of AI from static prediction to dynamic, goal-driven problem-solving. In medical imaging, the principles of agentic AI are beginning to shape how multiple models and tools can work together coherently to assist with complex clinical tasks [12,29]. Foundation models (FMs), including LLMs and large multimodal models (LMMs), provide a powerful foundation for building unified frameworks that integrate medical image-text reasoning, diagnostic analysis, and clinical decision support. Several recent agentic frameworks have advanced this direction by introducing structured reasoning, tool orchestration, and modality-aware workflows. MDAgents [13] presents a multi-agent system that dynamically configures collaboration between LLMs based on task complexity, achieving state-of-the-art (SOTA) performance on medical benchmarks. MMedAgent [16] introduces the first multi-modal medical AI agent capable of intelligently selecting and integrating specialized tools—such as segmentation, classification, and report generation—across five imaging modalities. Using instruction tuning and dynamic tool invocation, it outperforms both open- and closed-source models, including GPT-4o [9]. MedRAX [5] combines state-of-the-art analysis tools with multimodal LLMs to address complex queries involving visual question answering and report generation. While these frameworks represent major progress, many still lack explicit reasoning capabilities and robust image-grounded explanations—features critical for transparency, safety, and clinical trust. Together, these frameworks underscore the increasing role of agentic AI in developing more interactive and intelligent healthcare systems.

In this work, we present *AURA*, the first AI agent designed to analyze, generate visual-linguistic explanations (VLEs), and perform self-evaluations using a comprehensive suite of SOTA tools. Unlike conventional models that offer limited interpretability and operate in a static inference paradigm, *AURA* emphasizes dynamic VLE, introspective evaluation, and adaptive reasoning in data-scarce clinical settings. Our contributions include an agent capable of: (i) image segmentation with phase grounding (associating regions of medical image to corresponding text or clinical concepts), pathology segmentation, and anatomy segmentation to identify and attribute clinically relevant regions; (ii) CF image generation for probing understanding through controlled perturbations; and (iii) a self-evaluation toolkit incorporating difference map analysis, classification, and specialized tools on Chest X-ray (CXR) images such as RadEdit [20] and PRISM [15], enabling critical assessment of its own outputs. Notably, *AURA* performs robustly even in low-supervision scenarios (less textual prompt guidance) by autonomously identifying knowledge gaps, invoking report generation to gain context, and generating multiple candidate CFs. Finally, the best candidate is chosen using a self-evaluation scoring mechanism. This integration of explanation, self-assessment, and context-aware tool use marks a step toward more trustworthy, adaptable, and clinically actionable AI systems.

2   AURA: An Agent for Visual-Linguistic Explanation

We present *AURA*, an open-source, modular agent designed to provide interpretable, multimodal visual-linguistic explanations for medical imaging. Built to address the limitations of static and non-integrated AI pipelines, *AURA* flexibly integrates a diverse suite of expert medical tools—spanning report generation and grounding, visual question answering (VQA), segmentation, pathology detection, and counterfactual image editing—within a reasoning-driven loop powered by a large language model. *AURA* supports user-driven interactions and autonomous self-evaluation, allowing it to select and optimize among multiple

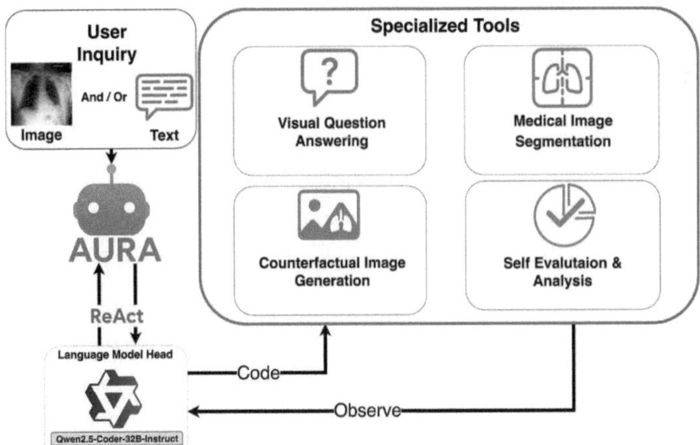

Fig. 1. *AURA architecture* integrates a reasoning loop that dynamically selects from a multimodal ecosystem of tools. The agent iteratively reasons over image-query pairs, generates *code* to invoke tools, and *observes* their resulting outputs to synthesize high-quality visual and textual medical explanations. ReAct framework enables visualization and control of the agent's reasoning process.

strategies to produce clinically meaningful outputs, Fig. 1. We conduct all evaluations and demonstrate the agent's capabilities in the context of chest X-rays, leveraging the CheXpert dataset [11].

Algorithm 1. Overview of *AURA*'s Dynamic Reasoning and Tool Orchestration

---

1: **Input:** Q: User query, I: Chest X-ray image (optional), $\mathcal{T}$: Set of available medical AI tools, M: LLM Head, t_{max}: Maximum allowed steps
2: **Initialize:** $step \leftarrow 0$; $memory \leftarrow \emptyset$; $current\_state \leftarrow$ OBSERVE$(Q, I, memory)$
3: **while** $step < t_{max}$ **do**
4:        ▷ Agent reasons about the next action based on current state and memory
5:     $thought \leftarrow$ M.REASON$(current_state, memory)$
6:     **if** M.CANGENERATEFINALANSWER(thought, memory) **then**   ▷ LLM determines if a final answer can be formed based on memory and the last "thought"
7:         **return** M.GENERATEFINALANSWER(thought, memory)
8:         ▷ Select the most appropriate tool based on reasoning and current state
9:     $tool \leftarrow$ SELECTTOOL$(thought, \mathcal{T}, current_state)$
10:     $tool\_input \leftarrow$ PREPARETOOLINPUT$(thought, current_state, I)$ ▷ Prepare input for the selected tool
11:   ▷ Execute the selected tool
12:     $tool\_result \leftarrow$ EXECUTE$(tool, tool_input)$
13:           ▷ Update memory and current state with the result of tool execution
14:     $memory \leftarrow memory \cup \{(thought, tool, tool\_result)\}$
15:     $current\_state \leftarrow$ M.OBSERVE$(current_state, tool_result, memory)$
16:     $step \leftarrow step + 1$
17: **return** M.GENERATETIMEOUTRESPONSE$(current_state, memory)$

2.1 Dynamic Reasoning with an Expert Toolkit

AURA is driven by a ReAct-style reasoning loop [27] powered by a code-instructed LLM, which is specifically fine-tuned and optimized for understanding and generating executable code. This enables the agent to break down a user's request into a sequence of logical steps and execute and perform each step by leveraging its integrated set of medical tools. (Algorithm 1). Unlike static pipelines, our dynamic approach enables *AURA* to infer chest CXR findings—crucial for visual verification and clinical interpretability—and to dynamically generate and select the most relevant visual evidence through a self-evaluation mechanism that effectively supports its textual responses. *AURA*'s analytical power comes from its ability to flexibly chain tools from its integrated toolkit:

- **Visual Question Answering**: Radiology specific dialogue using the Chex-Agent VQA [3] or generates pathology medical reports using MAIRA-2 [2].
- **Grounded Report Generation**: Employs MAIRA-2 [2] to align medical findings with bounding box or segmentation overlays for visual grounding.
- **Counterfactual Editing**: Utilizes *RadEdit* [20] for precision-guided image editing of pathologies and *PRISM* [15] to generate counterfactual images.

- **Segmentation and Detection**: Leverages MedSAM [17] and PSPNet [28] for anatomy localization and TorchXRayVision for pathology classification [4].
- **Analysis and Visualization**: Difference maps to quantify edits, generate subject-specific image variations, and manage the overall analysis session.

2.2 Modular Architecture for Self-Evaluation

AURA's architecture is implemented using modular segments where each tool is an independent component. This design supports parallel execution across multiple GPUs, robust fallback strategies, and makes the system easily extendable. More importantly, this modularity is the foundation for *AURA*'s most defining feature: self-evaluation. By treating its own tools as both actors and critics, *AURA* can autonomously assess and refine its work. For example, when tasked with removing a pathological structure from an image, the agent initiates the following multi-step, self-correcting workflow:

1. Generate several candidate CF images (offering varying explainability) using *RadEdit* [20] and *PRISM* [15] with varying hyper-parameters.
2. Leverage *TorchXRayVision* [4] to classify pathological structures in the original and for each CF, candidate explanation image.
3. Compare pathology and similarity scores, selecting CF that achieves the highest improvement in the target pathology's score while effectively preserving subject identity, ensuring a clinically relevant and high-quality edit.

This generate–test–select strategy is a core capability of *AURA*, enabling it to optimize parameters, compare outputs across models, and produce accurate, visually coherent explanations. The entire process is made transparent to the user through an interactive interface that displays step-by-step diagnostic evidence, including visual comparisons of counterfactual and factual images. This interface provides both qualitative explanations through difference maps and quantitative justification through corresponding pathology scores. For an overview of different scenarios that our model could handle, please refer to Fig. 2.

3 Experiments and Results

3.1 Dataset and Implementation Details

For evaluation, we utilize the held-out test set of the publicly available CheXpert dataset [11]. As *AURA* operates as an inference agent, leveraging off-the-shelf tools rather than being fine-tuned, this held-out set provides the sole relevant data for assessing its performance. To ensure a fair comparison and mitigate potential biases or distribution shifts, we adopt the identical data split used by PRISM [15]. *AURA* deploys Qwen2.5-Coder-32B-Instruct [8] as its backbone LLM. This leverages Qwen's capabilities in generating executable code actions and robust function calls for seamless tool interaction [25]. The agent orchestrates its integrated tools via the SmolAgents framework, allowing it to formulate requests with required arguments through programmatic Python function

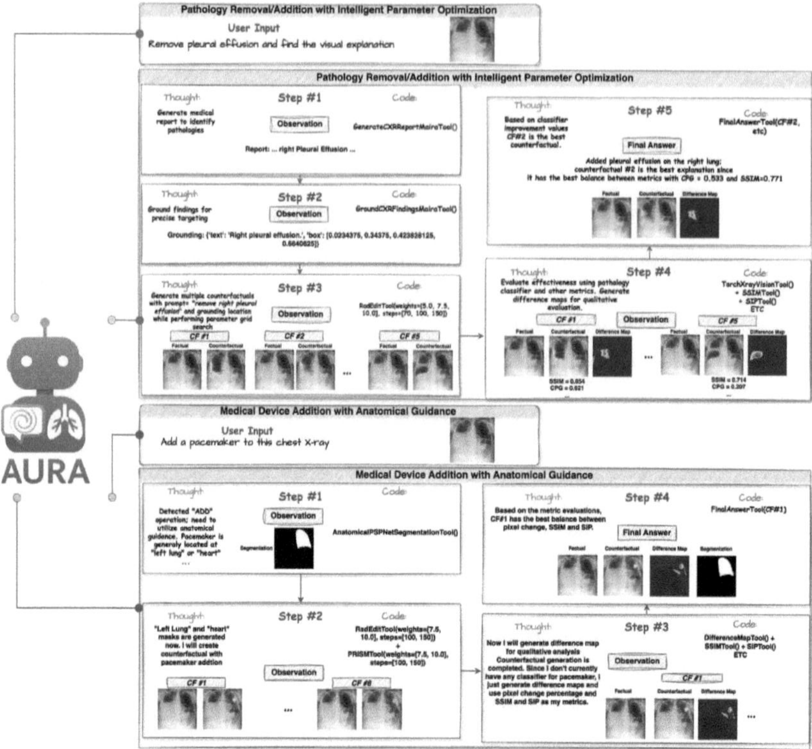

Fig. 2. AURA Explainability Flow. *AURA* handles an open-ended clinical question by analyzing the image, localizing findings, and producing both visual and linguistic explanations. The interface displays intermediate results and self-evaluations to ensure transparency. Note: The thought behind evaluation and selecting the final image is described in the image. Zoom in for better visibility.

calls [21]. *AURA* uses two NVIDIA A100 80GB GPUs as the backend for running inferences. In case multiple GPUs are detected on the system, our method is adaptive to parallelize on multiple GPUs. This on-premises deployment is particularly advantageous in medical settings, as it provides a secure and controlled environment for sensitive patient data and effectively mitigates the privacy risks associated with external, cloud-based LLM APIs.

3.2 Agentic Image-Based Explainability via CF Image Generation

We demonstrate the efficacy of *AURA*'s capabilities in CF image generation through a comparative experiment against its underlying image editing tools, RadEdit and PRISM. This setup highlights how *AURA*'s autonomous decision-making outperforms fixed-strategy baselines. For RadEdit and PRISM, we evaluate both single CF generation (one CF per instance) and an "Ensemble" approach, where five CFs are generated per instance by varying parameters such

Table 1. Comparison of counterfactual editing methods across evaluation metrics. Baselines (RadEdit, PRISM) include single CF generation (#CFs = 1) and an Ensemble (#CFs = 5). *AURA* (#CFs = 5) acts as an intelligent agent, internally self-evaluating and selecting the optimal CF.

Method	# CFs	CPG ↑	CFR ↑	SSIM ↑	SIP ↓
RadEdit	1	0.264	0.41	0.764	0.055
RadEdit-Ensemble	5	0.355	0.55	0.778	0.059
PRISM	1	0.418	0.67	0.648	0.081
PRISM-Ensemble	5	0.459	0.71	0.661	0.079
AURA	5	0.443	0.71	0.740	0.060

as guidance scale and inference steps (with RadEdit also including grounded and ungrounded conditions, where applicable). A post-processing script is then used to select the best-performing CF from the ensemble based on external metric evaluation. In contrast, *AURA*'s CF generation is entirely agent-driven: for each instance, AURA intelligently explores different generation settings and editing tools (RadEdit or PRISM), autonomously produces multiple candidate counterfactuals, and evaluates them using internal metrics. It then selects the optimal counterfactual as the final output—without any external tuning or post-processing. To ensure fairness, *AURA* is constrained to produce and present a maximum of five counterfactuals per instance, aligning with the ensemble conditions.

We quantitatively assess the CFs using four metrics: (i) *Subject Identity Preservation* (SIP) is the L1 distance between CF and factual images to assess preservation of identity [18]; [18]; (ii) *Counterfactual Prediction Gain* (CPG) measures the absolute difference in an off-the-shelf DenseNet121 [10] multi-head classifier's predictions (from TorchXRayVision [4]) between factual and CF images [19]. A higher CPG indicates a greater shift across the classifier's decision boundary; (iii) *Classifier Flip Rate* (CFR) is the number of samples with flipped predictions after CF generation [6]; (iv) *Structural Similarity Index Measure* (SSIM) to identify the visual similarity between factual and CF images [14].

Table 1 presents the quantitative comparison of these methods. We observe that generating multiple counterfactuals and selecting the best (Ensemble) significantly boosts performance over single CF generation across metrics. Specifically, *AURA* demonstrates a superior balance across metrics: its SSIM is notably higher than PRISM-Ensemble, indicating better identity preservation while achieving similar CPG/CFR, and its SIP is comparable to RadEdit-Ensemble.

3.3   Adaptive Explanations with Limited Pathological Knowledge

We evaluate *AURA*'s emergent ability to generate accurate explanations when provided with limited pathological details, a critical requirement for real-world clinical use where users' prior knowledge is limited. As illustrated in Fig. 3, ambiguous user inputs with no explicit pathological context typically result in poor-quality edits when standard image-editing tools rely solely on generic prompts such as *"Normal chest X-ray with no finding"*. These tools inherently depend on precise textual instructions to produce targeted edits. In contrast, *AURA* recognizes the lack of explicit information as a challenge and proactively engages a medical report generation tool to identify specific pathological findings. It then leverages this detailed information to create an accurate prompt for subsequent editing tasks. This adaptive, agent-based reasoning enables *AURA* to surpass predefined instructions, achieving edits that are both precise and contextually relevant. Quantitative results, including CPG and SIP scores in Fig. 3, further demonstrate *AURA*'s effectiveness in these demanding situations.

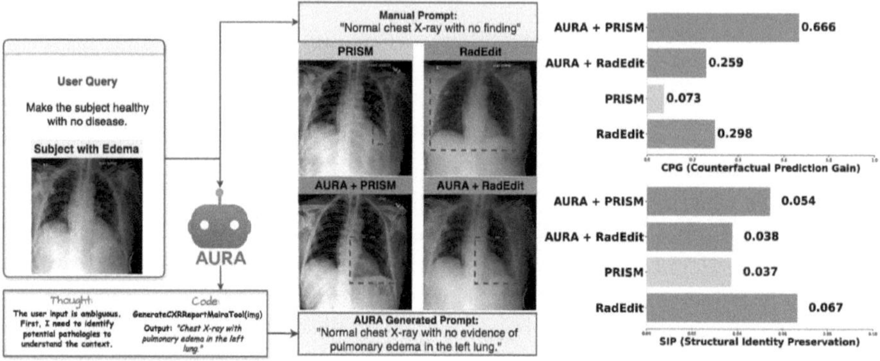

Fig. 3. Adaptive CF generation from user queries. This figure illustrates *AURA*'s agentic reasoning. Given an ambiguous query (top-left), *AURA* autonomously identifies limited pathology knowledge, calls a report tool, and generates a precise prompt (bottom-middle). Bounding boxes indicate the area of interest for each edit. Top: Baselines (PRISM/RadEdit) utilise generic prompts, resulting in over- or under-editing respectively. Bottom: AURA-guided edits show improved focus on correct pathology and visual quality. CPG and SIP plots (right) quantitatively confirm *AURA*'s superior performance.

4   Conclusion

In this work, we introduced *AURA*, an agentic AI system designed to analyze, explain, and evaluate chest X-ray images through a modular toolbox encompassing segmentation, counterfactual generation, and evaluation tools. Building upon recent works in LLMs, *AURA* brings explainability and autonomy to medical imaging agentic workflows. Our evaluations highlight *AURA*'s ability to operate effectively under limited supervision and incomplete pathological information, conditions that often reflect real-world clinical settings. Notably, *AURA* demonstrates intelligent decision-making by recognizing gaps in context, invoking report generation tools when necessary, and refining visual edits accordingly. This capacity for self-directed reasoning and adaptation represents a step forward in the development of clinically aligned AI systems. Looking ahead, we envision agentic frameworks like *AURA* playing a critical role in advancing trustworthy, transparent, and context-aware AI for medical diagnostics.

Acknowledgments. Funding was provided in part by the Natural Sciences and Engineering Research Council of Canada, the Canadian Institute for Advanced Research (CIFAR) Artificial Intelligence Chairs program, Mila - Quebec AI Institute, Google Research, Calcul Quebec, and the Digital Research Alliance of Canada.

Disclosure of Interests. The authors have no competing interests to declare that are relevant to the content of this article.

References

1. Acharya, D.B., Kuppan, K., Divya, B.: Agentic ai: autonomous intelligence for complex goals–a comprehensive survey. IEEE Access (2025)
2. Bannur, S., et al.: Maira-2: grounded radiology report generation. arXiv preprint arXiv:2406.04449 (2024)
3. Chen, Z., et al.: Chexagent: towards a foundation model for chest x-ray interpretation. arXiv preprint arXiv:2401.12208 (2024)
4. Cohen, J.P., et al.: Torchxrayvision: a library of chest x-ray datasets and models. In: International Conference on Medical Imaging with Deep Learning, pp. 231–249. PMLR (2022)
5. Fallahpour, A., Ma, J., Munim, A., Lyu, H., Wang, B.: Medrax: medical reasoning agent for chest x-ray. arXiv preprint arXiv:2502.02673 (2025)
6. Fathi, N., Kumar, A., Nichyporuk, B., Havaei, M., Arbel, T.: Decodex: confounder detector guidance for improved diffusion-based counterfactual explanations. arXiv preprint arXiv:2405.09288 (2024)
7. Hughes, L., et al.: Ai agents and agentic systems: a multi-expert analysis. J. Comput. Inf. Syst. 1–29 (2025)
8. Hui, B., et al.: Qwen2. 5-coder technical report. arXiv preprint arXiv:2409.12186 (2024)
9. Hurst, A., et al.: Gpt-4o system card. arXiv preprint arXiv:2410.21276 (2024)
10. Iandola, F., Moskewicz, M., Karayev, S., Girshick, R., Darrell, T., Keutzer, K.: Densenet: implementing efficient convnet descriptor pyramids. arXiv preprint arXiv:1404.1869 (2014)
11. Irvin, J., et al.: Chexpert: a large chest radiograph dataset with uncertainty labels and expert comparison. In: Proceedings of the AAAI Conference on Artificial Intelligence, vol. 33, pp. 590–597 (2019)
12. Karunanayake, N.: Next-generation agentic ai for transforming healthcare. Inf. Health **2**(2), 73–83 (2025)
13. Kim, Y., et al.: Mdagents: an adaptive collaboration of llms for medical decision-making. Adv. Neural. Inf. Process. Syst. **37**, 79410–79452 (2024)
14. Kumar, A., et al.: Counterfactual image synthesis for discovery of personalized predictive image markers. In: MICCAI Workshop on Medical Image Assisted Biomarkers' Discovery, pp. 113–124. Springer, Heidelberg (2022). https://doi.org/10.1007/978-3-031-19660-7_11

15. Kumar, A., Kriz, A., Havaei, M., Arbel, T.: Prism: high-resolution & precise counterfactual medical image generation using language-guided stable diffusion. MIDL (2025)
16. Li, B., et al.: Mmedagent: learning to use medical tools with multi-modal agent. arXiv preprint arXiv:2407.02483 (2024)
17. Ma, J., He, Y., Li, F., Han, L., You, C., Wang, B.: Segment anything in medical images. Nat. Commun. **15**(1), 654 (2024)
18. Mothilal, R.K., Sharma, A., Tan, C.: Explaining machine learning classifiers through diverse counterfactual explanations. In: Proceedings of the 2020 Conference on Fairness, Accountability, and Transparency, pp. 607–617 (2020)
19. Nemirovsky, D., Thiebaut, N., Xu, Y., Gupta, A.: Countergan: generating realistic counterfactuals with residual generative adversarial nets. arXiv preprint arXiv:2009.05199 (2020)
20. Pérez-García, F., et al.: Radedit: stress-testing biomedical vision models via diffusion image editing. In: European Conference on Computer Vision, pp. 358–376. Springer, Heidelberg (2025). https://doi.org/10.1007/978-3-031-73254-6_21
21. Roucher, A., del Moral, A.V., Wolf, T., von Werra, L., Kaunismäki, E.: 'smolagents': a smol library to build great agentic systems (2025). https://github.com/huggingface/smolagents
22. Sapkota, R., Roumeliotis, K.I., Karkee, M.: Ai agents vs. agentic ai: A conceptual taxonomy, applications and challenge. arXiv preprint arXiv:2505.10468 (2025)
23. Shavit, Y., Agarwal, S., Brundage, M., Adler, S., O'Keefe, C., Campbell, R., Lee, T., Mishkin, P., Eloundou, T., Hickey, A., et al.: Practices for governing agentic ai systems. Research Paper, OpenAI (2023)
24. Wang, G., et al.: Voyager: an open-ended embodied agent with large language models. arXiv preprint arXiv:2305.16291 (2023)
25. Wang, X., et al.: Executable code actions elicit better llm agents. In: Forty-First International Conference on Machine Learning (2024)
26. Yang, H., Yue, S., He, Y.: Auto-gpt for online decision making: benchmarks and additional opinions. arXiv preprint arXiv:2306.02224 (2023)
27. Yao, S., et al.: React: synergizing reasoning and acting in language models. In: International Conference on Learning Representations (ICLR) (2023)
28. Zhao, H., Shi, J., Qi, X., Wang, X., Jia, J.: Pyramid scene parsing network (2017). https://arxiv.org/abs/1612.01105
29. Zou, J., Topol, E.J.: The rise of agentic ai teammates in medicine. Lancet **405**(10477), 457 (2025)

M³Builder: A Multi-agent System for Automated Machine Learning in Medical Imaging

Jinghao Feng[1,2], Qiaoyu Zheng[1,2], Chaoyi Wu[1,2], Ziheng Zhao[1,2], Ya Zhang[1,2], Yanfeng Wang[1,2(✉)], and Weidi Xie[1,2(✉)]

[1] Shanghai Jiao Tong University, Shanghai, China
{wangyanfeng622,weidi}@sjtu.edu.cn
[2] Shanghai AI Laboratory, Shanghai, China

Abstract. In this paper, we aim to automate the labor-intensive process of developing machine learning (ML)-based tools for medical imaging, paving the way for the self-evolution of medical agentic systems. **(i)** We present **M³Builder**, a multi-agent collaboration framework designed to automate model training in medical imaging, that divide-and-conquers complex medical ML with four specialized agents. **(ii)** To better fit in the professional medical imaging domain, we build up a specialized ML context protocol, a structured environment designed to provide agents with comprehensive free-text descriptions of medical datasets, training code templates, and interaction tools. **(iii)** To monitor the progress, we propose **M³Bench**, spanning four medical imaging ML tasks across 14 datasets, covering both 2D and 3D data. Our experiments demonstrate that, when employing an identical agent core, M³Builder surpasses existing automated ML agentic architectures, achieving a superior task completion rate of 94.29% while maintaining satisfactory model performance. This highlights the potential of fully automated ML-based tool development in medical imaging. The source code is publicly available at https://github.com/MAGIC-AI4Med/M3Builder.

Keywords: Agentic System · Medical Imaging · Machine Learning

1 Introduction

Large Language Model (LLM)-powered agentic systems have demonstrated remarkable success across diverse domains. Leveraging their ability to orchestrate specialized tools, they can solve complex, multi-step tasks with precision. However, their application in medical remains challenging. Medical tasks are typically professional and complex, encompassing a wide range of diseases, imaging modalities, and task-specific requirements. This diversity makes it difficult for preparing a complete toolset to implement a specific medical agentic systems, often necessitating further labor-intensive development of specialized tools, *e.g.*, training a machine-learning-based model. However, clinicians often lack the coding skills or technical expertise to implement such processes, significantly hindering the tailored adaptation of agentic systems in medical practice.

J. Feng and Q. Zheng—Equal Contribution.

© The Author(s), under exclusive license to Springer Nature Switzerland AG 2026
J. Qiu et al. (Eds.): Agentic AI 2025/CMLLMs 2025/CREATE 2025, LNCS 16147, pp. 115–124, 2026.
https://doi.org/10.1007/978-3-032-06004-4_12

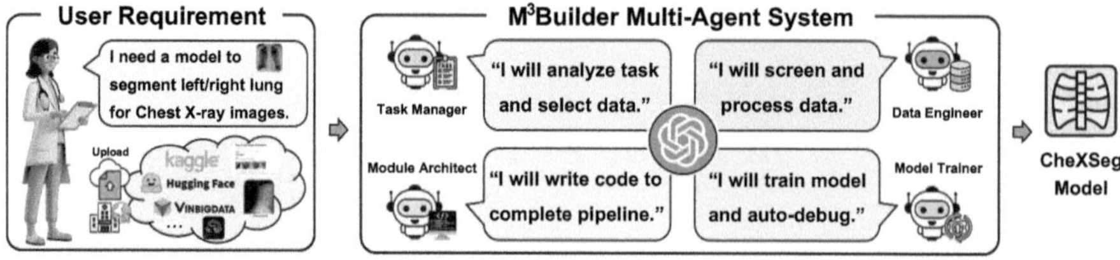

Fig. 1. General Pipeline of M³Builder. Based on user requirements and available raw clinical datasets, four agents play different roles and collaborate to complete the procedure from task analysis, data processing to the completion of model training.

In this paper, we propose **M³Builder** (M³ stands for Mutli-agent, Machine Learning, and Medical Imaging), a novel agentic framework for automating medical imaging machine learning (MIML) tasks. Given a MIML tool development demand and raw training data, **M³Builder** autonomously manages the entire process, from data preparation to model construction, training, and deployment (Fig. 1). When integrated with other medical agentic systems, the overall system gains the ability to self-evolve, automatically developing new ML-based medical imaging tools, significantly alleviating the burden of manual tool collection and creation.

Specifically, M³Builder employs a multi-agent collaboration framework, where four role-playing LLM agents work together to execute medical imaging ML tasks step-by-step. These include task management, data engineering, module architecture design, and model training. In each step, the system will self-refine its intermediate generation based on `Python` execution feedbacks.

Moreover, we design a new MIML-centered agentic context protocol. It includes: a free-text dataset description template emphasizing critical taxonomy tags, a collection of coding templates to structure agent outputs and reduce coding complexity, and a suite of supporting documentation tool functions. This protocol helps integrate MIML coding priors into the agentic workflow, simplifies system processes, emphasizes key MIML coding elements, and underlines core coding lines to enable seamless human-in-the-loop auditing.

Lastly, we introduce a benchmark, **M³Bench**, designed to evaluate the performance for automated MIML. **M³Bench** comprises four general tasks: *organ segmentation, anomaly detection, disease diagnosis,* and *report generation.* These tasks are associated with 14 distinct training datasets, spanning five anatomical regions and three primary imaging modalities, encompassing both 2D and 3D models. By covering a broad range of tasks and datasets, M³Bench offers a comprehensive quantitative assessment of the automated MIML capabilities of various agentic systems.

Experimentally, to validate the effectiveness of the proposed agentic system, we compare against other state-of-the-art (SoTA) ML agentic systems, including ML-AgentBench [9], Aider [5], MetaGPT [8], ToolMaker [24], Copilot Edits [6] and so on under fair settings. Our results demonstrate that **M³Builder** consistently achieves significant performance advancing across a range of metrics, achieving an average task execution success rate of 94.29% across four MIML tasks, where success is defined as producing a MIML model comparable (within 5% below baseline method on) to the official baseline.

2   Method

2.1   Problem Formulation

Given a task description on medical imaging analysis, denoted as $\mathcal{T}$, our objective is to automatically construct a functional AI model via multi-agents collaboration. As shown in Fig. 2, our proposed **M³Builder** comprises two key components: the Medical Imaging ML Workspace ($\mathcal{W}$), and the Multi-Agent Collaboration Framework ($\mathcal{A}$). Specifically, the workspace includes three elements: *dataset with datacard*, *toolset with descriptions*, and *code templates*. The data cards are represented in natural language, while the toolset descriptions and code templates are provided in `Python`. Together, these elements form a structured 'environment' that guides the automatic AI workflow in medical imaging.

Building on the workspace ($\mathcal{W}$), the multi-agent framework composes of four LLM agents with distinct roles, *i.e.*, $\mathcal{A} = \{a_1, a_2, a_3, a_4\}$. These agents adopt a divide-and-conquer strategy to collaboratively address the complexities of the AI task. The framework iteratively performs analysis and planning, code generation, editing and executing using tools defined by *toolset descriptions*, until a functional AI model is successfully produced. This process can be expressed as:

$$\{\mathcal{C}_i, \mathcal{R}_i\} = \mathcal{A}(\mathcal{C}_{i-1}, \mathcal{R}_{i-1}, \mathcal{T} \mid \mathcal{W}), \tag{1}$$

where $\mathcal{C}_i$ denotes code and scripts generated or edited in the i_{th} iteration, $\mathcal{R}_i$ denotes the compiler feedback in `Python` environment, with $\mathcal{C}_0 = \mathcal{R}_0 = \varnothing$.

2.2   Workspace Initialization

The ML Workspace for medical imaging analysis is designed to serve three key purposes: (i) provides the multi-agent framework with metadata of the available datasets, enabling informed dataset selection; (ii) supplies initial code templates, offering a standardized starting point for agents and demonstrating a typical AI model training pipeline; (iii) defines and describes all tools available to the agents, restricting their action space to a predefined, complete set of operations.

Dataset Preparation. The workspace includes a range of medical imaging datasets, each accompanied by a structured datacard for standardized descriptions. Each datacard contains key information such as the dataset name, a concise summary (covering data sources, categorical classifications, and scope), and detailed metadata. This flexible format ensures that any description meeting these criteria qualifies as a valid datacard, including documentation provided with the dataset itself. To enable seamless integration, users can add new datasets by completing corresponding datacards. Initially, the workspace provides 14 medical imaging datasets with their datacards as examples.

Code Template Design. To streamline AI model training while maintaining flexibility, we prepare standardized training pipeline code templates based on the Transformers Trainer [10] framework and nnU-Net [12]. These templates address four primary medical imaging tasks: *disease diagnosis, organ segmentation, anomaly detection,* and *report generation.* Each task is implemented as a modular package with configurable components, including forward architecture and network backbone options (2D/3D models). Templates share common features like loss functions, data augmentation, training utilities, and architectural frameworks, reducing coding complexity while preserving adaptability.

Interaction Toolset. To facilitate the agentic system to interact with the PC environment, we develop a comprehensive interaction toolset, comprising eight functions. Inspired by MLAgentBench, the toolset includes: *list_files, read_files, copy_files, write_files, edit_files,* and *run_script* functions. We also introduce two specialized tools: *preview_dirs* and *preview_files.* The former processes large dataset directories, while the latter extracts key segments from oversized metadata files. Together, this toolkit enables critical operations such as task understanding, dataset navigation, workspace management, and code execution.

2.3 Multi-agent Collaboration Framework

This section introduces our multi-agent collaboration framework ($\mathcal{A}$), which decomposes the AI task into four sub-tasks and assigns them to four specialized prompt-driven role-playing LLMs agents: Task Manager, Data Engineer, Module Architect, and Model Trainer, denoted as $\{a_1, a_2, a_3, a_4\}$, respectively, as illustrated in Fig. 2. System prompts defining working logic and implementation for each agent will be available in our GitHub repository upon publication.

Task Manager acts as the coordinator of the framework. Its primary responsibilities include selecting the most suitable dataset for the task, or alternatively, asking users to upload raw datasets with associated datacard to the workspace as the supplementary of pre-existing datasets, and generating a comprehensive planning document $\mathcal{P}$ to guide the collaboration among the other agents. Specifically, given a user-provided task description, as exemplified by "user requirements" in Fig. 2, the Task Manager will identify and select the optimal dataset ($\mathcal{D}$) for model training and generate the planning documents. This process can be formally represented as:

$$\{\mathcal{D}, \mathcal{P}\} = a_1(\mathcal{T} \mid \mathcal{W}_d), \tag{2}$$

where $\mathcal{W}_d$ denotes the data card in the pre-defined workspace.

Data Engineer is responsible for dataset preparation and processing. It transforms raw data into a format suitable for model training by performing tasks such as pre-screening the organizational structure of large-scale datasets, analyzing metadata files to extract relevant information, and splitting datasets into training and testing subsets. A critical aspect of the Data Engineer's role is its iterative interaction with the external compiler environment. It generates, edits,

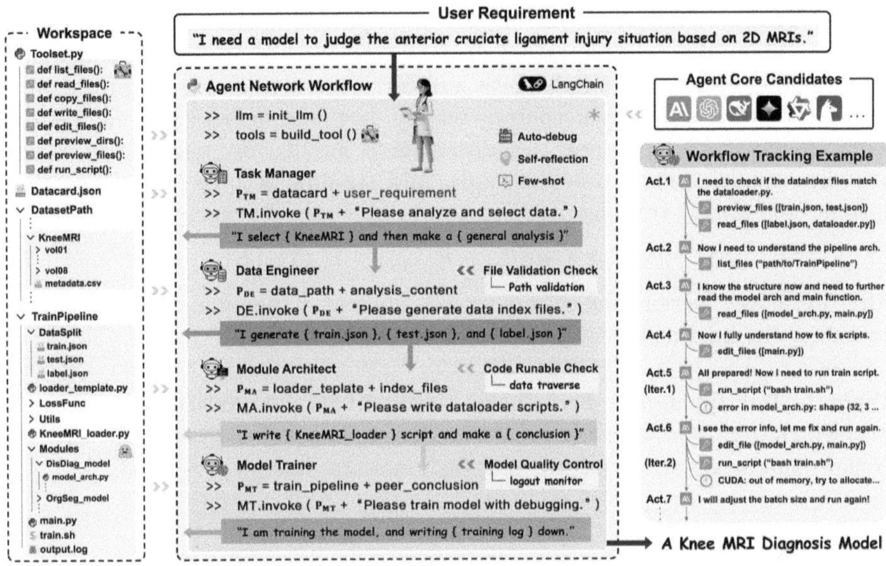

Fig. 2. Visualization of M³Builder workflow: From user's free-text request to model delivery. The system integrates user requirements, a workspace with candidate data, tools, and code templates, a network of four specialized collaborative agents. A sample log tracks the Model Trainer agent's activities during diagnosis model development.

and refines code, incorporating feedback from the compiler, until the code executes successfully. This iterative process ensures the dataset preparation code is both robust and functional. The process can be expressed as:

$$\{\mathcal{C}_i, \mathcal{R}_i\} = a_2(\mathcal{C}_{i-1}, \mathcal{R}_{i-1}, \mathcal{T} \mid \mathcal{D}, \mathcal{P}), \tag{3}$$

where $\mathcal{C}_i$ represents the ith version of the code generated by the Data Engineer, and $\mathcal{R}_i$ denotes the compiler feedback from the Python environment. Similarly as defined in **Problem Formulation**, $\mathcal{C}_0 = \mathcal{R}_0 = \varnothing$.

Module Architect. Upon the dataset preparation, the Module Architect integrates essential components into the training pipeline, including developing dataloader scripts, designing appropriate model architecture, initializing the entrance function for training, and selecting other components, such as loss functions, data augmentation strategies, and training utilities. Notably, the Module Architect will iteratively validates the dataloader to ensure it outputs batches with correct shapes and formats. The process can be formulated as:

$$\{\mathcal{C}_i, \mathcal{R}_i\} = a_3(\mathcal{C}_{i-1}, \mathcal{R}_{i-1}, \mathcal{T} \mid \mathcal{D}, \mathcal{W}_c), \tag{4}$$

where $\mathcal{W}_c$ represents the code templates in the workspace. Similar to the Data Engineer, the code generation process in the Module Architect is also iterative and we denote the final iteration output as $\mathcal{C}_{MA}$. After the integration of these modules, the architect finally synthesizes a summary $\mathcal{S}$ of all completed work.

120   J. Feng et al.

Model Trainer finalizes the debugging and optimizing the training procedure. Building upon the pipeline established by the Module Architect, the Model Trainer first verifies the completeness and correctness of the training framework thus far. It then selects hyperparameters to meet the model's specific training requirements, while retaining the authority to modify any part of the code— including model code, dataloader code, and training scripts—based on errors encountered during training. The entire workflow can be expressed as:

$$\{\mathcal{C}_i, \mathcal{R}_i\} = a_4(\mathcal{C}_{i-1}, \mathcal{R}_{i-1}, \mathcal{T} \mid \mathcal{D}, \mathcal{C}_{\mathrm{MA}}, \mathcal{S}). \tag{5}$$

After iteratively performing the above code generation pipeline until successfully executed, the final code will produce a desired AI model.

Table 1. Task Completion Performance Across LLMs. Results shown as successful completions over total rounds (a/b format). Green cells indicate that all runs passed, Yellow indicates partially passed, and Red indicates that all runs failed.

Task	Dataset	LLMs						
		Son.3.7	Son.3.5	GPT4o	QwMax	DSeekV3	Gemini2	Llama3.3
OrgSeg	BTCV [14]	5/5	4/5	5/5	3/5	2/5	0/5	1/5
	Verse [17]	4/5	4/5	5/5	2/5	0/5	0/5	0/5
	OASIS [21]	4/5	3/5	3/5	1/5	1/5	0/5	0/5
AnoDet	COV19 [20]	5/5	5/5	4/5	3/5	1/5	1/5	0/5
	INS22 [16]	5/5	5/5	4/5	2/5	2/5	0/5	0/5
	PancMSD [2]	5/5	5/5	5/5	4/5	2/5	0/5	1/5
	XDet10 [19]	5/5	5/5	5/5	3/5	1/5	1/5	0/5
DisDiag	ADNI [13]	5/5	4/5	2/5	1/5	1/5	0/5	0/5
	KneeMR [22]	4/5	5/5	4/5	2/5	1/5	0/5	0/5
	CC-CCII [26]	5/5	5/5	5/5	0/5	0/5	0/5	0/5
	KidCT [27]	5/5	5/5	4/5	2/5	1/5	0/5	0/5
RepGene	CT-RATE [7]	5/5	4/5	4/5	2/5	1/5	0/5	0/5
	GenBra [15]	5/5	5/5	4/5	3/5	1/5	0/5	1/5
	IU-Xray [3]	4/5	4/5	3/5	1/5	2/5	0/5	0/5
Average(%)		**94.29**	90.00	81.43	41.43	22.86	4.29	4.29

3   Benchmark and Experiments

3.1   Task Definition and Data Preparation

We define 14 tasks across *organ segmentation (Seg.)*, *anomaly detection (Det.)*, *disease diagnosis (Diag.)*, and *report generation (Gen.)*, each mapped to a specific dataset shown in Table 1. Tasks are categorized by anatomies (head & neck, chest, abdomen & pelvis, limb, spine), modalities (X-ray, CT, MRI), and 2D/3D. An example is: "Build a model for pneumonia classification on 3D chest CTs."

Table 2. Framework Comparison with SOTAs and Ablations on System Design using Sonnet. Results are averaged over two runs per task in dataset-level. "w/o Colab" represents single-agent execution, and "Iters" means the self-correction rounds.

Agentic System	Completion Runs (Total) ↑					Average Actions (Iters) ↓			
	Seg.	Det.	Diag.	Gen.	Avg(%)	Seg.	Det.	Diag.	Gen.
MLA-Bench	0(6)	0(8)	0(8)	0(6)	0.00	-(-)	-(-)	-(-)	-(-)
Aider	2(6)	3(8)	2(8)	3(6)	35.71	58.0(6.5)	51.3(7.0)	44.5(4.5)	56.7(5.7)
Cursor Comp	1(6)	3(8)	3(8)	2(6)	32.14	42.0(7.0)	35.7(4.3)	36.3(4.0)	51.5(5.5)
Wsurf Casc	1(6)	3(8)	4(8)	2(6)	35.71	39.0(5.0)	36.0(4.3)	35.3(4.8)	48.5(5.5)
Copilot Edits	2(6)	4(8)	3(8)	2(6)	39.29	48.0(3.5)	45.3(3.5)	44.7(4.0)	46.5(4.5)
Ours	4(6)	7(8)	8(8)	4(6)	82.14	34.5(1.8)	25.29(2.4)	35.0(1.9)	32.0(2.3)
w/o Colab	3(6)	3(8)	3(8)	2(6)	39.29	33.6(4.7)	37.5(4.8)	33.3(4.3)	33.5(3.5)
w/o Debug	2(6)	3(8)	1(8)	0(6)	21.43	30.5(1.0)	24.0(1.0)	33.0(1.0)	-(-)
w/o Reflect	4(6)	7(8)	7(8)	4(6)	78.57	28.3(2.3)	24.9(4.1)	35.4(3.9)	41.3(5.8)
w/o Fewshot	3(6)	4(8)	6(8)	3(6)	57.14	37.3(4.0)	23.5(4.3)	32.3(4.1)	34.0(3.0)

3.2   Task Completion Analysis

We design a specific model-building task for each dataset based on its unique characteristics, resulting in a total of 14 tasks. For each task, seven leading LLMs including GPT-4o [11], Claude-3.7-Sonnet [1], Claude-3.5-Sonnet [1], DeepSeek-v3 [18], Gemini-2.0-flash [23], Qwen-2.5-max [25], and Llama-3.3-70B [4] are ran independently five times, with a maximum of 100 actions (tool invocations) allowed per execution. The task completion is defined as successful training of a model with performance on test set falling within an acceptable range. As results in Table 1, the performance of different LLMs exhibit significant variation, with Sonnet-3.7 achieving the highest completion rate of 94.29%, while Gemini2.0 and Llama3.3 only reach 4.29%. The experimental results demonstrate that our M^{3}Builder posses impressive capabilities for medical imaging model training automation, particularly when using LLMs that excel in tool utilization and code generation, which significantly enhances the system's overall performance.

3.3   Comparison to State-of-the-Art Agentic Systems

We compare M^{3}Builder with other agentic systems including MLAgent-Bench, Aider, Cursor Composer, Windsurf Cascade, and Copilot Edits (all using Sonnet as the agent core). Each system performed each task twice on our workspace under their built-in framework. As shown in Table 2, across radiology tasks (Organ Segmentation, Anomaly Detection, Disease Diagnosis, and Report Generation), MLAgent-Bench performed poorly due to insufficient data structure understanding (tool limitation). Other frameworks achieved only moderate success rates (39.29% max) due to single-agent limitations, required human-in-the-loop confirmation, increasing operational complexity and max iteration constraints. Our M^{3}Builder demonstrated superior performance with a 42.85% higher average success rate while requiring fewer action steps and execution iterations.

Table 3. Role-specific agent performance on tasks using Sonnet. "Run", "Act", "Iter" and "Tkn" respectively denote "execution rounds", "actions", "iterations" and "tokens".

Task	Task Manager				Data Engineer				Module Architect				Model Trainer			
	Run	Act	Iter	Tkn	Run	Act	Iter	Tkn	Run	Act	Iter	Tkn	Run	Act	Iter	Tkn
Seg.	6/6	1.3	1.0	4.3k	5/6	10.2	1.3	72k	5/6	10.5	2.3	74k	4/6	10.7	3.3	115k
Det.	8/8	2.0	1.0	4.8k	8/8	10.0	1.4	92k	7/8	10.6	2.1	61k	7/8	9.3	2.1	67k
Diag.	8/8	2.0	1.0	4.4k	8/8	8.9	1.4	116k	7/8	11.3	2.0	84k	8/8	9.3	2.0	198k
Gen.	6/6	2.0	1.0	4.2k	6/6	8.7	1.3	66k	5/6	10.8	2.5	91k	5/6	11.2	2.2	70k

3.4 Analysis on Different Agent Roles

We evaluate $\mathbf{M^3Builder}$'s role-specific agents using distinct success criteria: Task Manager selects appropriate datasets, Data Engineer generates valid index files, Module Architect produces executable data loading scripts, and Model Trainer completes model training. As shown in Table 3, Task Manager demonstrated exceptional accuracy with stable token usage across tasks, while the other roles exhibited greater variability in token consumption and execution attempts due to strict requirements for code organization, pre-processing, and error-free training. Despite these unstable consumptions, most agents successfully complete their tasks, showcasing the multi-agent framework's robustness and adaptability.

3.5 Ablation Study

Here we present ablation studies on our system design, examining the impact of: single-agent versus multi-agent collaboration, auto-debugging capability, self-reflection mechanisms, and workflow few-shot examples. Results in Table 2 indicate that self-reflection has minimal influence on system performance, while auto-debugging proves crucial for successful training. Multi-agent collaboration and well-crafted example instructions also significantly impact performance, with their absence resulting in 42.85% and 25.00% performance gaps, respectively.

4 Conclusion

In this paper, we present M^3Builder, an agentic system for automating machine learning in medical imaging tasks. Our approach combines an efficient medical imaging ML workspace with free-text descriptions of datasets, code templates, and interaction tools. Additionally, we propose a multi-agent collaborative agent system designed specifically for AI model building, with four role-playing LLMs, Task Manager, Data Engineer, Module Architect, and Model Trainer. In benchmarking against 5 SOTA agentic systems across 14 radiology task-specific datasets, M^3Builder achieves a 94.29% model building success rate with Claude-3.7-Sonnet standing out among 7 SOTA LLMs. Future work will extend beyond medical imaging to broader medical tasks, develop more robust tool-building agent systems, implement automated dataset preparation capabilities, and incorporate visual processing to better approximate clinical expertise.

Acknowledgments. This study was funded by National Key R&D Program of China (No. 2022ZD0161400).

Disclosure of Interests. The authors have no competing interests to declare that are relevant to the content of this article.

References

1. Anthropic: The claude 3 model family: Opus, sonnet, haiku. https://api.semanticscholar.org/CorpusID:268232499
2. Antonelli, M.: The medical segmentation decathlon. Nature Commun. **13**(1), 4128 (2022)
3. Demner-Fushman, D., et al.: Preparing a collection of radiology examinations for distribution and retrieval. J. Am. Med. Inform. Assoc. **23**(2), 304–310 (2016)
4. Dubey, A., et al.: The llama 3 herd of models. arXiv preprint arXiv:2407.21783 (2024)
5. Gauthier, P.: Aider is ai pair programming in your terminal (2023). https://github.com/paul-gauthier/aider
6. GitHub: Copilot edits (2023). https://code.visualstudio.com/docs/copilot/copilot-edits
7. Hamamci, I.E., et al.: A foundation model utilizing chest ct volumes and radiology reports for supervised-level zero-shot detection of abnormalities. CoRR (2024)
8. Hong, S., et al.: MetaGPT: Meta programming for a multi-agent collaborative framework. In: The Twelfth International Conference on Learning Representations (2024). https://openreview.net/forum?id=VtmBAGCN7o
9. Huang, Q., Vora, J., Liang, P., Leskovec, J.: Mlagentbench: evaluating language agents on machine learning experimentation. arXiv preprint arXiv:2310.03302 (2023)
10. Huggingface: Transformers trainer (2022). https://huggingface.co/docs/transformers/main_classes/trainer
11. Hurst, A., et al.: Gpt-4o system card. arXiv preprint arXiv:2410.21276 (2024)
12. Isensee, F., Jaeger, P.F., Kohl, S.A., Petersen, J., Maier-Hein, K.H.: nnu-net: a self-configuring method for deep learning-based biomedical image segmentation. Nat. Methods **18**(2), 203–211 (2021)
13. Jack, C.R., et al.: The alzheimer's disease neuroimaging initiative (adni): Mri methods. J. Magnetic Resonance Imaging **27** (2008). https://api.semanticscholar.org/CorpusID:3272607
14. Landman, B., Xu, Z., Igelsias, J., Styner, M., Langerak, T., Klein, A.: Miccai multi-atlas labeling beyond the cranial vault–workshop and challenge. In: Proc. MICCAI Multi-Atlas Labeling Beyond Cranial Vault—Workshop Challenge, vol. 5, p. 12. Munich, Germany (2015)
15. Lei, J., et al.: Autorg-brain: Grounded report generation for brain mri. arXiv preprint arXiv:2407.16684 (2024)
16. Li, X., et al.: The state-of-the-art 3d anisotropic intracranial hemorrhage segmentation on non-contrast head ct: The instance challenge. arXiv preprint arXiv:2301.03281 (2023)

17. Liebl, H., et al.: A computed tomography vertebral segmentation dataset with anatomical variations and multi-vendor scanner data. Sci. Data **8**(1), 284 (2021)
18. Liu, A., et al.: Deepseek-v3 technical report. arXiv preprint arXiv:2412.19437 (2024)
19. Liu, J., Lian, J., Yu, Y.: Chestx-det10: Chest x-ray dataset on detection of thoracic abnormalities (2020)
20. Ma, J., et al.: Toward data-efficient learning: a benchmark for covid-19 ct lung and infection segmentation. Med. Phys. **48**(3), 1197–1210 (2021)
21. Marcus, D.S., Wang, T.H., Parker, J., Csernansky, J.G., Morris, J.C., Buckner, R.L.: Open access series of imaging studies (oasis): cross-sectional mri data in young, middle aged, nondemented, and demented older adults. J. Cogn. Neurosci. **19**(9), 1498–1507 (2007)
22. Štajduhar, I., Mamula, M., Miletić, D., Ünal, G.: Semi-automated detection of anterior cruciate ligament injury from MRI. Comput. Methods Programs Biomed. **140**, 151–164 (2017)
23. Team, G., et al.: Gemini: a family of highly capable multimodal models. arXiv preprint arXiv:2312.11805 (2023)
24. Wölflein, G., Ferber, D., Truhn, D., Arandjelović, O., Kather, J.N.: LLM agents making agent tools (2025). https://arxiv.org/abs/2502.11705
25. Yang, A., et al.: Qwen2. 5 technical report. arXiv preprint arXiv:2412.15115 (2024)
26. Zhang, K., et al.: Clinically applicable ai system for accurate diagnosis, quantitative measurements, and prognosis of covid-19 pneumonia using computed tomography. Cell **181**(6), 1423–1433 (2020)
27. Žukovec, M., Dular, L., Špiclin, Ž.: Modeling multi-annotator uncertainty as multi-class segmentation problem. In: International MICCAI Brainlesion Workshop, pp. 112–123. Springer (2021)

The Clinical-Driven Robotics and Embodied AI Technology (CREATE 2025)

Harnessing Foundation Models for Robust and Generalizable 6-DOF Bronchoscopy Localization

Qingyao Tian[1,2], Huai Liao[3], Xinyan Huang[3], Bingyu Yang[1,2],

and Hongbin Liu[1,4,5(✉)]

[1] State Key Laboratory of Multimodal Artificial Intelligence Systems, Institute of Automation, Chinese Academy of Sciences, Beijing, China
liuhongbin@ia.ac.cn

[2] School of Artificial Intelligence, University of Chinese Academy of Sciences, Beijing, China

[3] The First Affiliated Hospital, Sun Yat-sen University, Guangzhou, China

[4] Centre for Artificial Intelligence and Robotics, Chinese Academy of Sciences, Hong Kong, China

[5] School of Engineering and Imaging Sciences, King's College London, London, UK

Abstract. Vision-based 6-DOF bronchoscopy localization offers a promising solution for accurate and cost-effective interventional guidance. However, existing methods struggle with 1) *limited generalization* across patient cases due to scarce labeled data, and 2) *poor robustness* under visual degradation, as bronchoscopy procedures frequently involve artifacts such as occlusions and motion blur that impair visual information. To address these challenges, we propose **PANSv2**, a generalizable and robust bronchoscopy localization framework. Motivated by PANS [14] that leverages multiple visual cues for pose likelihood measurement, PANSv2 integrates depth estimation, landmark detection, and centerline constraints into a unified pose optimization framework that evaluates pose probability and solves for the optimal bronchoscope pose. To further enhance generalization capabilities, we leverage the endoscopic foundation model EndoOmni [13] for depth estimation and the video foundation model EndoMamba [16] for landmark detection, incorporating both spatial and temporal analyses. Pretrained on diverse endoscopic datasets, these models provide stable and transferable visual representations, enabling reliable performance across varied bronchoscopy scenarios. Additionally, to improve robustness to visual degradation, we introduce an automatic re-initialization module that detects tracking failures and re-establishes pose using landmark detections once clear views are available. Experimental results on bronchoscopy dataset encompassing 10 patient cases show that PANSv2 achieves the highest tracking success rate, with an 18.1% improvement in SR-5 (percentage of absolute trajectory error under 5 mm) compared to existing methods, showing potential towards real clinical usage.

Keywords: Surgical navigation · Foundation model · Bronchoscopy localization

1 Introduction

Bronchoscopy is widely used for visual inspection, diagnosis, and biopsy of pulmonary lesions [1]. During the procedure, clinicians navigate a camera-integrated flexible endoscope through the bronchial tree, guided by pre-operative CT scans to reach target

J. Qiu et al. (Eds.): Agentic AI 2025/CMLLMs 2025/CREATE 2025, LNCS 16147, pp. 127–135, 2026.
https://doi.org/10.1007/978-3-032-06004-4_13

regions such as peripheral airways and pulmonary nodules. However, the endoscope's limited field of view, coupled with the complex anatomy of the airway, makes accurate localization challenging and highly dependent on operator expertise. To address this, there is growing interest in developing automatic 6-DoF bronchoscope localization methods to enable robust and efficient navigation support [4,9].

Recent approaches for vision-based localization have explored a range of techniques, including image retrieval [11,18], visual odometry [3,5], landmark detection [8,17], and hybrid strategies incorporating rule-based logic [15] or probabilistic models [14]. While these methods show promise for cost-effective, image-based localization, their deployment in real-world clinical workflows is limited by two key challenges: (1) *poor generalization across patients*, and (2) *limited robustness to visual degradation*. The lack of generalization is primarily due to insufficient training data, as collecting large-scale, accurately labeled bronchoscopic pose annotations is labor-intensive and time-consuming. In terms of robustness, bronchoscopic videos are frequently degraded by visual occlusions such as fluids, bubbles, and motion blur. These artifacts severely impact visual feature extraction, often leading to failure cases. Consequently, many prior works filter out low-quality frames during evaluation [15,17], or require frequent re-calibration to reinitialize tracking [14].

In this work, we present PANSv2, a robust framework for 6-DoF bronchoscope localization. Building upon the original PANS [14], which fuses multiple visual cues to enhance localization robustness, PANSv2 integrates depth estimation, anatomical landmark detection, and centerline constraints to evaluate the probability of candidate poses and solve for the optimal pose that aligns the virtual bronchoscopic view with the real one. To overcome the generalization challenge, we construct the largest bronchoscopic localization dataset to date, comprising 66 procedures and over 30k labeled frames, providing a strong foundation for data-driven learning. To further enhance generalization, we incorporate foundation models pretrained on large-scale endoscopic datasets: EndoOmni [13] for depth estimation and EndoMamba [16] for landmark detection from video input. These models offer stable and transferable visual representations, enabling consistent performance across diverse patient anatomies and imaging conditions. To improve robustness under visual degradation, we introduce an automatic re-initialization module that detects tracking failures and recovers pose using landmark detections when clear views return. As a result, our method achieves significantly improved tracking continuity under intervention dataset captured under real clinical conditions. Evaluations on 10 full-length bronchoscopy procedures show that PANSv2 achieves the highest tracking success rate among existing methods, demonstrating strong potential for clinical deployment.

2   Methods

2.1   Overview

Our PANSv2 framework is illustrated in Fig. 1. The core of bronchoscopic localization is to align the view from a virtual bronchoscopic camera—rendered within the segmented airway from CT scans—with the real bronchoscopic image, thereby estimating the real camera's pose in the CT coordinate system. To achieve this, we formulate a

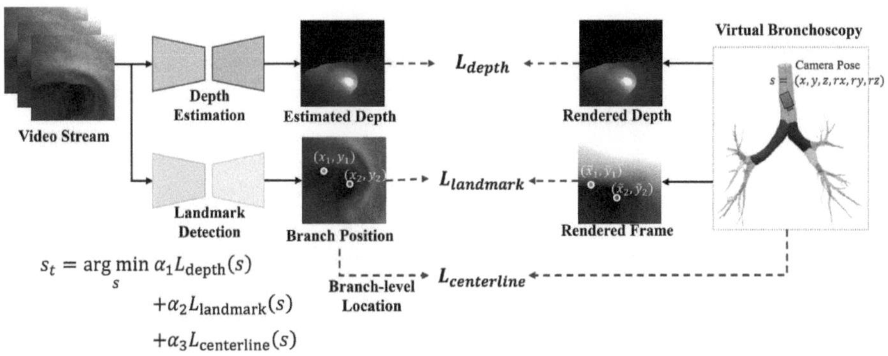

Fig. 1. Overview of the proposed PANSv2 for robust bronchoscope localization. The bronchoscope pose s_t at time t is estimated by optimizing the depth similarity, landmark alignment and centerline constraint.

probabilistic cost function that evaluates the likelihood of a given camera pose s. This cost L is computed based on three cues: depth similarity, alignment of landmark image coordinates, and a centerline constraint. Formally, the cost is formulated as:

$$L(s) = \alpha_1 L_{\mathrm{depth}}(s) + \alpha_2 L_{\mathrm{landmark}}(s) + \alpha_3 L_{\mathrm{centerline}}(s). \tag{1}$$

By solving for the camera pose that minimizes this cost, we obtain the estimated pose s_t of the bronchoscope at each time t:

$$s_t = \arg\min_s L(s). \tag{2}$$

Given the initial bronchoscopic pose at the first time step, at each following time step, our PANSv2 solves optimized bronchoscopic camera pose by Powell [6], using the previous estimated pose as initial value.

2.2 Depth Similarity

To align the virtual bronchoscopic view with the real one, we first estimate the cost of a candidate camera pose by measuring the similarity between the depth map rendered in the virtual airway and the depth estimated from the real bronchoscopic frame. By aligning depth rather than RGB images, we eliminate texture-specific appearance variations in the airway and focus solely on geometric structure, enabling more robust and interpretable alignment.

To estimate depth from bronchoscopic frames, we employ EndoOmni [13], a foundation model for endoscopic depth estimation trained on over 700,000 frames from a wide range of procedures and sources. Since EndoOmni predicts relative depth without an absolute scale, we use Normalized Cross-Correlation (NCC) to quantify similarity between the estimated and rendered depths. Formally:

$$L_{\mathrm{depth}}(s) = 1 - \mathrm{NCC}(z, \bar{z}(s)) = 1 - \frac{\sum_i (z_i - \mu_z)(\bar{z}_i(s) - \mu_{\bar{z}(s)})}{\sqrt{\sum_i (z_i - \mu_z)^2}\sqrt{\sum_i (\bar{z}_i(s) - \mu_{\bar{z}(s)})^2}}, \tag{3}$$

where z denotes the estimated depth map of the real bronchoscopic frame, $\bar{z}(s)$ is the depth rendered from the virtual camera at pose s, and $\mu_z, \mu_{\bar{z}(s)}$ are the mean depths of the respective maps. The cost L_{depth} is minimized when the two depth maps are maximally correlated, corresponding to high geometric alignment.

2.3 Landmark Alignment

Different sections of the human airway often share similar geometric structures, such as tubular shapes or bifurcations. As a result, relying solely on depth similarity can introduce ambiguity when optimizing the camera pose. To address this, PANSv2 leverages EndoMamba [16], an efficient endoscopic video foundation model, to detect anatomical landmarks within the bronchoscopic view. Built on a Mamba-based backbone for joint spatial and temporal modeling, EndoMamba leverages information from both the current frame and preceding frames to enhance landmark detection. This temporal context helps disambiguate visually similar airway regions, improving the accuracy and reliability of landmark-based pose estimation. Formally, given a bronchoscopic image I_t and hidden states from previous time points h_{t-1}, the landmark detection model outputs:

$$f_{\text{landmark}}(I_t, h_{t-1}) = (\mathbf{M}_t, h_t), \tag{4}$$

$$\mathbf{M}_t \in \mathbb{R}^{n \times 3}, \quad \mathbf{M}_{t,i} = (v_i, x_i, y_i), \tag{5}$$

where $v_i \in [0, 1]$ is the predicted visibility score, and (x_i, y_i) are the 2D image coordinates of landmark i, for $i = 1, \dots, n$, with n being the number of defined anatomical landmarks.

To evaluate how well a candidate camera pose aligns with observed anatomy, we compare the detected landmark positions from the model with the expected projections of known anatomical landmarks from the CT-based virtual model. Specifically, for a given camera pose s, we project the 3D landmark positions from the CT into image space, yielding the set of expected 2D coordinates $\{(\tilde{x}_i(s), \tilde{y}_i(s))\}_{i=1}^n$. The landmark alignment cost is then defined as the average L2 distance between the detected and projected positions of visible landmarks:

$$L_{\text{landmark}}(s) = \frac{1}{\sum_i v_i} \sum_{i=1}^n v_i \cdot \|(x_i, y_i) - (\bar{x}_i(s), \bar{y}_i(s))\|_2, \tag{6}$$

where (x_i, y_i) and v_i are the image coordinates and visibility predicted by the landmark detection model, and $(\bar{x}_i(s), \bar{y}_i(s))$ is the projection of landmark i given pose s. The loss is computed only over visible landmarks, as indicated by $v_i = 1$.

2.4 Centerline Constraint

Preliminary experiments show that using only depth and landmark alignment for pose evaluation often results in camera poses drifting outside the airway mesh. To address this issue, and to constrain the search space for improved convergence, PANSv2 incorporates a centerline-based constraint as part of the pose cost function. The centerline constraint is motivated by the assumption that the bronchoscopic camera should remain close to the airway centerline and generally align its viewing direction with the local orientation of the airway. To implement this, we first infer the branch-level location of the bronchoscope based on detected landmarks with a voting strategy following [17]. Then, given branch-level location $b = 1, \ldots, n$, where n being the number of defined anatomical branches, the centerline cost is defined as:

$$L_{\text{centerline}}(s) = \mathcal{N}(d; 0, \sigma_1^2) \cdot \mathcal{N}(\phi; 0, \sigma_2^2), \tag{7}$$

where d is the distance from camera pose s to branch b, and ϕ represents the angle between pose s and branch b. The variances σ_1 and σ_2 are set as $\frac{r}{2}$ and $\frac{\pi}{6}$, with r being the radius of the branch b.

2.5 Automatic Re-initialization

During bronchoscopy interventions, it is common for surgeons to navigate the bronchoscope into a new airway branch, even when the image is temporarily degraded by severe visual artifacts. In such cases, pose optimization often fails during the artifact period and cannot recover afterward without external intervention, since the previous estimated pose (used as the initialization for optimization) is far from the bronchoscope's true current location.

To address this, PANSv2 adopts a simple yet effective strategy for automatic pose re-initialization. We first detect optimization failure by thresholding the final pose cost:

$$\delta_{\text{fail}}(s) = \begin{cases} \text{True,} & \text{if } L(s) > \tau \\ \text{False,} & \text{otherwise} \end{cases}. \tag{8}$$

After the field of view recovers, EndoMamba recognizes visible anatomical landmarks and determines the current branch-level location b of the bronchoscope. We then reinitialize the pose optimization by setting the initial pose to the center point of branch b's centerline segment, and continue with the optimization process with Eq. 2.

3   Experiments

3.1   Implementation Details

Dataset. We constructed a bronchoscopy localization dataset consisting of 66 cases, with 56 cases for training and 10 for testing. Each case includes the patient-specific CT scan, camera checkerboard calibration, and a video clip containing approximately 1500 to 3500 frames. The camera poses for each bronchoscopic frame are annotated, except when the pose cannot be determined due to severe occlusions. Note that frames without pose annotations are still processed by PANSv2 but do not contribute to metric calculations. Localization annotations were generated using our OpenGL-based toolkit, which aligns virtual camera intrinsics with the actual bronchoscope. Three experts manually labeled the dataset by registering the virtual views to the real data. To assess annotation accuracy, two cases were labeled independently by the experts, yielding a group

Table 1. Comparison results across the 10-patient cases with 10,004 filtered frames.

Method	ATE (mm) ↓	SR-5 ↑	SR-10 ↑
EDM	35.68 ± 23.23	3.9%	8.8%
Depth-Reg	35.18 ± 28.37	13.9%	25.7%
DD-VNB	15.02 ± 11.68	22.9%	44.3%
PANS	**8.68 ± 5.97**	28.4%	70.0%
Ours	9.09 ± 11.86	**46.5%**	**73.0%**

Table 2. Ablation results across the 10-patient cases with 15441 unfiltered frames.

Method	ATE (mm) ↓	SR-5 ↑	SR-10 ↑
w/o L_{depth}	12.87 ± 17.99	34.4%	67.2%
w/o $L_{landmark}$	13.70 ± 18.00	34.1%	60.9%
w/o $L_{centerline}$	12.76 ± 17.87	39.5%	66.0%
w/o reinit	35.84 ± 32.13	17.5%	32.0%
Ours	**9.77 ± 12.15**	**41.8%**	**69.7%**

variance of 0.58 mm. After obtaining coordinate-level annotations, we generate branch landmark detection labels using the airway centerline. A branch is considered visible if any point along its centerline is visible from the current camera pose, with the 2D position of the branch defined as the furthest visible centerline point in the image.

Foundation Models. For depth estimation, we use the pre-trained EndoOmni-b without fine-tuning for zero-shot scale-and-shift-invariant depth estimation. Each video frame is resized to 378×378 as input to EndoOmni. To detect visible branches in each frame, we sample 16-frame input clips of spatial size 224×224 for fine-tuning EndoMamba. The learning rate is set to 1e-6 with 20 training epochs.

Evaluation Metrics. Following previous work [7, 12], we use Absolute Trajectory Error (ATE), SR-5 (percentage of ATE < 5 mm), and SR-10 (percentage of ATE < 10 mm) to evaluate our method. Among these, SR-5 and SR-10 serve as indicators of the success rate for endoscope tracking.

3.2   Comparison with State of the Art

We compare PANSv2 with existing state-of-the-art (SOTA) monocular endoscopic localization techniques. This includes depth registration (Depth-Reg) approaches [2, 12], the visual-odometry-based method Endo-Depth-and-Motion [10], the hybrid method DD-VNB [15], and the previous version, PANS [14]. Since prior methods exclude frames affected by visual degradation—caused by severe airway deformation or rapid endoscope withdrawal—we additionally evaluate PANSv2 on these filtered frames. Notably, PANSv2 processes the complete set of unfiltered video frames using only a single initialization at the trachea, whereas the compared methods require two separate initialization steps—one before entering each side of the airway.

Results are presented in Table 1. PANSv2 outperforms all compared SOTA methods, achieving an 18.1% improvement in SR-5, which reflects a substantial gain in localization success rate. However, we observe a slight increase in ATE compared to PANS. We attribute this to occasional misidentification of anatomical landmarks, which can result in large localization errors when the bronchoscope is estimated to be in an incorrect airway branch. We also illustrate the unfiltered localization error of PANSv2 compared to PANS on a representative case. As shown in Fig. 2, PANSv2 rapidly recovers from visual degradation once the field of view is restored, in contrast to PANS, which fails to recover. This further demonstrates the enhanced robustness of PANSv2.

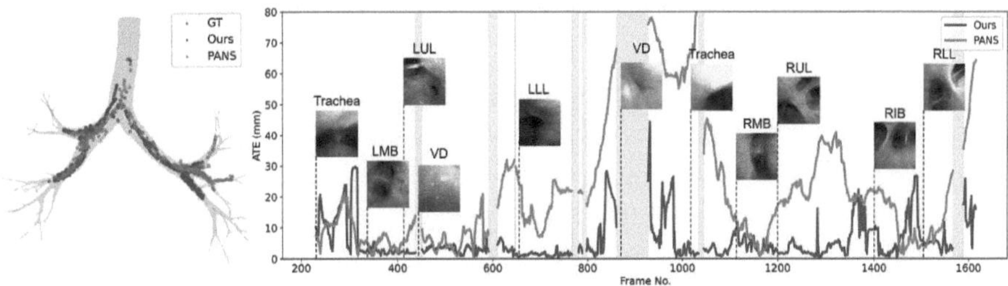

Fig. 2. Localization trajectory and error for a patient case. Timesteps lacking pose labels due to severe visual degradation are shown in light gray. Example frames are provided at anatomical landmarks, including the trachea, left main bronchus (LMB), left upper lobe (LUL), left lower lobe (LLL), right main bronchus (RMB), right upper lobe (RUL), right intermediate bronchus (RIB), and right lower lobe (RLL), as well as frames affected by visual degradation (VD). (Color figure online)

3.3 Ablation Studies

We evaluate the localization performance of PANS on test cases by sequentially removing the depth similarity cost, landmark alignment cost, centerline constraint, and the re-initialization module. In contrast to comparisons with SOTA methods, we present results on complete bronchoscopic videos without any filtering. As shown in Table 2, the automatic re-initialization module significantly contributes to robust bronchoscope tracking. Figure 2 illustrates the consistent visual degradation caused by occlusion, motion blur, and airway deformation, highlighted by the gray temporal-axis blocks. These visual artifacts pose significant challenges to airway localization methods, as the bronchoscope often continues to move despite poor visual information. By detecting tracking failures and re-initializing at identified landmarks, PANSv2 exhibits strong robustness in complex bronchoscopy scenes, quickly recovering once clear views are available. Furthermore, PANSv2 estimates pose likelihood through an integrated visual modality that combines depth similarity, landmark alignment, and centerline regularization. This approach is enhanced by foundation models with strong generalization capabilities, enabling the system to adapt to diverse anatomical variations and unseen conditions. As a result, PANSv2 achieves higher localization accuracy and stability compared to prior methods. These results underscore the effectiveness of PANSv2 framework and demonstrate its potential as a reliable tool for real-world bronchoscopic navigation.

Furthermore, we evaluate the runtime of PANSv2 and its individual components on a workstation equipped with a NVIDIA RTX 3090 GPU. The overall inference time per frame is approximately 200 ms, comprising 28 ms for depth estimation, 57 ms for landmark detection, and 111 ms for optimization. This processing speed is slower than the typical bronchoscope video capture rate (10 ~ 20 Hz). The primary computational bottleneck arises from the repeated rendering given candidate bronchoscopic poses during the optimization process, which is used to assess the similarity between the rendered depth and landmark positions with the real bronchoscopic view. In future work, we aim to explore advanced optimization techniques and faster rendering engines to move the PANSv2 framework closer to real-time inference capabilities.

4   Conclusion

In this work, we present PANSv2, a 6-DoF bronchoscopy localization framework that leverages foundation models to enable generalizable multi-visual representation learning for improved robustness. Localization is achieved by minimizing the visual alignment cost between virtual and real bronchoscopic views using three complementary visual cues: depth similarity, landmark alignment, and centerline constraint. Combined with an automatic re-initialization module that recovers tracking during challenging scenarios, PANSv2 demonstrates state-of-the-art performance on a bronchoscopy dataset comprising 10 patient cases captured in real clinical workflow. These results highlight the effectiveness and robustness of PANSv2 in realistic and complex clinical environments, paving the way for more reliable autonomous navigation in bronchoscopy.

Acknowledgments. This work was supported by the Centre of AI and Robotics, Hong Kong Institute of Science and Innovation, Chinese Academy of Sciences, sponsored by InnoHK Funding, HKSAR, and partially supported by Institute of Automation, Chinese Academy of Sciences.

Disclosure of Interests. The authors have no competing interests to declare that are relevant to the content of this article.

References

1. Andolfi, M., Potenza, R., Capozzi, R., Liparulo, V., Puma, F., Yasufuku, K.: The role of bronchoscopy in the diagnosis of early lung cancer: a review. J. Thorac. Dis. **8**(11), 3329 (2016)
2. Banach, A., King, F., Masaki, F., Tsukada, H., Hata, N.: Visually navigated bronchoscopy using three cycle-consistent generative adversarial network for depth estimation. Med. Image Anal. **73**, 1361–8415 (2021)
3. Borrego-Carazo, J., Sanchez, C., Castells-Rufas, D., Carrabina, J., Gil, D.: Bronchopose: an analysis of data and model configuration for vision-based bronchoscopy pose estimation. Comput. Methods Programs Biomed. **228**, 107241 (2023)
4. Cold, K.M., Xie, S., Nielsen, A.O., Clementsen, P.F., Konge, L.: Artificial intelligence improves novices' bronchoscopy performance: a randomized controlled trial in a simulated setting. Chest **165**(2), 405–413 (2024)
5. Deng, J., Li, P., Dhaliwal, K., Lu, C.X., Khadem, M.: Feature-based visual odometry for bronchoscopy: a dataset and benchmark. In: 2023 IEEE/RSJ International Conference on Intelligent Robots and Systems (IROS), pp. 6557–6564. IEEE (2023)
6. Fletcher, R., Powell, M.J.: A rapidly convergent descent method for minimization. Comput. J. **6**(2), 163–168 (1963)
7. Gu, Y., Gu, C., Yang, J., Sun, J., Yang, G.Z.: Vision-kinematics interaction for robotic-assisted bronchoscopy navigation. IEEE Trans. Med. Imaging **41**(12), 3600–3610 (2022)
8. Keuth, R., Heinrich, M., Eichenlaub, M., Himstedt, M.: Airway label prediction in video bronchoscopy: capturing temporal dependencies utilizing anatomical knowledge. Int. J. Comput. Assist. Radiol. Surg. **19**(4), 713–721 (2024)
9. Kops, S.E., et al.: Diagnostic yield and safety of navigation bronchoscopy: a systematic review and meta-analysis. Lung Cancer **180**, 107196 (2023)

10. Recasens, D., Lamarca, J., Fácil, J.M., Montiel, J., Civera, J.: Endo-depth-and-motion: reconstruction and tracking in endoscopic videos using depth networks and photometric constraints. IEEE Robot. Automation Lett. **6**(4), 7225–7232 (2021)
11. Sganga, J., Eng, D., Graetzel, C., Camarillo, D.B.: Autonomous driving in the lung using deep learning for localization. arXiv preprint arXiv:1907.08136 (2019)
12. Shen, M., Gu, Y., Liu, N., Yang, G.Z.: Context-aware depth and pose estimation for bronchoscopic navigation. IEEE Robot. Automation Lett. **4**(2), 732–739 (2019)
13. Tian, Q., et al.: EndoOmni: zero-shot cross-dataset depth estimation in endoscopy by robust self-learning from noisy labels. arXiv preprint arXiv:2409.05442 (2024)
14. Tian, Q., et al.: PANS: probabilistic airway navigation system for real-time robust bronchoscope localization. In: MICCAI (2024)
15. Tian, Q., et al.: DD-VNB: a depth-based dual-loop framework for real-time visually navigated bronchoscopy. In: IROS (2024)
16. Tian, Q., et al.: EndoMamba: an efficient foundation model for endoscopic videos. arXiv preprint arXiv:2502.19090 (2025)
17. Tian, Q., et al.: BronchoTrack: airway lumen tracking for branch-level bronchoscopic localization. IEEE Trans. Med. Imaging (2024)
18. Zhao, C., Shen, M., Sun, L., Yang, G.Z.: Generative localization with uncertainty estimation through video-ct data for bronchoscopic biopsy. IEEE Robot. Automation Lett. **5**(1), 258–265 (2019)

Embodied Surgical Intelligence via Digital Twins: Autonomous Trocar Insertion

Duy Ho[1], Ahmed Alanazi[2], Saeed Alqarni[2], Chi Lee[2], Gary Sutkin[2], and Yugyung Lee[2(✉)]

[1] California State University, Fullerton, CA, USA
duyho@fullerton.edu
[2] University of Missouri–Kansas City, Kansas City, MO, USA
{aha85b,saacfb,leech,sutking,leeyu}@umkc.edu

Abstract. Robotic-assisted surgery (RAS) has improved the precision and consistency of minimally invasive procedures, yet most systems remain teleoperated and lack adaptability. We present a clinically grounded framework that integrates digital twins, virtual reality (VR), and embodied AI to enable autonomous trocar insertion in Mid-Urethral Sling (MUS) surgery—a representative task involving blind anatomical navigation. Our approach combines reinforcement learning (RL), behavioral cloning (BC), and generative adversarial imitation learning (GAIL) to train robotic agents from expert demonstrations and synthetic interactions. A high-fidelity 3D digital twin environment supports safe policy training and sim-to-real transfer to a physical robot (UFactory Lite6). Evaluation across simulation and deployment shows a 23% improvement in procedural fidelity, 0.78 mm average deviation, and superior performance compared to manual and gesture-based control. Three operational prototypes, VR-controlled, hand-tracked, and fully autonomous, demonstrate the system's flexibility and clinical relevance. This work underscores the potential of embodied AI and digital twins to enhance surgical autonomy, reduce variability, and scale safe, adaptive training systems for intelligent operating rooms.

Keywords: Embodied AI · Surgical Robotics · Digital Twin · Reinforcement Learning · Mid-Urethral Sling · Surgical Autonomy

1 Introduction

Robotic-assisted surgery (RAS) is transforming healthcare by enhancing precision, dexterity, and reproducibility in minimally invasive procedures. A key application is *Mid-Urethral Sling* (MUS) surgery, a widely adopted intervention for treating *urinary stress incontinence* (USI) [11]. While current RAS platforms offer mechanical advantages, they are predominantly teleoperated, limiting adaptability and slowing progress toward intelligent, semi-autonomous systems.

© The Author(s), under exclusive license to Springer Nature Switzerland AG 2026
J. Qiu et al. (Eds.): Agentic AI 2025/CMLLMs 2025/CREATE 2025, LNCS 16147, pp. 136–145, 2026.
https://doi.org/10.1007/978-3-032-06004-4_14

Recent developments in artificial intelligence (AI)—notably *reinforcement learning* (RL), *behavioral cloning* (BC), and *generative adversarial imitation learning* (GAIL)—have enabled robots to learn context-sensitive surgical behaviors and adapt to intraoperative variability. These learning paradigms support the shift toward *conditional autonomy*, which promises to reduce surgeon workload, increase consistency, and enhance procedural safety [7,13,19].

To bridge the gap between simulation and real-world execution, we propose a unified framework that integrates *virtual reality* (VR), *augmented reality* (AR), and a real-time digital twin with a physical robotic platform (UFactory Lite6). The system enables immersive skill acquisition, procedural rehearsal, and seamless policy deployment. The digital twin mirrors surgical environments via multimodal sensing—visual tracking, force feedback, and imaging—allowing AI agents to generalize across anatomical and situational variability [3,6,9].

Using MUS surgery as a clinically grounded use case, we demonstrate how embodied AI and digital twin technologies enable autonomous trocar insertion and surgical decision support. Our main contributions include:

- **Hybrid Learning for Embodied Agents:** We combine reinforcement learning (RL), generative adversarial imitation learning (GAIL), and behavioral cloning (BC) to train agents under real-time safety constraints, enabling adaptive, context-aware motor behavior.
- **Sim-to-Real Policy Transfer via Digital Twin:** A tightly coupled VR/AR simulator and physical robot allow seamless transfer of trained policies. Our ML agent outperforms manual and gesture-based control in fidelity, precision, and task efficiency, achieving a 23% improvement in procedural fidelity and robust real-world execution.
- **Safety-Conscious Evaluation:** Imaging-guided reward shaping and trajectory supervision ensure anatomically constrained, compliant execution, addressing critical safety requirements for autonomous surgical systems.

By embedding interpretable, feedback-driven AI policies within a digital twin-augmented environment, our system offers a scalable foundation for developing semi-autonomous surgical platforms suited for high-volume, anatomically complex procedures.

2   Clinical Motivation and System Overview

2.1   From Surgical Autonomy to Learning-Driven Embodied AI

Surgical robotics has evolved from fully teleoperated platforms to systems supporting higher autonomy levels [2,18]. *Level 0* involves full human control; *Level 1* adds assistance through constrained guidance or enhanced visualization. *Level 2* introduces *task-level autonomy* (e.g., autonomous suturing [2]), and *Level 3* enables *conditional autonomy*, where robots propose plans for human approval [17].

Learning-based methods facilitate this transition. *Reinforcement Learning (RL)* refines actions through trial-and-error [14]; *GAIL* adversarially mimics expert behavior [8]; and *Behavioral Cloning (BC)* maps demonstrations to actions via supervised learning [13]. These enhance adaptability across anatomies and procedures. Digital twins and VR simulation environments also play a key role in pre-deployment policy validation. Immersive training platforms have successfully onboarded healthcare providers using digital twin simulations [20], highlighting their importance in bridging virtual and real-world execution.

2.2   Mid-Urethral Sling (MUS) Procedure and Trocar Challenges

Mid-Urethral Sling (MUS) surgery is a common, evidence-based treatment for *Urinary Stress Incontinence (USI)*, which involves involuntary urine leakage during exertion, coughing, or sneezing [11]. The procedure places a synthetic mesh under the mid-urethra to restore continence. A high-risk step involves blindly inserting a sharp trocar through the pelvic cavity to create a dissection tunnel for sling passage. This must be done with millimeter-level precision to avoid damaging the bladder, iliac arteries, or obturator veins. Misplacement can lead to serious complications, including vascular injury, urethral trauma, or hematoma [11].

Trocar Insertion Workflow. The process includes five coordinated stages: (1) proper hand placement, (2) alignment with the dissection tunnel, (3) angling toward the ipsilateral shoulder, (4) passage behind the pubic bone while avoiding the symphysis, and (5) controlled force to avoid overshooting or tissue damage. Due to its difficulty, this step remains a key focus for training and AI assistance.

Motivation for Autonomy. Robotic systems with real-time sensing, force feedback, and trajectory optimization can help: (i) detect resistance, (ii) maintain safe force thresholds, (iii) adjust paths via learned models, and (iv) offer feedback during training. We propose a digital twin-enhanced platform to support immersive simulation, rehearsal, and real-time execution.

2.3   Digital Twin Framework

Our digital twin architecture bridges simulation and real-world execution via integrated modules for virtual reality (VR), augmented reality (AR), and robotic deployment. This bidirectional pipeline enables immersive rehearsal, policy training, and seamless transition to clinical settings.

Pipeline Stages. The platform comprises four stages:

- *Stage 1: Simulation Setup* – 3D pelvic, trocar, and phantom models are built in Unity and paired with the UFactory Lite6 robot.
- *Stage 2: Workflow Modeling* – AI models extract spatiotemporal patterns from motion-force data to learn phase transitions and control policies.
- *Stage 3: AR-Based Interaction* – Users engage via HoloLens or Oculus Quest for gesture-based rehearsal and feedback.
- *Stage 4: Deployment and Feedback* – Trained policies run on the physical robot; real-world data is looped back for continuous refinement.

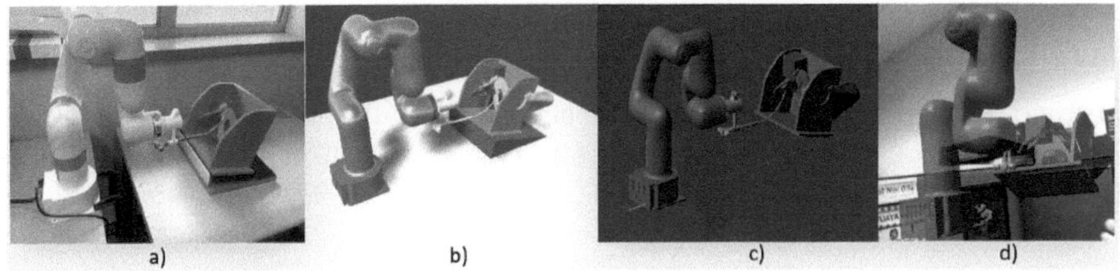

Fig. 1. Learning-Integrated Robotic Training: (a) Physical robot interaction; (b) 3D virtual coordination; (c) Virtual robotic maneuvers. Demonstration videos: Demo 1 (VR controller-based simulation),   Demo 2 (hand-tracked VR procedure), Demo 3 (ML-driven autonomous execution).

Trocar Force-Time Tracking. A 6-axis force-torque sensor and stereo vision track trocar insertion in both simulated and cadaveric contexts. These multi-modal inputs guide trajectory estimation, detect resistance shifts, and provide real-time feedback. Characteristic force patterns signal stages like vaginal entry, bone contact, and posterior navigation—used as cues for learning and validation. To ensure reliable deployment, we model the robot's mechanical constraints during simulation. Figure 2 illustrates the UFactory Lite6's kinematics, workspace, and end-effector mount for accurate policy transfer (Fig. 1).

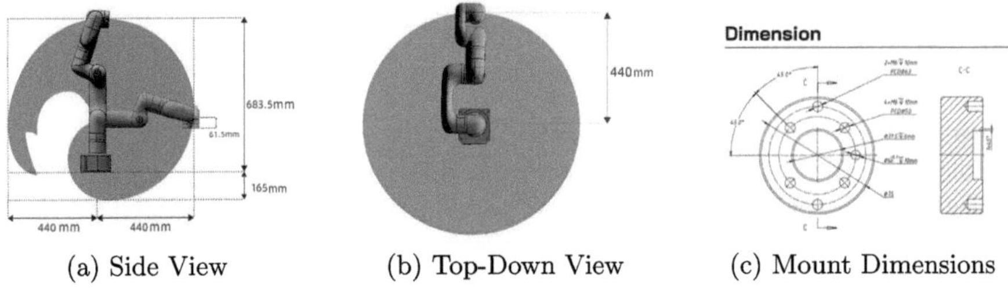

(a) Side View   (b) Top-Down View   (c) Mount Dimensions

Fig. 2. UFactory Lite6 specifications modeled in simulation to ensure fidelity in planning and safe execution.

Integrated System Benefits. This architecture supports safe skill development, real-time feedback on motion and force, and continuous VR-to-robot policy updates. Its modular design ensures generalizability to other procedures. The closed-loop workflow enables scalable, reliable deployment of embodied AI in safety-critical surgical settings [1,4,20].

3   Learning and Trajectory Modeling Framework

To enable autonomous trocar insertion with real-time safety, we adopt a hybrid learning architecture combining *Reinforcement Learning (RL)*, *Behavioral Cloning (BC)*, and *Generative Adversarial Imitation Learning (GAIL)*. BC enables fast bootstrapping from expert demonstrations, RL refines policies via trial-and-error, and GAIL improves generalization to novel anatomies by aligning agent behavior with expert trajectories [10,13,14].

Hybrid Learning Integration. The agent is initially pre-trained using BC on 23 expert demonstrations to accelerate policy convergence. The supervised loss is defined as:

$$\mathcal{L}_{\mathrm{BC}} = \sum_{i=1}^{N} \left\| \pi(a_i \mid s_i) - a_i^{\mathrm{expert}} \right\|^2$$

This BC-initialized policy is fine-tuned using Proximal Policy Optimization (PPO), a robust, sample-efficient RL algorithm that maximizes expected return.

$$\max_{\pi} \mathbb{E}_{\tau \sim \pi} \left[\sum_{t} \gamma^t R(s_t, a_t) \right]$$

To further align the learned policy with expert strategies, GAIL introduces a discriminator $D(s, a)$ trained to differentiate expert from agent behaviors:

$$\mathcal{L}_{\mathrm{GAIL}} = \log D(s, a) + \log(1 - D(s_{\mathrm{expert}}, a_{\mathrm{expert}}))$$

The agent's policy π is optimized adversarially to generate trajectories that are indistinguishable from expert demonstrations, effectively capturing domain-specific nuances of trocar insertion.

State, Action, and Reward Formulation. The agent's state space integrates sensory and contextual cues for safe, precise trocar manipulation.

1. 6-DoF pose of the robotic end-effector,
2. Real-time 6-axis force-torque data from the trocar handle,
3. Estimated proximity to anatomical boundaries (e.g., bone, vessels),
4. Phase label indicating procedural stage (e.g., entry, tunneling, alignment).

The action space defines continuous control over linear and angular movements of the trocar, along with grip modulation. These actions drive trajectory planning and low-level compliance for tissue interaction. The reward function encodes surgical safety and precision, as detailed in Table 1:

Table 1. Reward Components for AI-Guided Trocar Insertion

Term	Description
R_{entry}	Reward for anatomically accurate trocar entry point.
$R_{\mathrm{deviation}}$	Penalty for deviation from expert-demonstrated trajectories.
R_{force}	Penalty when applied force exceeds safety thresholds.
R_{progress}	Incremental reward for progressing through each surgical phase.
R_{terminal}	High reward for successful and safe completion.

Trajectory Estimation via Visual Segmentation. To support trajectory modeling, we use a two-stage vision-based segmentation pipeline that estimates

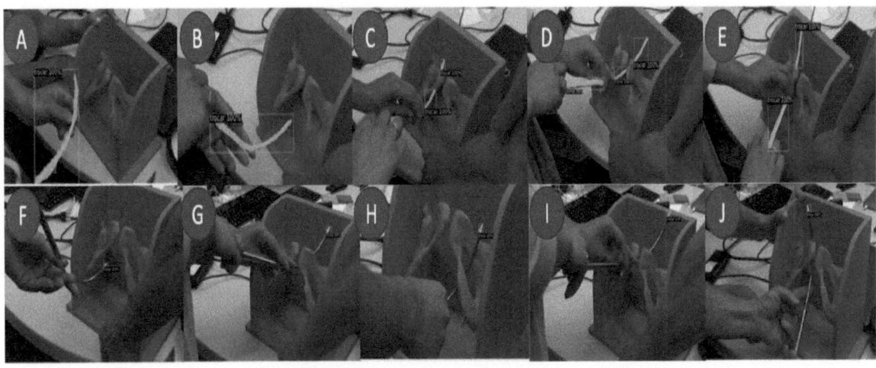

Fig. 3. Trocar Segmentation Pipeline: Mask R-CNN produces coarse object segmentation; U-Net enhances edge fidelity for accurate pose estimation.

the trocar pose in real time. As illustrated in Fig. 3, we employ Mask R-CNN [5] for initial object detection and segmentation, followed by U-Net [15] to refine boundaries and improve spatial resolution. This hierarchical process ensures accurate visual cues for adaptive policy updates.

Trajectory Modeling and Force Constraints. The trocar trajectory is modeled as a damped parabolic curve under Newtonian dynamics:

$$p(t) = p_0 + v_0 t + \frac{1}{2}at^2$$

where p_0 is the starting position, v_0 the insertion velocity, and a the net acceleration combining robot actuation and tissue response (Fig. 4).

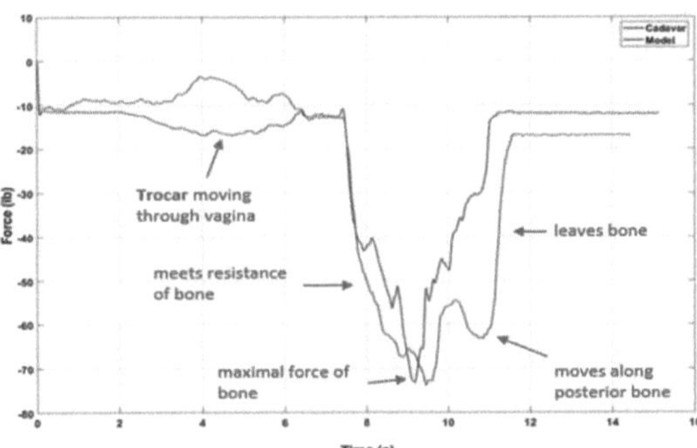

Fig. 4. Force-time profile of trocar insertion, showing key transitions: (1) vaginal entry, (2) bone resistance, and (3) posterior maneuvering. These stages inform state transitions and reward design.

Force safety is maintained using a 6-axis force-torque sensor, enabling real-time adjustment to insertion trajectory and velocity. This ensures protection during transitions like pubic bone contact or posterior redirection.

Curiosity-Driven Exploration and Transfer. To improve robustness in sparse-reward environments, we include an intrinsic curiosity module based on prediction error [12]:

$$r_{\text{curiosity}}(s, a, s') = \|f_\phi(s, a) - s'\|^2$$

where f_ϕ is a learned forward dynamics model. High error incentivizes exploration of unfamiliar state transitions.

Unity-based domain randomization (lighting, tissue stiffness, anatomy) [16] supports clinical generalization. Final deployment on UFactory Lite6 uses threshold and trajectory tuning for safe execution.

4   Experimental Results, Evaluation, and Deployment

We evaluate our digital twin-augmented framework in both simulated and real-world settings using a Unity-based surgical simulator and the UFactory Lite6 robot. Evaluation focuses on trajectory accuracy, generalization across variable conditions, and robustness of learned policies under clinical constraints.

Simulation Fidelity and Learning Performance. Table 2 summarizes the performance of three control modes: (1) manual VR teleoperation, (2) gesture-based hand tracking, and (3) autonomous execution via an ML agent trained with hybrid reinforcement and imitation learning (PPO + BC + GAIL). Metrics include procedural fidelity score (PFS), task completion rate (TCR), and real-time deviation from motion logs (RTDML). The ML agent achieved the highest fidelity and precision with perfect task completion, validating the hybrid learning framework for clinically relevant motor behavior in simulation.

Table 2. Simulation Performance Across Learning Modes

Control Mode	PFS (%)	TCR (%)	RTDML (mm)
VR Controller (Teleop)	91.2	100	1.7
VR Hand Tracking	88.5	94.2	2.1
ML Agent (PPO + BC + GAIL)	**94.3**	**100**	**1.2**

Real-World Robotic Deployment. To evaluate sim-to-real transfer, we deployed the trained policy on the UFactory Lite6 robot for repeated trocar insertion trials. Table 3 presents path consistency (PC), entry coordinate stability (ECS), and task execution time (TET). The robot achieved sub-millimeter ECS and consistent trajectories, confirming the robustness and transferability of simulation-trained policies.

Skill Transfer and Generalization. Compared to rule-based and BC-only baselines, the hybrid model achieved a 23% improvement in procedural fidelity and a 29% reduction in RTDML. Domain randomization and curriculum learning were essential for generalizing across variations in lighting, anatomy, and sensor noise—factors typical in clinical settings.

Table 3. Performance Metrics in Physical Robotic Deployment

Trial	PC (%)	ECS (mm)	TET (s)
Trial 1	92.5	0.78	31.2
Trial 2	94.1	0.82	28.5
Trial 3	93.7	0.75	29.9
Average	**93.4**	**0.78**	**29.9**

Qualitative Observations. Video analysis revealed that ML-driven agents produced smoother motion, maintained trocar alignment across varied conditions, and adjusted velocity based on contact. These behaviors help reduce tissue trauma and improve insertion safety.

Prototype Demonstrations. We developed three digital twin-enabled prototypes for mid-urethral sling (MUS) training, each with a distinct control mode to support flexible surgical training: (1) *Prototype 1: VR Controller-Based Training* Manual control via Oculus Quest VR controllers supports real-time simulation with haptic feedback. Motion and force data are logged for motor planning and spatial reasoning. (2) *Prototype 2: VR Hand-Tracking Interface* Bare-hand 3D tracking enables controller-free tool manipulation. This immersive setup facilitates intuitive input and early-stage rehearsal. (3) *Prototype 3: Autonomous Execution with ML Policies* Autonomous control using GAIL + BC + Curiosity + PPO allows real-time trajectory adaptation. This hands-free system supports precision execution and skill assessment.

5   Discussion and Conclusion

Our digital twin platform demonstrates the feasibility of integrating immersive simulation, AI autonomy, and real-time robotic control for procedures such as mid-urethral sling (MUS) insertion. All three prototypes share a unified backend, enabling consistent evaluation and transfer learning across stages. The ML agent outperformed manual and gesture modes in entry accuracy, force control, and task time, while earlier prototypes supported pre-training and rehearsal. Clinically, the system reduces variability and improves consistency in anatomically constrained, high-volume surgeries. Although developed for MUS, the modular architecture generalizes to laparoscopic procedures, including cholecystectomy and hernia repair, through domain randomization and hybrid learning.

Limitations include limited expert demonstration diversity, visual sensitivity to occlusion and lighting, and latency across Unity, HoloLens, and robotic systems. Despite these challenges, the AI-augmented digital twin framework offers a scalable foundation for intelligent surgical robotics. By integrating autonomy with real-time sensing and simulation, it supports safer, more consistent workflows. Future directions include: (1) surgeon-in-the-loop reinforcement learning, (2) tactile sensing for tissue feedback, (3) multi-agent coordination, and (4) LLM-based natural language interfaces for hands-free control—advancing toward collaborative AI-assisted surgery.

Disclosure of Interests. The authors have no competing interests to declare that are relevant to the content of this article.

References

1. Asciak, L., et al.: Digital twin assisted surgery, concept, opportunities, and challenges. npj Digit. Med. **8**(1), 32 (2025)
2. Attanasio, A., Bianchi, M., Pino, G., Formica, D., Guglielmelli, E., Vitiello, N.: Autonomy in surgical robotics. Annu. Rev. Control. Robotics Auton. Syst. **4**, 651–679 (2021)
3. Diniz, P., et al.: Digital twin systems for musculoskeletal applications: a current concepts review. Knee Surg. Sports Traumatol. Arthrosc. (2025)
4. Han, J.M., Kwon, D.S., Kyung, K.U.: A novel hybrid ureteroscope tracking for robotic-assisted retrograde intrarenal surgery via recognition of pathway with lumen identification. IEEE Robot. Autom. Lett. (2025)
5. He, K., Gkioxari, G., Dollár, P., Girshick, R.: Mask R-CNN. In: Proceedings of the IEEE International Conference on Computer Vision (ICCV), pp. 2961–2969 (2017)
6. Jiang, P., Zhang, D.: A digital twin-driven immersive teleoperation framework for robot-assisted microsurgery. In: 2024 IEEE/RSJ International Conference on Intelligent Robots and Systems (IROS), pp. 13495–13501. IEEE (2024)
7. Li, Z., Shi, L., Wang, J., Cristea, A.I., Zhou, Y.: Sim-GAIL: a generative adversarial imitation learning approach of student modelling for intelligent tutoring systems. Neural Comput. Appl. **35**(34), 24369–24388 (2023)
8. Liu, M., Guo, M., Fu, Y., O'Neill, Z., Gao, Y.: Expert-guided imitation learning for energy management: evaluating Gail's performance in building control applications. Appl. Energy **372**, 123753 (2024)
9. Liu, R., Wang, J., Chen, Y., Liu, Y., Wang, Y., Gu, J.: Proximal policy optimization with time-varying muscle synergy for the control of an upper limb musculoskeletal system. IEEE Trans. Autom. Sci. Eng. (2023)
10. Luo, H., Lee, C.H., Li, C., He, S.: Generative adversarial imitation learning-based continuous learning computational guidance. IEEE Trans. Aerosp. Electron. Syst. (2025)
11. Okui, N., Okui, M.A., Okui, M.: Mesh extraction surgery and laser treatment for pain after mid-urethral sling surgery: a case series. Cureus **16**(1) (2024)
12. Pathak, D., Agrawal, P., Efros, A.A., Darrell, T.: Curiosity-driven exploration by self-supervised prediction. In: Proceedings of the IEEE Conference on Computer Vision and Pattern Recognition Workshops, pp. 16–17 (2017)
13. Peloso, A., Damiano, R., Zhang, X., Bicchi, A., Votta, E., De Momi, E.: Imitation learning for path planning in cardiac percutaneous interventions. IEEE Trans. Biomed. Eng. (2025)
14. Qian, C., Ren, H.: Deep reinforcement learning in surgical robotics: enhancing the automation level. In: Handbook of Robotic Surgery, pp. 89–102 (2025)
15. Ronneberger, O., Fischer, P., Brox, T.: U-Net: convolutional networks for biomedical image segmentation. In: Navab, N., Hornegger, J., Wells, W.M., Frangi, A.F. (eds.) MICCAI 2015. LNCS, vol. 9351, pp. 234–241. Springer, Cham (2015). https://doi.org/10.1007/978-3-319-24574-4_28

16. Tobin, J., Fong, R., Ray, A., Schneider, J., Zaremba, W., Abbeel, P.: Domain randomization for transferring deep neural networks from simulation to the real world. In: 2017 IEEE/RSJ International Conference on Intelligent Robots and Systems (IROS), pp. 23–30. IEEE (2017)
17. Wagner, M., et al.: The importance of machine learning in autonomous actions for surgical decision making. Artif. Intell. Surg. **2**(2), 64–79 (2022)
18. Yip, M., Das, N.: Robot autonomy for surgery. In: The Encyclopedia of MEDICAL ROBOTICS: Volume 1 Minimally Invasive Surgical Robotics, pp. 281–313. World Scientific (2019)
19. Yip, M., et al.: Artificial intelligence meets medical robotics. Science **381**(6654), 141–146 (2023)
20. Zackoff, M.W., et al.: Immersive virtual reality onboarding using a digital twin for a new clinical space expansion: a novel approach to large-scale training for health care providers. J. Pediatr. **252**, 7–10 (2023)

Neural Ultrasound Shape Reconstruction via Gated Fusion and Sampson Distance

Zhinuo Zhou[1], Jiuan Chen[1], Guanglin Cao[1], Xue Li[3], Mingyang Zhao[1,2(✉)], and Gaofeng Meng[1,4(✉)]

[1] Centre for Artificial Intelligence and Robotics, HK Institute of Science and Innovation, Chinese Academy of Sciences, Hong Kong SAR, China
gaofeng.meng@cair-cas.org.hk
[2] State Key Laboratory of Mathematical Sciences, Academy of Mathematics and Systems Science, Chinese Academy of Sciences, Beijing, China
[3] Beijing Huairou District Hospital of Traditional Chinese Medicine, Beijing, China
[4] State Key Laboratory of Multimodal Artificial Intelligence Systems, Institute of Automation, Chinese Academy of Sciences, Beijing, China

Abstract. Recent advancements demonstrate that implicit neural representations of the *Signed Distance Field* (SDF) yield remarkable results in reconstructing shapes from freehand 3D ultrasound imaging. Typically, multi-view data is arranged in rows and columns, but fusing ultrasound data or networks remains a significant challenge. To address this issue, we propose a novel neural signed distance function for ultrasound shape reconstruction, guided by a newly introduced *gated fusion* and *Sampson distance*, termed GS-SDF. The core of our method lies in combining row-SDF and column-SDF networks using an *adaptive gating mechanism* and optimizing a fusion SDF network with a *Sampson loss*. Specifically, for row- and column-scanned data, we first employ implicit neural representations to model their SDFs. We then design a gating mechanism to dynamically assign weights to the row-SDF and column-SDF networks, significantly enhancing the fusion process and enabling more accurate fitting during training. Additionally, we introduce the Sampson distance to improve the accuracy of distance computation between points and the surface, compared to conventional algebraic methods. This approach provides a more faithful evaluation of the loss between scanned data and model predictions. We demonstrate the efficacy of GS-SDF on four benchmark datasets acquired using ultrasound transducer probes and computed tomography, achieving state-of-the-art performance compared to the competing reconstruction approach. The code for our method will be made publicly available.

Keywords: Neural shape representation · Freehand 3D ultrasound

1 Introduction

Freehand three-dimensional (3D) ultrasound is a technique that captures 3D ultrasound data by tracking the trajectory of a conventional 2D ultrasound

© The Author(s), under exclusive license to Springer Nature Switzerland AG 2026
J. Qiu et al. (Eds.): Agentic AI 2025/CMLLMs 2025/CREATE 2025, LNCS 16147, pp. 146–155, 2026.
https://doi.org/10.1007/978-3-032-06004-4_15

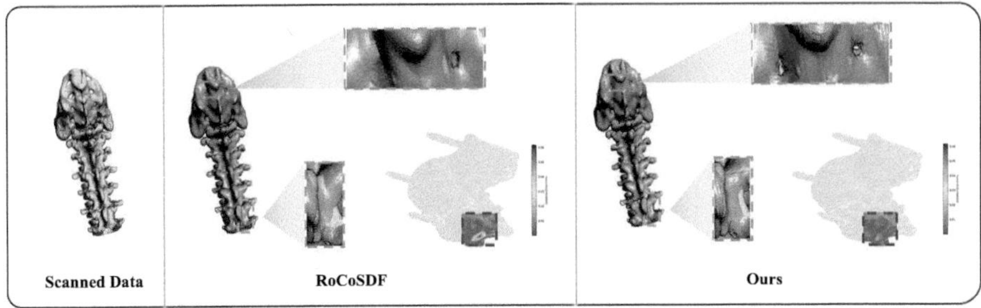

Fig. 1. Comparison between RoCoSDF and GS-SDF. Grey represents the reconstruction results, while yellow represents the scanned data. (Color figure online)

probe as a clinician moves it across an object of interest [6]. Compared to traditional ultrasound equipment, freehand 3D ultrasound offers numerous advantages, including a wider range of applications, compact size, and improved portability [5]. Its ease of use and simplicity make it a promising tool for advancing hierarchical diagnosis and treatment [11–13]. Recent research on 3D reconstruction in freehand 3D ultrasound imaging has primarily focused on the accurate reconstruction of geometric shapes or surfaces [15], as ultrasound signals cannot penetrate the boundaries of hard tissues [2,7].

In the field of 3D reconstruction, SDF [14,20] plays a crucial role. It defines the signed distance from any point in space to the surface of an object, providing an accurate representation of its geometric boundary [8,9,17,19]. In multi-view imaging, images captured from different perspectives offer complementary information about the object. SDF is particularly effective in this context, as it accurately represents the object's geometric boundary and remains unaffected by variations in viewing angles or lighting conditions. Previous work on multi-view scanning for 3D reconstruction has typically addressed the limitations of single-view data by fusing multi-view data to achieve comprehensive 3D reconstruction [18,23,24]. By integrating geometric information from multiple views, a more stable and precise 3D structure of the object can be reconstructed [4,25].

The seminar work RoCoSDF [2] encodes ultrasound data from different views into their corresponding neural SDFs and introduces two regularizers to refine shapes by constraining the SDFs near the surface. RoCoSDF utilizes max pooling to fuse row- and column-SDF networks. Nevertheless, this coarse fusion of multi-view data can result in *information redundancy* and cause the network to *miss critical geometric details*. In Fig. 1, the yellow region represents the scanned data, while the grey region corresponds to the model reconstruction. A higher degree of overlap between the grey and yellow regions indicates better reconstruction performance. For the Spine (entire) dataset, the overlap region of RoCoSDF is notably smaller in the bottom-right corner compared to GS-SDF, indicating the incompleteness of structure reconstruction in RoCoSDF. Furthermore, in the top-left region of the reconstruction, aside from obvious structural errors (*e.g.*, the absence of holes), RoCoSDF tends to produce overly smooth surfaces,

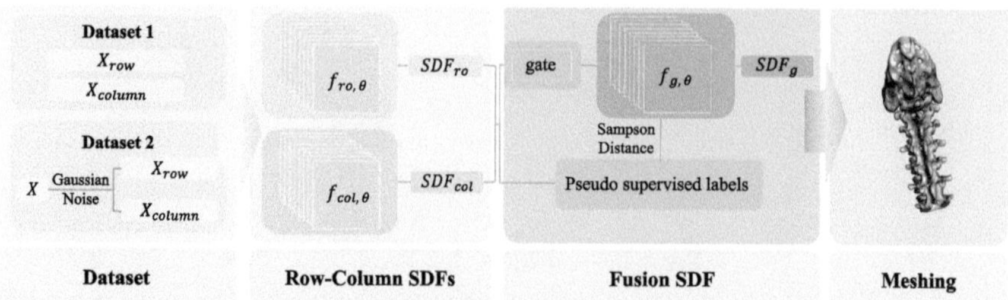

Fig. 2. Framework of GS-SDF. For datasets, we add Gaussian noise to stimulate row and column data. For Row-Column SDFs, we train two neural SDF networks separately. In fusion SDF, we optimize the network by introducing the gate and train it with the Sampson distance. In meshing, we utilize Marching Cubes to extract the 3D mesh.

leading to discrepancies between the reconstructed models and the scanned data regarding geometric details.

To address these limitations, specifically for the structure errors and the lack of geometric details, in this work, we propose a novel framework termed GS-SDF using gated fusion and Sampson distance for 3D neural ultrasound shape reconstruction. GS-SDF avoids the use of max pooling or averaging operations for multi-view networks fusion and instead introduces a novel gating mechanism to distribute weights more effectively. Also, GS-SDF incorporates the adaptive fusion scheme with the Sampson distance [16,22], which enables more realistic and geometrically accurate reconstructions. Specifically, row-SDF and column-SDF networks are first trained on the scanned data. Then, query points are sampled randomly, and pseudo-supervised labels are generated using row-SDFs and column-SDFs. The designed gate and the fusion SDF network are then trained jointly with pseudo labels. In the second training phase, the Sampson distance is introduced to reduce computational errors during loss computing and improve the accuracy of the model [22]. We validate our approach on the benchmark datasets, with extensive experimental results demonstrating its effectiveness. To summarize, the contributions of this work are as follows:

- We present a novel pipeline for ultrasound shape reconstruction that enhances the fusion process within the network for ultrasound-scanned data.
- We introduce a novel gating mechanism to enable adaptive weight distribution during the fusion of multi-view data in network training.
- We introduce the Sampson distance as an effective loss function, significantly reducing computational errors when calculating the loss between scanned data and model predictions.

2 The Proposed Method

The pipeline of our proposed gated and Sampson distance-guided SDF network for ultrasound reconstruction is presented in Fig. 2. First, row and column-scanned SDF networks are trained separately to coarsely predict SDFs from different views. To reduce information redundancy and extract key features from both row and column views, a gating network is introduced and trained jointly with the supervised fusion network. In this setup, the predictions of row- and column-SDF networks are used as pseudo training labels. Also, Sampson distance is added into the loss function to reduce errors in loss computation. Finally, the 3D mesh is extracted from the gated fusion network using the Marching Cubes algorithm [10].

2.1 Gated SDF Prediction

Gated Neural SDF Network. To address the issue of insufficient accuracy in ultrasound reconstruction caused by the use of max pooling in network fusion, we propose integrating a gating mechanism into the reconstruction framework. Three separate multilayer perceptron (MLP) networks are trained for row-column scanned data and fusion-supervised prediction. For row-column SDF prediction, we use a basic MLP, denoted as f_θ. Given a set of scanned data $X = \{\mathbf{x}_i = [x_i, y_i, z_i] \in \mathbb{R}^3\}_{i=1}^N$, the network aims to predict the value of the corresponding signed distance field $s = f_\theta(X) \in \mathbb{R}$ for X, where the zero-level set $f_\theta(\cdot) = 0$ implicitly represents the surface of the object, $i.e.$,

$$f_\theta(\mathbf{x}_i) = \begin{cases} d(\mathbf{x}_i, S), & \mathbf{x}_i \in \mathbb{R}^3 \setminus \mathrm{int}(S) \\ 0, & \mathbf{x}_i \in S \\ -d(\mathbf{x}_i, S), & \mathbf{x}_i \in \mathrm{int}(S) \end{cases} \tag{1}$$

where S represents the surface of the prediction object, $d(\cdot)$ represents the distance between the point $\mathbf{x}_i$ and S. According to Eq. 1, when $\mathbf{x}_i$ is outside the surface S, the value of $f_\theta(\mathbf{x}_i)$ is positive. If the data is inside S, $f_\theta(\mathbf{x}_i)$ is negative. Trained separately on row and column data, the corresponding row and column SDF networks are denoted as $f_{\mathrm{ro},\theta}$ and $f_{\mathrm{co},\theta}$, respectively.

In contrast to the row and column SDF networks, our proposed fusion network for fusion-supervised prediction incorporates an additional gating mechanism, as shown in Eq. 2. With the help of the gate, the features learned in previous networks are fused better. Specifically, the gate is a single-layer linear network designed for weight distribution:

$$s_1 = f_{\mathrm{ro},\theta}(X), \quad s_2 = f_{\mathrm{co},\theta}(X), \quad f_{\mathrm{g},\theta}(X) = f_\theta(\mathrm{gate}(s_1, s_2)), \tag{2}$$

where s_1 and s_2 represent the SDF predicted by networks, $\mathrm{gate}(\cdot)$ denotes the gate mechanism, and $f_{\mathrm{g},\theta}(\cdot)$ denotes the fusion network.

Row and Column SDFs Training. For the neural row- and column-SDF networks, like RoCoSDF [2], our training objective function is defined as follows:

$$L_{\mathrm{ro,co}} = D_1(X, f_\theta) + \lambda_1 D_2(X, f_\theta) + \lambda_2 R_{\mathrm{nonmfd}} D_3(X, f_\theta), \tag{3}$$

where $D_1 = (X, f_\theta) = ||X - \hat{X}||_2^2$, $D_2(X, f_\theta) = \frac{1}{2}(C(X) - 1)^2$, $D_3(X, f_\theta) = 1 - \cos(\nabla f_\theta(X), \frac{X - \hat{X}}{||X - \hat{X}||_2})$, $\hat{X} = X - f_\theta(X)\nabla f_\theta(X)$ represents the projected points, $D(\cdot)$ denotes the distance, $C(\cdot)$ denotes the confidence of the model prediction, $R_{\mathrm{nonmfd}} = e^{-\alpha_{\mathrm{nonmfd}}|f_\theta(X)|}$ represents the non-manifold regularizer, λ_1 and λ_2 are the hyper-parameters. For each scanned data point $\mathbf{x_i}$, it is first projected onto the surface S at $\hat{X}$. D_1 measures the Euclidean distance between X and $\hat{X}$. D_2 follows the adversarial learning strategy in [3].

Fusing SDF Training with the Sampson Distance. The fusion SDF training strategy inherits D_1 and D_3 from the row and column SDFs training, as proposed in RoCoSDF [2]. However, to constrain the influence of the gate and better facilitate supervised learning, we incorporate the Sampson distance into the loss function. As defined in Eq. 4, the Sampson distance measures the error of data scanned from different angles, enhancing the accuracy of distance computation between points and the surface while reducing computational errors in the loss:

$$d_{\mathrm{Sampson}}(\mathbf{x_i}) = \frac{|\mathbf{x_i}^T C_{\mathbf{x_i}}|}{\|\nabla \mathbf{x_i}^T C_{\mathbf{x_i}}\|} = \frac{|(\hat{\mathbf{x}}_i + f_\theta(\mathbf{x_i})\nabla f_\theta(\mathbf{x_i}))^T C_{\mathbf{x_i}}|}{\|\nabla(\hat{\mathbf{x}}_i + f_\theta(\mathbf{x_i})\nabla f_\theta(\mathbf{x_i}))^T C_{\mathbf{x_i}}\|}$$
$$= \frac{|f_\theta(\mathbf{x_i})|}{\|\nabla f_\theta(\mathbf{x_i})\| + c_1} + \frac{|\nabla f_\theta(\mathbf{x_i})|}{\|\nabla^2 f_\theta(\mathbf{x_i})\| + c_2} \approx \frac{|f_\theta(\mathbf{x_i})|}{\|\nabla f_\theta(\mathbf{x_i})\| + c_1}, \tag{4}$$

where C is the reconstruction shape, $C_{\mathbf{x_i}}$ is the polar line of $\mathbf{x_i}$ related to C [21] and c_1, c_2 are small constants to avoid the denominator equating to zero. With the introduced training objective function designed for the gate, the final loss function for the fusion SDF network is given in the following:

$$L_{\mathrm{fusion}} = \|f_{\mathrm{pseudo}}, f_\theta\| + \lambda_2 R_{\mathrm{nonmfd}} D_3(X, f_\theta) + \lambda_{\mathrm{Samp}} D_{\mathrm{Samp}}(X), \tag{5}$$

where $f_{\mathrm{pseudo}} = max(f_{ro,\theta}, f_{co,\theta})$ is the pseudo labels, $D_{\mathrm{Samp}}(X) = \Sigma \frac{|f_\theta(X)|}{\|\nabla f_\theta(X)\|}$, λ_{Samp} is the hyper-parameter.

3   Experiments and Discussions

In this section, we perform extensive experiments to test the proposed method for ultrasound reconstruction and compare it with the SOTA method ReCoSDF [2].

3.1   Dataset

We evaluate the reconstruction efficacy of the proposed method using three ultrasound datasets: Spine (partial), Thyroid, and Vessel, each representing different body parts. The Spine (partial) dataset is sourced from RoCoSDF [2] and includes both row- and column-scanned data. The Thyroid dataset, acquired using an ultrasound transducer (UT) from a patient, consists of 430 frames. The Vessel dataset, also collected using UT, is derived from a silicone model and contains 1,200 frames. To simulate row and column data, we add Gaussian noise with a mean of 0 and a variance of 1 to these datasets. A robotic arm is attached to the transducers to ensure precise localization of images in the tracking space. Moreover, to assess the generalization capability of the proposed method, we test it on a Spine (entire) dataset captured using computed tomography (CT).

3.2   Evaluation Metrics

Following prior work [1,2], we adopt two evaluation metrics to quantitatively assess the reconstruction quality of our model: the symmetric Chamfer distance D_{Chamfer} and Hausdorff distance $D_{\text{Hausdorff}}$, which are defined in the following:

$$
\begin{aligned}
D_{\text{Chamfer}}(X_1, X_2) &= \frac{1}{|X_1|} \sum_{\mathbf{x} \in X_1} \min_{\mathbf{y} \in X_2} \|\mathbf{x} - \mathbf{y}\|_2^2 + \frac{1}{|X_2|} \sum_{\mathbf{y} \in X_2} \min_{\mathbf{x} \in X_1} \|\mathbf{y} - \mathbf{x}\|_2^2, \\
D_{\text{Hausdorff}}(X_1, X_2) &= \max_{\mathbf{x} \in X_1} \{ \min_{\mathbf{y} \in X_2} \|\mathbf{x} - \mathbf{y}\| \},
\end{aligned}
\tag{6}
$$

where $\mathbf{x}$ and $\mathbf{y}$ are sampled points. For both D_{Chamfer} and $D_{\text{Hausdorff}}$, we report the average performance across raw row- and column-scanned data.

3.3   Implementations

The adaptive gating mechanism is a single-layer linear network with Softmax as the activation function. The MLP consists of 8 layers with hidden channels of 256. We utilize the Adam optimizer with a learning rate of 0.001. The iteration, batch size, α_{nonmfd}, λ_1, λ_2 and λ_{Samp} are set as 10,000, 5,000, 100, 0.01, 0.01 and 1.0, respectively. Our model is trained on two NVIDIA A100-80 GPUs. The mesh extraction resolution is 2,563 and the threshold of Marching Cubes is 0.

Table 1. Quantitative comparisons on the four benchmark reconstruction datasets. Bont fonts indicate the top performer.

Method	Data	$D_{\text{Chamfer}}(\times 10^{-2}\text{m})$	$D_{\text{Hausdorff}}(\times 10^{-2}\text{m})$
RoCoSDF [2]	Thyroid	6.655	23.878
GS-SDF (Ours)	Thyroid	**4.162**(2.493 ↓)	**14.224**(9.654 ↓)
RoCoSDF [2]	Spine (partial)	1.922	**8.873**
GS-SDF (Ours)	Spine (partial)	**1.559**(0.363 ↓)	9.064(0.191 ↑)
RoCoSDF [2]	Spine (entire)	1.357	11.145
GS-SDF (Ours)	Spine (entire)	**1.316**(0.041 ↓)	**10.587**(0.558 ↓)
RoCoSDF [2]	Vessel	1.311	**7.948**
GS-SDF (Ours)	Vessel	**1.295**(0.016 ↓)	8.106(0.158 ↑)
RoCoSDF [2]	Average	2.811	12.961
GS-SDF (Ours)	Average	**2.083**(0.728 ↓)	**10.495**(2.466 ↓)

3.4   Results

The quantitative evaluation results are reported in Table 1. As observed, our method achieves overall higher reconstruction accuracy than the competitor across all datasets. Specifically, our method delivers superior results on all

Table 2. Ablation experiments on the proposed Sampson loss and gating mechanism.

Method	Data	$D_{\text{Chamfer}}(\times 10^{-2}\text{m})$	$D_{\text{Hausdorff}}(\times 10^{-2}\text{m})$
GS-SDF w/o D_{Samp}	Thyroid	4.478(0.316 ↑)	14.994(0.770 ↑)
GS-SDF w/o gate	Thyroid	6.646(2.484 ↑)	23.988(9.764 ↑)
GS-SDF (ours)	Thyroid	**4.162**	**14.224**
GS-SDF w/o D_{Samp}	Spine (partial)	1.560(0.001 ↑)	9.244(0.180 ↑)
GS-SDF w/o gate	Spine (partial)	1.926(0.367 ↑)	**8.768**(0.296 ↓)
GS-SDF (ours)	Spine (partial)	**1.559**	9.064
GS-SDF w/o D_{Samp}	Spine (entire)	1.324(0.008 ↑)	11.008(0.421 ↑)
GS-SDF w/o gate	Spine (entire)	1.358(0.042 ↑)	11.101(0.514 ↑)
GS-SDF (ours)	Spine (entire)	**1.316**	**10.587**
GS-SDF w/o D_{Samp}	Vessel	1.295(0.000)	8.621(0.515 ↑)
GS-SDF w/o gate	Vessel	1.357(0.062 ↑)	8.288(0.182 ↑)
GS-SDF (ours)	Vessel	**1.295**	**8.106**
GS-SDF w/o D_{Samp}	Average	2.164(0.081 ↑)	10.967(0.472 ↑)
GS-SDF w/o gate	Average	2.822(0.739 ↑)	13.036(2.541 ↑)
GS-SDF (ours)	Average	**2.083**	**10.495**

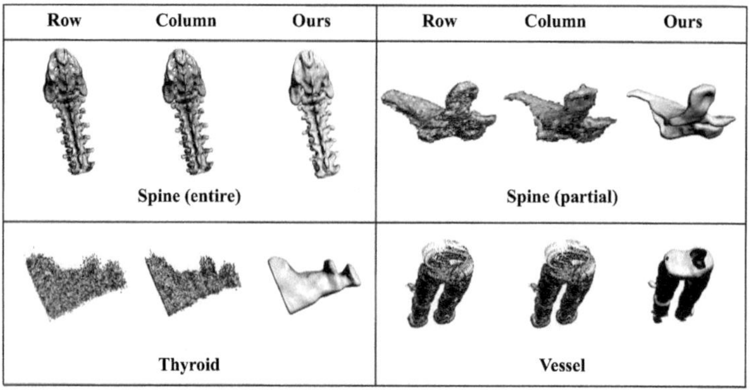

Row	Column	Ours	Row	Column	Ours
	Spine (entire)			Spine (partial)	
	Thyroid			Vessel	

Fig. 3. Visualization of the reconstruction results by our proposed method. Various perspectives of the reconstructed models are presented in the *Supplementary Video*.

datasets in terms of the D_{Chamfer} metric. For the $D_{\text{Hausdorff}}$ metric, GS-SDF outperforms RoCoSDF by a significant margin in Thyroid reconstruction, reducing the reconstruction error from 23.878 to 14.224. Although our method performs slightly lower than RoCoSDF in Spine (entire) and Vessel reconstruction for $D_{\text{Hausdorff}}$, it still achieves the best average reconstruction accuracy, demonstrating its stability and superiority.

We present visualizations to better illustrate our reconstruction performance in Figs. 1 and 3. In Fig. 1, it can be observed that the reconstruction result

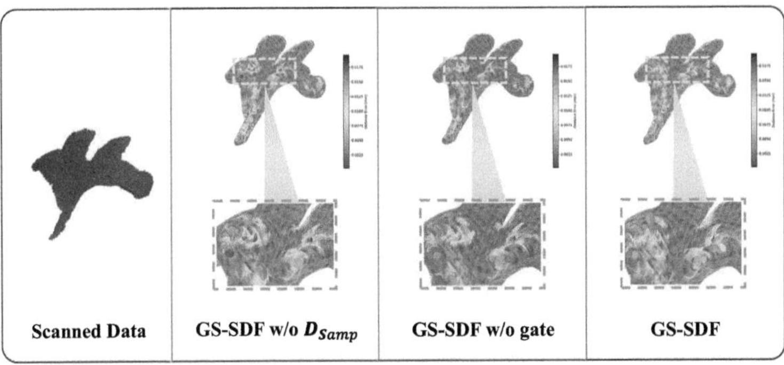

Fig. 4. Reconstruction errors for ablation studies indicated by the heatmap.

of our method accurately restores the overall 3D shape of the Spine data and outperforms RoCoSDF, particularly in capturing shape details. In the magnified upper part of the reconstructed result, our method successfully reconstructs two holes, while RoCoSDF produces only an ambiguous one. Additionally, the details in the bottom region demonstrate that RoCoSDF fails to reconstruct the shape as accurately as our method. The heat map showing the distance between sampled query points and the SDFs in Fig. 1 further highlights the significant lack of reconstruction in the bottom region of the Spine (entire) data for RoCoSDF. To better illustrate the reconstruction performance of the GS-SDF model, we provide visualizations of four benchmark datasets in Fig. 3.

Additionally, we conduct additional ablation experiments to validate the effectiveness of the introduced gate scheme and the Sampson loss. On average, our method demonstrates superior performance compared to the other two ablation methods, achieving reductions of 0.081 and 0.739 in D_{Chamfer}, as well as 0.472 and 2.541 in $D_{\text{Hausdorff}}$. From Table 2, we can see that without the Sampson loss, the method's performance has a decrease of 0.316, 0.001, 0.008, 0.000 on D_{Chamfer}. And without the gate, compared to GS-SDF, most of the performance decreases. Both the gate and the Sampson loss contribute to the effectiveness of GS-SDF. The distance between sampled query points and SDF visualized in Fig. 4 further demonstrates the contribution of these two modules.

4   Discussion and Conclusions

The technical advancements presented in GS-SDF could benefit clinical applications of freehand 3D ultrasound imaging in several aspects: (1) The method's sub-millimeter reconstruction accuracy enables precise visualization of critical anatomical features like thyroid nodules and vascular structures, directly supporting clinical diagnosis. (2) With 2.466×10^{-2} m Hausdorff and 0.728×10^{-2} m Chamfer distance reduction, the system provides more reliable intraoperative guidance for procedures, minimizing registration errors.

We presented GS-SDF, a novel gated-guided SDF method for reconstructing the shape of 3D images from freehand ultrasound data using row-column scanning. Our approach adaptively distributes weights between the row-SDF and column-SDF networks, effectively reducing information redundancy through max pooling and significantly improving reconstruction precision, especially for intricate geometric structures and fine details. Additionally, we integrate the Sampson distance into our loss computation to further refine the accuracy of distance measurements between the pseudo labels and predictions. Extensive experiments conducted on four benchmark datasets, complemented by ablation studies, validate the effectiveness of our method.

Limitations and Future Work. While our method has demonstrated impressive performance on diverse datasets, it is important to note that the SDF intrinsically relies on sign information, which inherently favors closed shapes. Furthermore, the observed discrepancy between the improvements in Hausdorff and Chamfer distances warrants further investigation. In future research, we aim to investigate the potential of replacing the SDF with an *Unsigned Distance Field* (UDF) to improve robustness and accuracy when dealing with open surfaces.

References

1. Ben-Shabat, Y., Koneputugodage, C.H., Gould, S.: DiGS: divergence guided shape implicit neural representation for unoriented point clouds. In: Proceedings of the IEEE/CVF Conference on Computer Vision and Pattern Recognition, pp. 19323–19332 (2022)
2. Chen, H., Gao, Y., Zhang, S., Wu, J., Ma, Y., Zheng, R.: RoCoSDF: row-column scanned neural signed distance fields for freehand 3D ultrasound imaging shape reconstruction. In: International Conference on Medical Image Computing and Computer-Assisted Intervention, pp. 721–731. Springer (2024)
3. Chen, H., et al.: Neural implicit surface reconstruction of freehand 3d ultrasound volume with geometric constraints. Med. Image Anal. **98**, 103305 (2024)
4. Chen, Z., et al.: UW-SDF: exploiting hybrid geometric priors for neural SDF reconstruction from underwater multi-view monocular images. In: 2024 IEEE/RSJ International Conference on Intelligent Robots and Systems (IROS), pp. 14248–14255. IEEE (2024)
5. Chen, Z., Huang, Q.: Real-time freehand 3D ultrasound imaging. Comput. Meth. Biomech. Biomed. Eng. Imaging Vis. **6**(1), 74–83 (2018)
6. Hsu, P.W., Prager, R.W., Gee, A.H., Treece, G.M.: Freehand 3D ultrasound calibration: a review. In: Advanced Imaging in Biology and Medicine: Technology, Software Environments, Applications, pp. 47–84 (2009)
7. Huang, Q., et al.: Anatomical prior based vertebra modelling for reappearance of human spines. Neurocomputing **500**, 750–760 (2022)
8. Jones, M.W., Baerentzen, J.A., Sramek, M.: 3d distance fields: a survey of techniques and applications. IEEE Trans. Visual Comput. Graphics **12**(4), 581–599 (2006)
9. Lipman, Y.: Phase transitions, distance functions, and implicit neural representations. arXiv preprint arXiv:2106.07689 (2021)

10. Lorensen, W.E., Cline, H.E.: Marching cubes: a high resolution 3d surface construction algorithm. In: Seminal Graphics: Pioneering Efforts that Shaped the Field, pp. 347–353 (1998)
11. Luo, M., et al.: Multi-IMU with online self-consistency for freehand 3D ultrasound reconstruction. In: International Conference on Medical Image Computing and Computer-Assisted Intervention, pp. 342–351. Springer (2023)
12. Mohamed, F., Siang, C.V.: A survey on 3D ultrasound reconstruction techniques. In: Artificial Intelligence—Applications in Medicine and Biology, pp. 73–92 (2019)
13. Mozaffari, M.H., Lee, W.S.: Freehand 3-D ultrasound imaging: a systematic review. Ultrasound Med. Biol. **43**(10), 2099–2124 (2017)
14. Park, J.J., Florence, P., Straub, J., Newcombe, R., Lovegrove, S.: DeepSDF: learning continuous signed distance functions for shape representation. In: Proceedings of the IEEE/CVF Conference on Computer Vision and Pattern Recognition, pp. 165–174 (2019)
15. Solberg, O.V., Lindseth, F., Torp, H., Blake, R.E., Hernes, T.A.N.: Freehand 3d ultrasound reconstruction algorithms—a review. Ultrasound Med. Biol. **33**(7), 991–1009 (2007)
16. Szpak, Z.L., Chojnacki, W., Eriksson, A., Van Den Hengel, A.: Sampson distance based joint estimation of multiple homographies with uncalibrated cameras. Comput. Vis. Image Underst. **125**, 200–213 (2014)
17. Tancik, M., et al.: Fourier features let networks learn high frequency functions in low dimensional domains. Adv. Neural. Inf. Process. Syst. **33**, 7537–7547 (2020)
18. Wang, D., et al.: Multi-view 3D reconstruction with transformers. In: Proceedings of the IEEE/CVF International Conference on Computer Vision, pp. 5722–5731 (2021)
19. Wang, Y., Rahmann, L., Sorkine-Hornung, O.: Geometry-consistent neural shape representation with implicit displacement fields. In: The Tenth International Conference on Learning Representations. OpenReview (2022)
20. Wiesner, D., et al.: Generative modeling of living cells with SO (3)-equivariant implicit neural representations. Med. Image Anal. **91**, 102991 (2024)
21. Wu, Y., Wang, H., Tang, F.: Conic fitting: new easy geometric method and revisiting Sampson distance. In: 2017 4th IAPR Asian Conference on Pattern Recognition (ACPR), pp. 37–42. IEEE (2017)
22. Wu, Y., Wang, H., Tang, F., Wang, Z.: Efficient conic fitting with an analytical polar-n-direction geometric distance. Pattern Recogn. **90**, 415–423 (2019)
23. Xie, H., Yao, H., Sun, X., Zhou, S., Zhang, S.: Pix2Vox: context-aware 3D reconstruction from single and multi-view images. In: Proceedings of the IEEE/CVF International Conference on Computer Vision, pp. 2690–2698 (2019)
24. Zhao, M., Huang, X., Jiang, J., Mou, L., Yan, D.M., Ma, L.: Accurate registration of cross-modality geometry via consistent clustering. IEEE Trans. Visual Comput. Graphics **30**(7), 4055–4067 (2023)
25. Zhao, M., Jiang, J., Ma, L., Xin, S., Meng, G., Yan, D.M.: Correspondence-free non-rigid point set registration using unsupervised clustering analysis. In: Proceedings of the IEEE/CVF Conference on Computer Vision and Pattern Recognition, pp. 21199–21208 (2024)

ReCAP: Recursive Cross Attention Network for Pseudo-label Generation in Robotic Surgical Skill Assessment

Julien Quarez[1]([✉]), Marc Modat[1], Sebastien Ourselin[1], Jonathan Shapey[1,2], and Alejandro Granados[1]

[1] School of Biomedical Engineering and Imaging Sciences, King's College London, London, UK
`julien.quarez@kcl.ac.uk`
[2] Neurosurgery, King's College Hospital, London, UK

Abstract. In surgical skill assessment, the Objective Structured Assessments of Technical Skills (OSATS) and Global Rating Scale (GRS) are well-established tools for evaluating surgeons during training. These metrics, along with performance feedback, help surgeons improve and reach practice standards. Recent research on the open-source JIGSAWS dataset, which includes both GRS and OSATS labels, has focused on regressing GRS scores from kinematic data, video, or their combination. However, we argue that regressing GRS alone is limiting, as it aggregates OSATS scores and overlooks clinically meaningful variations during a surgical trial.

To address this, we developed a weakly-supervised recurrent transformer model that tracks a surgeon's performance throughout a session by mapping hidden states to six OSATS, derived from kinematic data. These OSATS scores are averaged to predict GRS, allowing us to compare our model's performance against state-of-the-art (SOTA) methods. We report Spearman's Correlation Coefficients (SCC) demonstrating that our model outperforms SOTA using kinematic data (SCC 0.83–0.88), and matches performance with video-based models.

Our model also surpasses SOTA in most tasks for average OSATS predictions (SCC 0.46–0.70) and specific OSATS (SCC 0.56–0.95). The generation of pseudo-labels at the segment level translates quantitative predictions into qualitative feedback, vital for automated surgical skill assessment pipelines. A senior surgeon validated our model's outputs, agreeing with 77% of the weakly-supervised predictions $p = 0.006$.

Keywords: Surgical Skill · Kinematic Data · Automated Assessment · Robotic Assisted Surgery

1 Introduction

Robotic surgery has rapidly expanded across specialties and autonomy levels. Although evidence on the benefits of robotic-assisted surgery (RAS) is

© The Author(s), under exclusive license to Springer Nature Switzerland AG 2026
J. Qiu et al. (Eds.): Agentic AI 2025/CMLLMs 2025/CREATE 2025, LNCS 16147, pp. 156–166, 2026.
https://doi.org/10.1007/978-3-032-06004-4_16

mixed [20], its use is increasing—for example, in the UK, RAS for radical prosta-tectomy rose from 5% in 2006 to 88% in 2018 [20], though adoption varies widely (e.g., 1.8% in ENT). A significant barrier to wider adoption is the variability in training across systems and institutions [3,23]. Skill assessment is central to surgical training, enabling evaluation of trainee progress. However, reliance on senior surgeons for feedback limits junior doctors' opportunities [10]. Automated, system-agnostic assessment could overcome this.

The Objective Structured Assessment of Technical Skills (OSATS) [19] pro-vides a standardized, Likert-scale framework widely used in RAS. It yields a Global Rating Score (GRS) summarizing multiple skill components. Despite its usefulness, OSATS depends on expert assessors [6,25] and remains time-consuming and subjective. Machine learning (ML) and deep learning (DL) offer promising paths to automate and scale assessment.

Kinematic data is particularly suited for this task, offering standardization, lower computational cost, and system-agnostic features compared to video data [12,17,29,30]. Recent works focus on regressing GRS or expertise levels [1,7,14, 27,31], but these high-level scores provide limited clinical insight and still require expert feedback. Efforts to model score changes during procedures [1,7,11,15, 27,28,31] face challenges such as increased labeling burden [15,27] or insufficient validation [1,28,31].

For instance, Wang *et al.* [27] used supervised recurrent networks on inter-mediate GRS scores but depended on granular labels and lacked interpretabil-ity [9,18]. Anastasiou *et al.* [1] employed contrastive learning for GRS regression, yet their method lacks actionable feedback. Zia *et al.* [31] explored hand-crafted features to link input segments to OSATS, improving interpretability but with-out directly translating predictions into clinical feedback.

Our work addresses this gap by predicting intermediate OSATS scores during a surgical trial in a weakly-supervised manner, without additional labels. Using a recurrent cross-attention model on kinematic data, we provide segment-level OSATS predictions that offer detailed, actionable insights, advancing automated surgical skill assessment.

We summarize our contributions as follows:

1. An objective function enabling recurrent cross-attention models to predict trial-level GRS and OSATS scores alongside granular, segment-level OSATS scores in a weakly-supervised way, outperforming existing kinematics-based methods.
2. Revisiting kinematic data for task-agnostic modeling that links segment-level OSATS predictions to qualitative feedback.

2   Methods

We propose a recurrent model called ReCAP, Recursive Cross-Attention for Pseudo-label generation, where segments of kinematic data are processed into intermediate OSATS scores (Fig. 1). Those scores are then averaged into trial-level OSATS predictions. Our multi-task model is trained in an end-to-end fashion to output all six OSATS. We assess the model's performance on the GRS label by aggregating the individual OSATS predicted scores.

Problem Formulation. An input signal $X_i \in \mathbb{R}^{D \times T_i}$, of feature size D and length T_i, is divided into equal segments x_i^s of size L ($x_i^s \in \mathbb{R}^{D \times L}$): $\{x_i^1, x_i^2, \ldots, x_i^s\} \in X_i$ where S_i is the total number of segments, i.e. $S_i = \frac{T_i}{L}$ for a given signal i. For simplicity, in the rest of this paper, we omit i and use s as a subscript to refer to different segments within a trial i, i.e. $x_i^s \to x_s$. We fit a function F to map X to the label space $\mathcal{Y} : (F : X \to \mathbf{Y})$:

$$\mathbf{Y} = F(x_1, x_2, \ldots, x_s, \ldots x_S) \tag{1}$$

where $\mathbf{Y} \in \mathcal{Y}$ is a vector composed of all OSATS. The GRS is the aggregate of OSATS scores: $Y = \sum y_n$ where y_n is the nth OSATS.

Considering clinical practice where a given score is representative of their average performance through the trial i.e.: $y_n = \frac{1}{S} \sum y_s^n$, we rewrite Eq. 1 into Eq. 2, where f_n maps a segment s to the nth OSATS intermediate label ($x_s \to y_s^n$), similarly to many-to-many training in recursive training. Note that there is no ground truth for y_s^n and we learn $\hat{y}_s^n$ in a weakly-supervised manner.

$$y_n = \frac{1}{S} \sum_{s=1}^{S} f_n(x_s) \tag{2}$$

Model Overview: Our model recurrently processes segments of a kinematic signal by taking two inputs: the previous hidden state of the recurrent network, $z_{s-1} \in \mathbb{R}^{D \times L}$, and the current segment-level kinematic signal, x_s (Fig. 1). The two inputs are fused into the current hidden state, $z_s = h(x_s, z_{s-1})$, through the model backbone h (Fig. 1). We initialise z_0 as a zero-filled tensor. Each hidden state is then passed to six classification heads, c_n, giving $f_n(x_s) = c_n(h(x_s, z_{s-1}))$. The output of our model is a final n^{th} OSATS score, the average of all segment-level OSATS predictions:

$$\hat{y}_n = \frac{1}{S} \sum_{s=0}^{S} c_n[h(x_s, z_{s-1})] \tag{3}$$

The backbone h is composed of one fusion module (Fig. 1), where previous temporal information is fused with the current input through a series of multi-head self- and cross-attention blocks.

The classification heads c_n are five multilayer perceptron (MLPs) classifying the hidden state z_s into segment-level OSATS predictions $\hat{y}_{ns} = c_n(z_s)$. Each MLP layer consists of batch normalisation, ReLU activation function, and fully connected layers.

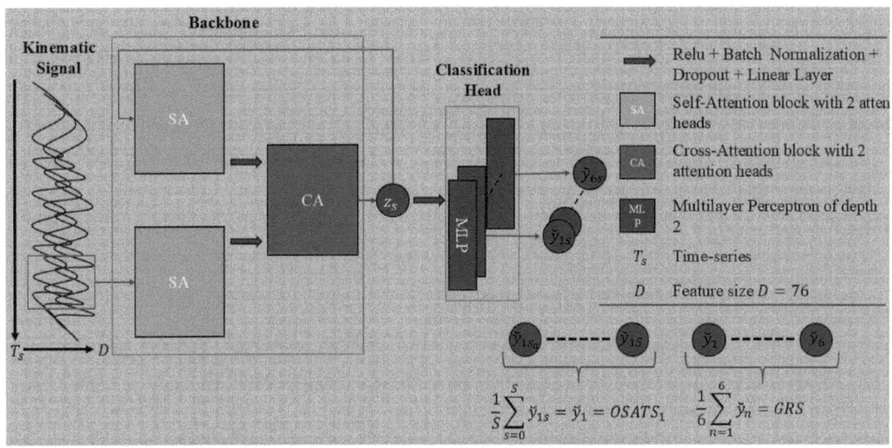

Fig. 1. ReCAP Architecture Overview: A kinematic signal is split into segments x_s of size L and used as inputs to our backbone h recurrently. The previous state z_{s-1} is also passed as an input and fused to x_s to produce z_s. This fusion is performed by a fusion module consisting of self- and cross-attention blocks. The current hidden state z_s is given as an input to six classification heads c_n to predict the respective OSATS score. $\hat{y}_{1s}$ to $\hat{y}_{6s}$ corresponding to the respective OSATS category prediction at segment position s

Loss: ReCAP is trained end-to-end using a cross-entropy loss. The loss is applied to the average of the classification head segment predictions $\hat{y}_n = \frac{1}{S}\sum \hat{y}_{ns}$ for a given OSATS category label y_n. An $L2$ penalty term weighted by λ is added to regularize the network to help with generalisation. Our final loss is expressed as:

$$\mathcal{L} = \sum_{n=0}^{N} CE(\hat{y}_n, y_n) + \lambda * L2 \tag{4}$$

Experimental Design: We evaluated our model on the JIGSAWS dataset [5]. This dataset consists of video and kinematics data generated by eight clinicians evaluated on three distinct tasks, namely needle passing (NP), suturing (SU), and knot-tying (KT). Altogether, there are 39, 28, and 36 labelled data samples for SU, NP, and KT, respectively. The labels are comprised of six OSATS score (1–5) and one GRS (6–35) the aggregate of all OSATS. It is worth noticing that the dataset is biased towards lower OSATS as the score 5 only appears in the suturing task. The OSATS include 1) respect for tissue, 2) suture/needle handling, 3) time and motion, 4) flow of operation, 5) overall performance, and 6) quality of the final product.

Cross-validation Scheme: Following the JIGSAWS cross-validation framework, we evaluate our method using Leave-One-Supertrial-Out (LOSO) whereby the i-th trial performed by the surgeons are left out as the validation set. The cross-validation scheme Leave-One-User-Out (LOUO) is not considered in this work since there is a lack of literature reporting OSATS for kinematic data. Literature [1,2] on the JIGSAWS dataset seems to indicate that no testing fold is used to assess performance.

Evaluation: Similar to relevant work [4,16,21,31], we evaluate our method using Spearman's Correlation Coefficient (SCC) ρ to compare the predicted ranked GRS score with the ground truth:

$$\rho = 1 - \frac{6 \sum d_i^2}{n(n^2 - 1)} = 1 - \frac{6 \sum \left(Y - \sum_{n=0}^{6} \hat{y}_n\right)^2}{n(n^2 - 1)} \tag{5}$$

We report SCC averaged across folds for the LOSO cross-validation scheme. The intermediate OSATS scores are averaged into a signal-level OSATS score after processing the whole kinematic sample. The final six OSATS scores are then summed to give a video-level GRS. Note that the predicted GRS, $\hat{Y} = \sum_{n=0}^{6} \hat{y}_n$, is only used to assess model performance, but is not directly learned from our model.

The mean average error (MAE) is also reported. To our knowledge, Benmansour *et al.* [2] is the only recent work reporting OSATS-specific performance under the LOSO validation scheme. OSATS performance is reported at the same epoch used to report GRS performance under the same training parameters. OSATS SCC was averaged across 10 epochs.

Data Augmentation: Two augmentation techniques were added to the kinematic signals to improve generalisation: 1) Gaussian noise based on the standard deviation of the signal, and 2) flipping, i.e. reversing the signal. Augmentations were done at a rate of 50%. Label smoothing and dropout was performed at 30%.

Implementation: Our model is trained with the Adam optimizer for 5000 epochs with a learning rate of 10^{-6}. Kinematic data was pre-processed by normalising across time and feature dimensions [24]. The kinematics from the slave and master device were used ($D = 76$). The sequence length of 75 was chosen. It corresponds to 2.5 s in time (force data acquired at 30 hz) and is consistent with the minimal time required for our clinician to rate a gesture. A lambda of 0.01 for L2 regularization and a batch size of 25 was used. In line with existing literature, design decisions and hyperparameter adjustments were experimentally conducted using the averaged cross-validation test fold [1]. ReCAP was implemented in Pytroch and trained on an Nvidia A100 GPU.

Validation of Model Behaviour: To validate our model's ability to generate interim OSATS scores, we asked a consultant surgeon in endoscopic interventions to agree or disagree with the model's intermediate predictions for the OSATS of Overall Performance. Every 75 frames was assigned a generated pseudo-label. The label is shown on the screen during viewing. Similar to Wang *et al.*'s framework [27], the surgeon was aware of the ground truth i.e. the trial level OSATS label. The predicted OSATS scores were divided into three categories: poor (1-2), average (3), and good (4-5). We randomly generate some segment predictions in two videos to mitigate potential bias without informing the surgeon. We then present these predictions and capture agreement or disagreement at the segment level when playing the video sequentially (Table 3).

Table 1. Performance for OSATS scores, where the $\rho Osats$ is the average across the 6 scores under LOSO scheme. *: results from training across the 3 tasks. Across Tasks (AT)

	KT	NP	SU	AT
Apen [31]	0.66	0.45	0.59	0.57
FCN [7]	0.65	**0.57**	0.60	**0.61**
ReCAP	**0.70**	0.46	**0.62**	0.59/0.58*

Table 2. Ablation of ReCAP components for GRS under LOSO scheme.

	KT	NP	SU
ReCAP no augmentation	0.86	0.85	0.83
ReCAP no pseudo-label	0.85	0.54	0.28
ReCAP	**0.88**	**0.85**	**0.83**

Table 3. OSATS performance, $\rho Osats$ is reported for RT: *Respect for tissue*, TM: *Time and Motion*, OP: *Overall Performance*, SNH: *Suture and Needle Handling*, FO: *Flow of Operation*, QFP: *Quality of Final Product* for the best/average fold performance. Across tasks(AT) is trained on the three tasks. [2] only report best results

	CNN+Bilstm [2]			ReCAP			
	KT	NP	SU	KT	NP	SU	AT
RT	0.83	0.49	0.46	**0.92**/0.78	**0.75**/0.43	**0.78**/0.52	0.56
TM	0.87	0.85	0.68	**0.95**/0.8	**0.91**/0.72	**0.84**/0.60	0.62
OP	0.89	**0.58**	0.71	**0.9**/0.79	0.42/0.23	**0.69**/0.5	0.65
SNH	0.82	0.79	0.75	**0.84**/0.61	**0.91**/0.69	**0.88**/0.78	0.65
FO	0.76	0.58	0.62	**0.78**/0.63	**0.66**/0.45	**0.89**/0.66	0.64
QFP	0.75	0.31	0.67	**0.85**/0.59	**0.56**/0.22	**0.91**/0.64	0.62
AVG	0.82	0.60	0.65	**0.87**/0.70	**0.70**/0.46	**0.83**/0.62	0.62

3   Results

We report the performance of our model against previous work that uses kinematic data or video data and report GRS performance (Table 4). Although we don't regress the GRS's most recent work only report on it. To allow for easier comparison we use the GRS as a performance proxy. The model outperforms all methods using kinematic data and achieves competitive performance against models using video (Table 4). When looking at the performance of our model in predicting OSATS under the LOSO validation scheme, we underperform only in NP (Table 1). The CNN+Bilstm [2] only reports the best-performing fold, whereas we also report the average across the 5 folds. As can be seen in Table 2 the introduced Gaussian noise and flipping had very little effect on the performance of the model. However the flipping does allow for the model to be

Table 4. Performance comparison of GRS score on JIGSAWS trained on independent and across tasks. **K**: Kinematic, **V**: Video. *: Results from training on the three domains. GRS is used as a performance proxy and not regressed directly in this paper.

Input	Method	Task			
		KT	NP	SU	AT
		Spearman's Correlation Coef (SCC)			
V	C3D-MTL-VF [26]	_0.89_	0.75	0.77	0.80
V	Contra-Sformer [1]	_0.89_	0.71	**0.86**	0.82
V	ViSA [14]	**0.92**	**0.93**	_0.84_	**0.90**
K	SMT-DCT-DFT [31]	0.70	0.38	0.64	0.59
K	DCT-DFT-ApEn [31]	0.63	0.46	0.75	0.63
K	**ReCAP**	0.88	_0.85_	0.83	_0.85/0.79*_
		Mean Average Error (MAE)			
V	Contra-Sformer [1]	**1.75**	3.15	2.74	2.55
V	ViSa [14]	2.16	**1.66**	**2.58**	**2.13**
K	**ReCAP**	2.04	3.12	2.89	2.68/2.71*

time invariant. We see that the pseudo-label drastically improves performance, especially for the two tasks, NP and SU, with the most class imbalance [13].

To validate the weakly supervised outputs, 9 videos were reviewed by a consultant surgeon, where each 75 frames (the segment length) had an assigned OSATS pseudo-label. The selected videos regrouped three levels of expertise (novice, intermediate, expert) across the three tasks. Two of those videos were shown with the randomly generated predictions. We found that the clinician agreed 69% of the time when shown random predictions while agreeing 77% of the time when shown our model's predictions. A one-tailed binomial test between the two distributions indicates a statistically significant difference between the agreements (p = 0.006).

4   Discussion

To our knowledge we are the first to report task-agnostic metrics on JIGSAWS. While video-based models dominate recent works; kinematic data, though underexplored, remains promising. Our model outperforms prior kinematic-only methods and rivals video-only approaches, despite a simple architecture. The high feature-to-parameter ratio (238,644:1,440) contributes to overfitting, common in deep learning.

Our model shows strong OSATS prediction overall, though underperforms on Needle Passing and Quality of Final Product. This likely stems from kinematic data's inability to capture visual nuances. As noted by Kasa _et al._ [8], quality of final product is very video subjective. Similar kinematic profiles may

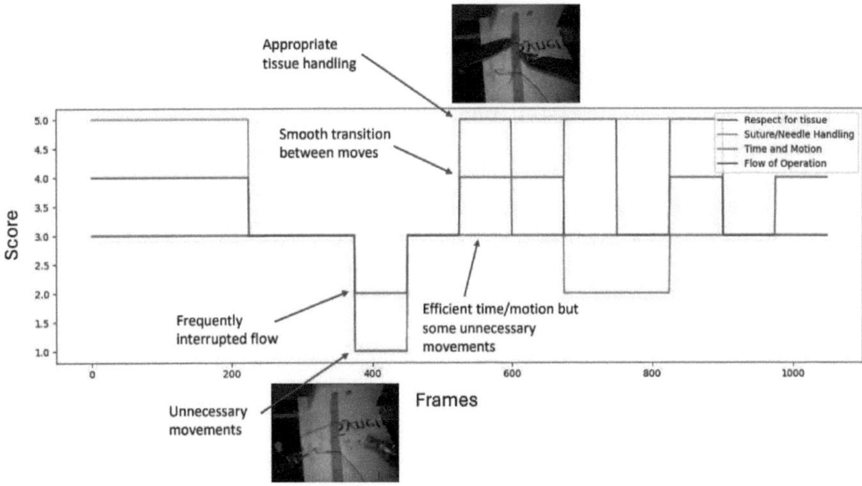

Fig. 2. Variations of OSATS from model for a Knot Tying task performed by a self-designated expert. Qualitative descriptions are taken from [19]. In this example, the senior clinician disagreed with the model's intermediate scores in 3 instances.

yield different outcomes—e.g., a depth misjudgment in Needle Passing affects performance but not kinematics.

As Lefor *et al.* [13] noted, JIGSAWS is imbalanced, with some OSATS scores inversely correlated with skill level. Also, using Spearman's ρ on only 3 samples for certain folds is very biased. Thus, model generalisation to real-world practice remains limited.

Our model benefits from a formulation that segments input with temporal context, enabling sparse, flexible learning. Ablations show that adding pseudo-labels improves performance, likely by regularising via intermediate predictions. This mirrors human raters, who assess cumulatively. However, the current objective doesn't capture catastrophic errors well; allowing segment-wise weighting could address this.

Pseudo-label loss aids both performance and interpretability. We visualise this in Fig. 2. It supports clinician feedback and online use due to the model's recurrent nature. Our work is limited by validation of fine-grained OSATS scores, which are difficult to distinguish by experts, as seen in the 69% agreement with random noise rather than the expected 33%. Rater variability complicates ground truth extraction. While more data and raters would help, it's often impractical. Weak supervision offers a scalable alternative. Our model's performance vs. noise suggests promise in this direction.

5   Conclusion and Future Work

We present a novel formulation for skill assessment, extensible to recurrent models and other domains. Our competitive results on JIGSAWS support its potential for granular feedback. Future work will improve validation, incorporate more OR time-series data (audio, bodytracking, ...) [22], and target longer, complex procedures. Since annotating multi-hour tasks is laborious, weakly-supervised methods extracting gesture/step/phase-level labels may enable scalable, automated assessment.

Acknowledgments. This work was supported by core funding from the Wellcome/EPSRC Centre for Medical Engineering [WT203148/Z/16/Z]. For the purpose of Open Access, the author has applied a CC BY public copyright license to any Author Accepted Manuscript version arising from this submission.

Disclosure of Interests. The authors have no competing interests to declare that are relevant to the content of this article.

References

1. Anastasiou, D., Jin, Y., Stoyanov, D., Mazomenos, E.: Keep your eye on the best: contrastive regression transformer for skill assessment in robotic surgery. IEEE Robot. Autom. Lett. **8**(3), 1755–1762 (2023)
2. Benmansour, M., Malti, A., Jannin, P.: Deep neural network architecture for automated soft surgical skills evaluation using objective structured assessment of technical skills criteria. Int. J. Comput. Assist. Radiol. Surg. **18**, 929–937 (2023)
3. Chen, R., Rodrigues Armijo, P., Krause, C., Siu, K.C., Oleynikov, D.: A comprehensive review of robotic surgery curriculum and training for residents, fellows, and postgraduate surgical education. Surg. Endosc. **34**(1), 361–367 (2019)
4. Gao, J., et al.: An asymmetric modeling for action assessment. In: Vedaldi, A., Bischof, H., Brox, T., Frahm, J.-M. (eds.) ECCV 2020. LNCS, vol. 12375, pp. 222–238. Springer, Cham (2020). https://doi.org/10.1007/978-3-030-58577-8_14
5. Gao, Y., et al.: Jhu-isi gesture and skill assessment working set (jigsaws): a surgical activity dataset for human motion modeling. In: MICCAI Workshop: M2CAI, vol. 3, p. 3 (2014)
6. Hackney, L., O'Neill, S., O'Donnell, M., Spence, R.: A scoping review of assessment methods of competence of general surgical trainees. The Surgeon **21**(1), 60–69 (2023)
7. Ismail Fawaz, H., Forestier, G., Weber, J., Idoumghar, L., Muller, P.A.: Accurate and interpretable evaluation of surgical skills from kinematic data using fully convolutional neural networks. Int. J. Comput. Assist. Radiol. Surg. **14**(9), 1611–1617 (2019)
8. Kasa, K., Burns, D., Goldenberg, M.G., Selim, O., Whyne, C., Hardisty, M.: Multimodal deep learning for assessing surgeon technical skill. Sensors **22**(19), 7328 (2022)
9. Kelly, C.J., Karthikesalingam, A., Suleyman, M., Corrado, G., King, D.: Key challenges for delivering clinical impact with artificial intelligence. BMC Med. **17**(1) (2019)
10. Khairy, G.: Challenges in creating the educated surgeon in the 21st century: where do we stand? Ann. Saudi Med. **24**(3), 218–220 (2004)
11. Khalid, S., Goldenberg, M., Grantcharov, T., Taati, B., Rudzicz, F.: Evaluation of deep learning models for identifying surgical actions and measuring performance. JAMA Netw. Open **3**(3), e201664 (2020)

12. Kulik, D., Bell, C.R., Holden, M.S.: Fast skill assessment from kinematics data using convolutional neural networks. Int. J. Comput. Assist. Radiol. Surg. **19**(1), 43–49 (2023)
13. Lefor, A.K., Harada, K., Dosis, A., Mitsuishi, M.: Motion analysis of the JHU-ISI gesture and skill assessment working set using robotics video and motion assessment software. Int. J. Comput. Assist. Radiol. Surg. **15**(12), 2017–2025 (2020)
14. Li, Z., Gu, L., Wang, W., Nakamura, R., Sato, Y.: Surgical skill assessment via video semantic aggregation (2022)
15. Liu, D., Jiang, T., Wang, Y., Miao, R., Shan, F., Li, Z.: Surgical skill assessment on in-vivo clinical data via the clearness of operating field (2020)
16. Liu, D., et al.: Towards unified surgical skill assessment (2021)
17. Liu, R., Holden, M.S.: Kinematics data representations for skills assessment in ultrasound-guided needle insertion. In: Hu, Y., Torrents Barrena, J. (eds.) ASMUS/PIPPI -2020. LNCS, vol. 12437, pp. 189–198. Springer, Cham (2020). https://doi.org/10.1007/978-3-030-60334-2_19
18. Markus, A.F., Kors, J.A., Rijnbeek, P.R.: The role of explainability in creating trustworthy artificial intelligence for health care: a comprehensive survey of the terminology, design choices, and evaluation strategies. J. Biomed. Inform. **113**, 103655 (2021)
19. Martin, J.A., et al.: Objective structured assessment of technical skill (OSATS) for surgical residents: objective structured assessment of technical skill. Br. J. Surg. **84**(2), 273–278 (1997)
20. Maynou, L., Pearson, G., McGuire, A., Serra-Sastre, V.: The diffusion of robotic surgery: examining technology use in the English NHS. Health Policy **126**(4), 325–336 (2022)
21. Pan, J.H., Gao, J., Zheng, W.S.: Action assessment by joint relation graphs. In: 2019 IEEE/CVF International Conference on Computer Vision (ICCV), pp. 6330–6339 (2019)
22. Quarez, J., et al.: Mutual: towards holistic sensing and inference in the operating room. In: International Conference on Medical Image Computing and Computer-Assisted Intervention, pp. 178–188. Springer (2024)
23. Ravi, K., Anyamele, U.A., Korch, M., Badwi, N., Daoud, H.A., Shah, S.S.N.H.: Undergraduate surgical education: a global perspective. Indian J. Surg. **84**(S1), 153–161 (2021)
24. Reyzabal, M.D.I., Chen, M., Huang, W., Ourselin, S., Liu, H.: DaFoES: mixing datasets towards the generalization of vision-state deep-learning force estimation in minimally invasive robotic surgery (2024)
25. Singh, P., Aggarwal, R., Pucher, P., Duisberg, A., Arora, S., Darzi, A.: Defining quality in surgical training: perceptions of the profession. Am. J. Surg. **207**, 628–636 (2014)
26. Wang, T., Wang, Y., Li, M.: Towards accurate and interpretable surgical skill assessment: a video-based method incorporating recognized surgical gestures and skill levels. In: Martel, A.L., et al. (eds.) MICCAI 2020. LNCS, vol. 12263, pp. 668–678. Springer, Cham (2020). https://doi.org/10.1007/978-3-030-59716-0_64
27. Wang, T., Wang, Y., Li, M.: Towards accurate and interpretable surgical skill assessment: a video-based method incorporating recognized surgical gestures and skill levels. In: Martel, A.L., et al. (eds.) MICCAI 2020. LNCS, vol. 12263, pp. 668–678. Springer, Cham (2020). https://doi.org/10.1007/978-3-030-59716-0_64
28. Wang, Y., et al.: Evaluating robotic-assisted surgery training videos with multi-task convolutional neural networks. J. Robot. Surg. **16**(4), 917–925 (2021)

29. Yanik, E., et al.: Deep neural networks for the assessment of surgical skills: a systematic review (2021)
30. Zago, M., et al.: Educational impact of hand motion analysis in the evaluation of fast examination skills. Eur. J. Trauma Emerg. Surg. **46**(6), 1421–1428 (2019)
31. Zia, A., Essa, I.: Automated surgical skill assessment in RMIS training. Int. J. Comput. Assist. Radiol. Surg. **13**(5), 731–739 (2018)

MM2CT: MR-to-CT Translation for Multi-modal Image Fusion with Mamba

Chaohui Gong, Zhiying Wu[✉], Zisheng Huang, Gaofeng Meng, Zhen Lei, and Hongbin Liu

Centre for Artificial Intelligence and Robotics, Hong Kong Institute of Science and Innovation, Chinese Academy of Sciences, Hong Kong, China
`zhiying.wu@cair-cas.org.hk`

Abstract. Magnetic resonance (MR)-to-computed tomography (CT) translation offers significant advantages, including the elimination of radiation exposure associated with CT scans and the mitigation of imaging artifacts caused by patient motion. The existing approaches are based on single-modality MR-to-CT translation, with limited research exploring multimodal fusion. To address this limitation, we introduce Multimodal MR to CT (MM2CT) translation method by leveraging multimodal T1- and T2-weighted MRI data, an innovative Mamba-based framework for multi-modal medical image synthesis. Mamba effectively overcomes the limited local receptive field in CNNs and the high computational complexity issues in Transformers. MM2CT leverages this advantage to maintain long-range dependencies modeling capabilities while achieving multi-modal MR feature integration. Additionally, we incorporate a dynamic local convolution module and a dynamic enhancement module to improve MRI-to-CT synthesis. The experiments on a public pelvis dataset demonstrate that MM2CT achieves state-of-the-art performance in terms of Structural Similarity Index Measure (SSIM) and Peak Signal-to-Noise Ratio (PSNR). Our code is publicly available at https://github.com/Gots-ch/MM2CT.

Keywords: MR-to-CT Translation · Fusion · Mamba

1 Introduction

In clinical practice, medical image such as Magnetic Resonance Imaging (MRI), Computed Tomography (CT) are crucial as it provides information about the human body [5]. The radiation exposure during CT scanning [3] has prompted researchers to explore image synthesis technology as an alternative solution. Medical image synthesis technology aims to predict imaging manifestations of one modality using data from another modality. However, different imaging devices operate on distinct physical principles [1]. This leads to significant non-linear differences in tissue contrast. Consequently, cross-modal synthesis methods still facer challenges in feature space mapping. Currently, research is largely

© The Author(s), under exclusive license to Springer Nature Switzerland AG 2026
J. Qiu et al. (Eds.): Agentic AI 2025/CMLLMs 2025/CREATE 2025, LNCS 16147, pp. 167–176, 2026.
https://doi.org/10.1007/978-3-032-06004-4_17

limited to image synthesis within a single modality. However, complementary features carried by different modality images can provide significant synergistic enhancement for anatomical structure analysis and pathological feature recognition [22]. MRI datasets usually consist of images from multiple imaging protocols [11], which provides conditions for multi-modal medical image fusion.

Image fusion is a specific algorithm that combines two or more images into a new image [12]. The current mainstream methods mainly rely on basic operations like feature superposition and channel concatenation for multi-modal information integration. For example, WavTrans [10] and MMHCA [6] both concatenate the input tensors in the medical image super-resolution task. However, these operations have obvious limitations. First, WavTrans [10] and MMHCA [6] struggle to deeply explore and fully utilize the inherent correlational features between cross-modal data. Second, even with the introduction of attention mechanisms or convolutional neural networks, technical bottlenecks remain. These include inefficient inter-modal interactions and insufficient modeling of long-range dependencies.

To address these challenges, we propose MM2CT (Multi-Modal MR-to-CT), a novel deep learning framework based on state space modeling for MRI-to-CT cross-modal synthesis. The framework innovatively integrates Mamba for multimodal fusion. First, shallow features are extracted using convolutional layers and Mamba blocks. Subsequently, we employ channel swapping and a Mamba fusion module to effectively extract local detail features from different modalities. Following this processing, the features undergo further refinement through Mamba blocks and convolutional layers to generate the final synthesized image. The main contributions are summarized as follows:

1. We first introduce the MM2CT model, a CT translation framework capable of simultaneously processing multimodal MR images. This model integrates complementary features from T1- and T2-weighted modalities to achieve high-quality unsupervised conversion between MR and CT images.
2. A Mamba-based fusion module is designed to effectively capture image features. Combined with a dynamic enhancement module, the proposed MM2CT efficiently handles inter-modality differences to enhance texture information perception.
3. Our method outperforms existing state-of-the-art approaches across all evaluation metrics in the pelvic dataset.

2   Related Work

2.1   Medical Image Translation

Image-to-image translation technology based on deep learning provides a new paradigm for medical imaging modality conversion. Early research mainly focused on the generative adversarial network (GAN) framework. These methods capture distribution features of target modalities through adversarial learning. This significantly improves the preservation of structural details [1]. For example, CycleGAN uses cycle consistency loss to deal with the problem of unpaired data and has been successfully applied to MRI-to-CT conversion [2]. MaskGAN [18] innovatively introduces automatically extracted anatomical structure masks as prior knowledge. This enhances the anatomical structure accuracy and consistency of generated images. However, GAN-based methods generally suffer from unstable adversarial training [14]. This may lead to artifacts or local detail distortion in generated images.

In recent years, diffusion models have made significant breakthroughs in the field of medical image generation. Notably, Syndiff [17] pioneered an innovative approach by combining adversarial training with the diffusion framework, demonstrating exceptional performance in MRI-to-CT translation tasks. However, existing research has primarily focused on single-modality translation, which somewhat limits the full utilization of medical imaging information. Image fusion techniques thus present a promising solution to address this challenge.

2.2 Multi-modal Image Translation

Multi-modal image fusion aims to integrate complementary information from different modalities to generate high-quality fused images with rich textural details [2,4]. With the advancement of deep learning technologies, neural network-based multi-modal fusion methods have achieved efficient cross-modal information integration through their powerful nonlinear modeling capabilities. For instance, Wu et al. [20] proposed an innovative framework combining convolutional operations and attention mechanisms, which was successfully applied to prostate cancer classification using multi-modal transrectal ultrasound images.

Although Transformer models excel at capturing long-range dependencies, their practical application is limited by computational complexity that grows quadratically with sequence length [19]. Recent studies have shown that architectures based on State Space Models (SSM) maintain model performance while significantly improving computational efficiency due to their lower computational complexity. In particular, the Visual State Model (Vmamba) [15] has been successfully applied to computer vision tasks, opening new technical pathways for medical image fusion.

Building on this foundation, the Mamba-based dual-phase fusion model (Mambadfuse) [13] innovatively proposes a dual-phase feature fusion module that effectively captures and integrates complementary information between different modalities, advancing multi-modal medical image fusion technology. Additionally, FusionMamba [21] enhances multi-modal image fusion performance through a novel dynamic feature enhancement method, demonstrating significant advantages in biomedical image processing.

3   Method

The detailed workflow of our MM2CT model is illustrated in Fig. 1, showcasing how these components interact to achieve seamless image translation and enhanced feature extraction. The translation module is responsible for converting

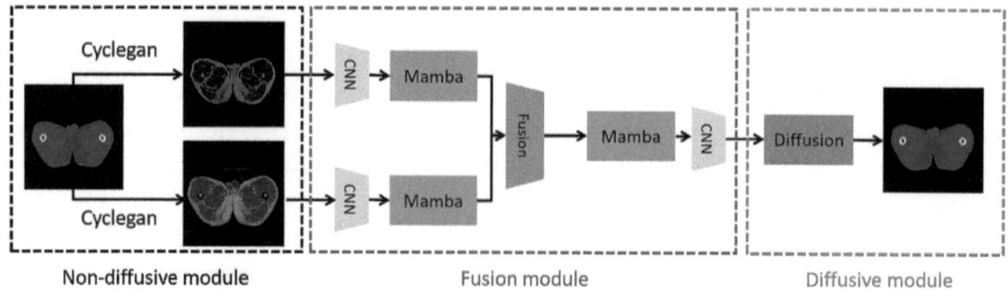

Fig. 1. The network architecture. It includes three key components: GAN-based generation module, feature fusion module and diffusive module. The process of translating CT into MR is similar.

multi-contrast MR images into corresponding CT images, ensuring high-quality and accurate transformations. The multi-modal feature fusion module aims to integrate information from multiple MRI modalities.

3.1   Translation Module

The conversion module consists of two key components: a non-diffusive module and a diffusive module. The non-diffusive module initially processes unpaired data from the training set by estimating source images corresponding to target images using ResNet-based generators [9], thereby establishing a foundation for subsequent transformation. To fully utilize information from both imaging modalities, we introduce a feature fusion module prior to the diffusive module input. The diffusive module based on the UNet backbone then uses the fused multi-modal MRI features as conditional information to guide the generation process of CT images. The diffusive module plays a central role in the conversion process, where the forward diffusion process acts as a low-pass filter, effectively extracting low-frequency information from images, which serves as a crucial starting point for the reverse diffusion process.

3.2   Fusion Module

We introduce a feature fusion module based on the Mamba architecture, which effectively integrates information from T1- and T2-weighted modalities. The fusion module's architecture can be described as follows. Initially, for low-level feature extraction, convolutional layers and Mamba blocks are employed for preliminary processing. Subsequently, we implement shallow channel swapping and deep fusion modules to effectively integrate local detail features from different modalities, ensuring comprehensive extraction of modality specific information. Following these processing steps, the features undergo further processing through Mamba blocks and convolutional layers before generating the synthesized image, which is then forwarded to the subsequent diffusive module.

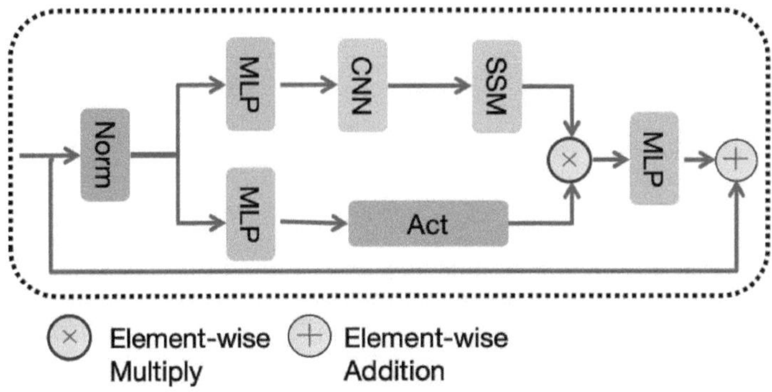

Fig. 2. Mamba block used in Fusion module.

The convolutional layers provide a direct and effective approach for capturing local semantic details. We employ two convolutional layers, each with a 3×3 kernel size and stride of 1. The internal structure of the Mamba block, as illustrated in Fig. 2, begins with a normalization layer for input processing, followed by a dual-path structure. In the first path, the signal sequentially passes through a linear layer, convolution operation, and SSM module. The second path consists of a linear layer and activation function. The outputs from both paths are merged through multiplication, then processed by a linear layer and added to the original input, forming a residual structure output.

The channel swapping Mamba module represents shallow feature fusion. Inspired by Mambadfuse [13], we implement alternating channel swapping between the two input feature to achieve lightweight exchange. The channel swapping module facilitates feature interaction between T1 and T2 modalities, establishing inter-modal correlations.

The cross-modal Mamba module is responsible for deep feature fusion. The current Mamba architecture faces challenges when processing multi-modal image information, primarily due to its lack of cross-modal feature fusion capabilities similar to cross-attention mechanisms. To address this limitation, inspired by the concept of cross-attention, we leverage a cross-modal Mamba block [13] designed to facilitate cross-modal feature interaction and fusion. In this approach, we project features from both modalities into a shared space and employ a gating mechanism to encourage complementary feature learning while suppressing redundant features.

Moreover, to better handle inter-modality differences, the fusion feature maps are adjusted based on the disparities between the two modality feature maps to enhance the differences between distinct features. First, a dynamic local convolution mechanism [7] restores similarities between neighborhoods and enhances texture information perception. Second, a dynamic difference-aware attention mechanism is described as Fig. 3a. It amplifies subtle differences between input feature maps. The final fusion module is shown in Fig. 3b.

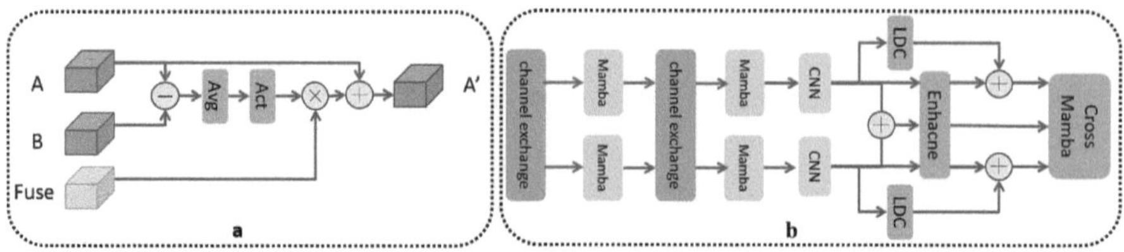

Fig. 3. Enhance module architecture and Fusion module.

3.3 Objective Loss

In this study, our model is trained by incorporating both traditional adversarial loss and cycle consistency loss, with the discriminator network implemented using a PatchGAN-like architecture [8]. These loss functions jointly constitute the model's overall objective function, ensuring the quality of generated images and maintaining consistent feature mapping between the source and target domains. The overall objective loss function is formulated as follows:

$$Loss_{all} = \lambda_{l1}Loss_{l1} + \lambda_{gan}Loss_{GAN} + Loss_{Dis} \tag{1}$$

where λ_{l1} and λ_{gan} are the weighting of $Loss_{l1}$ and $Loss_{GAN}$, respectively. All parameters used in the loss function are the same as the setting in [17].

4 Experiments and Result

4.1 Dataset

We demonstrate our model on a multi-modal pelvic MRI-CT dataset [16]. While all unsupervised medical image translation models are trained on unpaired images, performance assessments evaluate on the paired and registered images.

Pelvic from 15 subjects were analyzed, with a split of (9,2,4) subjects. Elastic registration was performed using the SimpleITK library to register T1 and CT volumes onto T2 volumes in validation and test sets. For T1 scans, $TE = 7.2$ ms, $TR = 500$–600 ms, $0.10 \times 0.10 \times 3$ mm^3 resolution. For T2 scans, $TE = 97$ ms, $TR = 6000$–6600 ms, $0.88 \times 0.88 \times 2.50$ mm^3 resolution. For CT scans, $0.10 \times 0.10 \times 3$ mm^3 resolution, Kernel = B30f, or $0.10 \times 0.10 \times 2$ mm^3 resolution, Kernel = FC17 were prescribed. We apply the following pre-processing steps: resampling of all volumes and corresponding labels to $1.0 \times 1.0 \times 1.0$ mm^3.

Table 1. Quantitative Comparison Between On The Pelvis Dataset.

Modality	Model	PSNR	SSIM
$T1 \rightarrow CT$	CycleGAN [23]	22.69	81.47
$T2 \rightarrow CT$	CycleGAN [23]	21.57	79.56
$T1 \rightarrow CT$	Syndiff [17]	25.36	88.97
$T2 \rightarrow CT$	Syndiff [17]	23.99	87.76
$T1\&T2 \rightarrow CT$	MM2CT	**25.72**	**89.54**

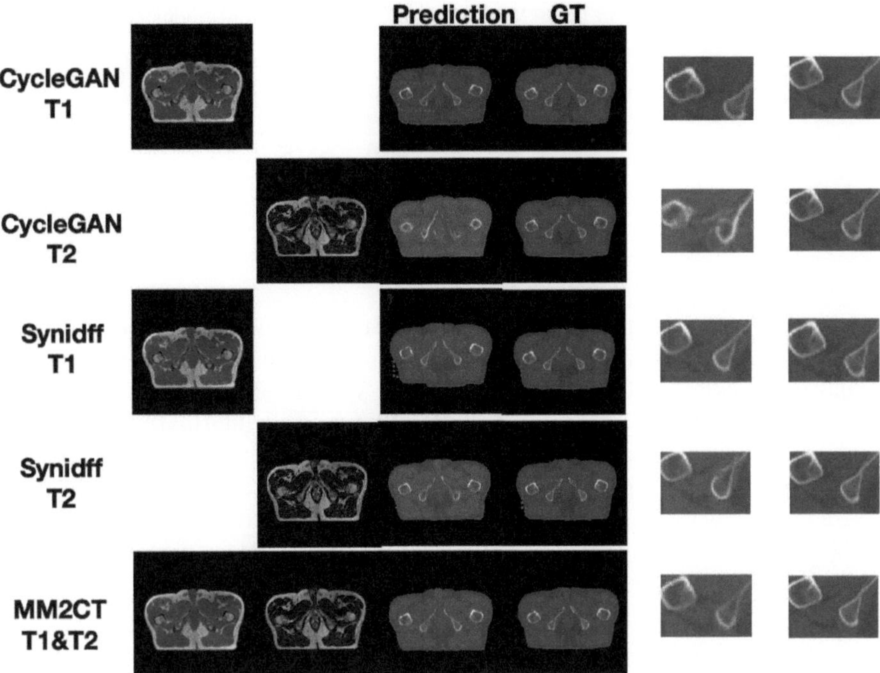

Fig. 4. The MR-to-CT translation results of different models are shown along with the source image (Source) and the true target (GT). The result demonstrated on the pelvic dataset for multi-modal MRI-to-CT translation. From left to right, input T1, input T2, synthetic image CT and ground truth image CT.

4.2 Implement Details

The networks are trained using an Adam optimizer with a batch size of 4 and 100 epochs. The experiments are implemented using Pytorch and are trained on an A100 GPU with 80 GB of memory. For faithfulness, we employ two widely recognized metrics for quantitative evaluations, namely Peak Signal-to-Noise Ratio (PSNR) and Structural Similarity Index Measure (SSIM). PSNR assesses whether the synthesized images is a uniform projection of the image, and SSIM concentrates on the visible structure of the images.

4.3 Results

Compare with State-of-the-Art Approaches. The model performance was quantitatively analyzed by comparing the synthesized images with real CT images. Table 1 presents representative test sample results. Initially, we compared the performance of two single-modality models. The performance of the model is obtained by retraining and testing based on the divided dataset. Syndiff demonstrated the best baseline performance. Further analysis revealed that our proposed model MM2CT significantly outperformed the Syndiff model in both

Table 2. Ablation study on cross Mamba and Dynamic Enhancement (DE) modules.

Configure		Metrics	
Mamba module	DE	PSNR ↑	SSIM ↑
–	–	25.36	88.97
✓	–	25.49	89.45
✓	✓	25.72	89.54

metrics. This superior performance can be attributed to the effective utilization of features from both modalities. The performance of our MM2CT is improved by 0.36dB PSNR and 0.57% SSIM compared to Syndiff. These results confirm that our model better preserves high-frequency detail information in images, producing synthetic images that more closely resemble the quality of real CT images. As shown in Fig. 4, the images generated by MM2CT demonstrate superior detail preservation and more natural texture features. The effectiveness of MM2CT is validated not only through quantitative metrics but also through direct visual quality assessment.

4.4   Ablation Studies

To validate the effectiveness of core components in the MM2CT model, we conducted a series of ablation experiments. Our analysis focused on two key modules: the Mamba fusion module and the Dynamic Enhancement (DE) module, examining their impact on the model's overall performance. Since the DE module is integrated within the Fusion block, conducting isolated ablation experiments for this component presented certain challenges.

As demonstrated by the experimental results in Table 2, the complete MM2CT model significantly outperformed its ablated variants across all evaluation metrics, substantiating the necessity of each component. Specifically, compared to the version without the DE module, the fully-equipped MM2CT achieved improved image quality metrics, with a 0.23 dB increase in PSNR and 0.09% improvement in SSIM. These experimental findings strongly validate that our proposed complete MM2CT architecture better preserves and reconstructs image details, thereby achieving superior performance in the cross-modal image translation task from MR to CT.

5   Conclusion

The proposed MM2CT framework presents an innovative solution for cross-modal synthesis in medical imaging, significantly improving the accuracy of MRI-to-CT image translation. By developing a state-space-based multi-modal feature fusion mechanism, this work establishes novel theoretical foundations for multi-modal image synthesis. Experimental results demonstrate that the framework outperforms current state-of-the-art methods across multiple quantitative metrics, including SSIM and PSNR, validating its potential value and prospects for clinical applications. Future work will focus on extending validation to multi-center datasets to furtheßr enhance generalization and clinical applicability of the proposed model.

Acknowledgments. The research in this paper was funded by Inno HK program.

Disclosure of Interests. The authors have no competing interests to declare relevant to this article's content.

References

1. Armanious, K., et al.: MedGAN: medical image translation using GANs. Comput. Med. Imaging Graph. **79**, 101684 (2020)
2. Atli, O.F., et al.: I2I-Mamba: multi-modal medical image synthesis via selective state space modeling. arXiv preprint arXiv:2405.14022 (2024)
3. Bosch de Basea Gomez, M., et al.: Risk of hematological malignancies from CT radiation exposure in children, adolescents and young adults. Nat. Med. **29**(12), 3111–3119 (2023)
4. Dalmaz, O., Yurt, M., Çukur, T.: ResViT: residual vision transformers for multimodal medical image synthesis. IEEE Trans. Med. Imaging **41**(10), 2598–2614 (2022)
5. Dayarathna, S., Islam, K.T., Uribe, S., Yang, G., Hayat, M., Chen, Z.: Deep learning based synthesis of MRI, CT and PET: review and analysis. Med. Image Anal., 103046 (2023)
6. Georgescu, M.I., et al.: Multimodal multi-head convolutional attention with various kernel sizes for medical image super-resolution. In: Proceedings of the IEEE/CVF Winter Conference on Applications of Computer Vision, pp. 2195–2205 (2023)
7. Huang, P.K., Ni, H.Y., Ni, Y., Hsu, C.T.: Learnable descriptive convolutional network for face anti-spoofing. In: BMVC, vol. 2, p. 7 (2022)
8. Isola, P., Zhu, J.Y., Zhou, T., Efros, A.A.: Image-to-image translation with conditional adversarial networks. In: Proceedings of the IEEE Conference on Computer Vision and Pattern Recognition, pp. 1125–1134 (2017)
9. Johnson, J., Alahi, A., Fei-Fei, L.: Perceptual losses for real-time style transfer and super-resolution. In: European Conference on Computer Vision, pp. 694–711. Springer (2016)
10. Li, G., Lyu, J., Wang, C., Dou, Q., Qin, J.: WavTrans: synergizing wavelet and cross-attention transformer for multi-contrast MRI super-resolution. In: International Conference on Medical Image Computing and Computer-Assisted Intervention, pp. 463–473. Springer (2022)
11. Li, H., et al.: DiamondGAN: unified multi-modal generative adversarial networks for MRI sequences synthesis. In: Shen, D., et al. (eds.) MICCAI 2019. LNCS, vol. 11767, pp. 795–803. Springer, Cham (2019). https://doi.org/10.1007/978-3-030-32251-9_87
12. Li, Y., Zhao, J., Lv, Z., Li, J.: Medical image fusion method by deep learning. Int. J. Cogn. Comput. Eng. **2**, 21–29 (2021)
13. Li, Z., Pan, H., Zhang, K., Wang, Y., Yu, F.: MambaDFuse: a mamba-based dual-phase model for multi-modality image fusion. arXiv preprint arXiv:2404.08406 (2024)

14. Li, Z., Xia, P., Tao, R., Niu, H., Li, B.: A new perspective on stabilizing GANs training: direct adversarial training. IEEE Trans. Emerg. Top. Comput. Intell. **7**(1), 178–189 (2022)
15. Liu, Y., et al.: Vmamba: visual state space model. Adv. Neural. Inf. Process. Syst. **37**, 103031–103063 (2025)
16. Nyholm, T., et al.: MR and CT data with multiobserver delineations of organs in the pelvic area–part of the gold atlas project. Med. Phys. **45**(3), 1295–1300 (2018)
17. Özbey, M., et al.: Unsupervised medical image translation with adversarial diffusion models. IEEE Trans. Med. Imag. (2023)
18. Phan, V.M.H., Liao, Z., Verjans, J.W., To, M.S.: Structure-preserving synthesis: MaskGAN for unpaired MR-CT translation. In: International Conference on Medical Image Computing and Computer-Assisted Intervention, pp. 56–65. Springer (2023)
19. Qu, H., et al.: A survey of mamba. arXiv preprint arXiv:2408.01129 (2024)
20. Wu, H., et al.: Towards multi-modality fusion and prototype-based feature refinement for clinically significant prostate cancer classification in transrectal ultrasound. In: International Conference on Medical Image Computing and Computer-Assisted Intervention, pp. 724–733. Springer (2024)
21. Xie, X., Cui, Y., Tan, T., Zheng, X., Yu, Z.: FusionMamba: dynamic feature enhancement for multimodal image fusion with mamba. Vis. Intell. **2**(1), 37 (2024)
22. Zhang, J., Zhang, S., Shen, X., Lukasiewicz, T., Xu, Z.: Multi-condos: multimodal contrastive domain sharing generative adversarial networks for self-supervised medical image segmentation. IEEE Trans. Med. Imaging **43**(1), 76–95 (2023)
23. Zhu, J.Y., Park, T., Isola, P., Efros, A.A.: Unpaired image-to-image translation using cycle-consistent adversarial networks. In: Proceedings of the IEEE International Conference on Computer Vision, pp. 2223–2232 (2017)

Can DeepSeek Reason Like a Surgeon? An Empirical Evaluation for Vision-Language Understanding in Robotic-Assisted Surgery

Boyi Ma[1,3], Yanguang Zhao[1], Jie Wang[1], Guankun Wang[1], Kun Yuan[2], Tong Chen[1], Long Bai[1], and Hongliang Ren[1,3(✉)]

[1] Department of Electronic Engineering, The Chinese University of Hong Kong, Sha Tin, Hong Kong
`b.long@link.cuhk.edu.hk, hlren@ee.cuhk.edu.hk`
[2] University of Strasbourg, CNRS, INSERM, ICube & IHU Strasbourg, Strasbourg, France
[3] Shenzhen Loop Area Institute (SLAI), Shenzhen, China

Abstract. The DeepSeek models have shown exceptional performance in general scene understanding, question-answering (QA), and text generation tasks, owing to their efficient training paradigm and strong reasoning capabilities. In this study, we investigate the dialogue capabilities of the DeepSeek model in robotic surgery scenarios, focusing on tasks such as Single Phrase QA, Visual QA, and Detailed Description. The Single Phrase QA tasks further include sub-tasks such as surgical instrument recognition, action understanding, and spatial position analysis. We conduct extensive evaluations using publicly available datasets, including EndoVis18 and CholecT50, along with their corresponding dialogue data. Our empirical study shows that, compared to existing general-purpose multimodal large language models, DeepSeek-VL2 performs better on complex understanding tasks in surgical scenes. Additionally, although DeepSeek-V3 is purely a language model, we find that when image tokens are directly inputted, the model demonstrates better performance on single-sentence QA tasks. However, overall, the DeepSeek models still fall short of meeting the clinical requirements for understanding surgical scenes. Under general prompts, DeepSeek models lack the ability to effectively analyze global surgical concepts and fail to provide detailed insights into surgical scenarios. Based on our observations, we argue that the DeepSeek models are not ready for vision-language tasks in surgical contexts without fine-tuning on surgery-specific datasets.

1 Introduction

Robotic-assisted surgery (RAS) is a pivotal advancement in modern healthcare, aiming to assist surgery procedures and enable remote operation [8]. With

---

B. Ma and Y. Zhao—Equal contribution.
L. Bai—Project lead.

© The Author(s), under exclusive license to Springer Nature Switzerland AG 2026
J. Qiu et al. (Eds.): Agentic AI 2025/CMLLMs 2025/CREATE 2025, LNCS 16147, pp. 177–186, 2026.
https://doi.org/10.1007/978-3-032-06004-4_18

the advancements in artificial intelligence and robotic control systems, contemporary Robotic-assisted surgery (RAS) technology has demonstrated superior precision in control [12,31]. Robotic-assisted surgery (RAS) is a pivotal advancement in modern healthcare, aiming to assist surgery procedures and enable remote operation [6,22]. Therefore, it is significant to develop a powerful AI model capable of comprehending and analyzing surgical scenarios for the potential substitution of expert surgeons [7,18]. Recent Multimodal Large Language Models (MLLMs) [14,26,28] have demonstrated significant advancements in their ability to interpret both visual and linguistic information. Given their remarkable success in general-purpose applications, MLLMs present new opportunities for enhancing contextual understanding and question-answering (QA) in the surgical domain [2–4,23,24,27,30]. Recent studies have begun exploring the potential of MLLMs in surgical AI through supervised fine-tuning of open-source models [6,11,16,24,25], aiming to adapt their capabilities to domain-specific challenges. The recently emerging open-source large language model, DeepSeek [5,17,29] has gradually demonstrated superior performance compared to other open-source models, while utilizing fewer activated parameters. Through comprehensive optimizations in the training framework, DeepSeek models achieve cost-efficient training and high inference efficiency. Moreover, leveraging reinforcement learning [10], it exhibits exceptional reasoning capabilities. Given its remarkable performance alongside cost-effective training and inference, DeepSeek models are poised to become a promising solution in surgical vision-language learning [15,20].

Although DeepSeek models show exceptional performance in general tasks, their effectiveness in contextual understanding and reasoning within the RAS domain remains uncertain. To bridge this gap, we systematically evaluate its potential using two publicly available robotic surgery datasets, EndoVis18 [1] and CholecT50 datasets [19]. Specifically, we compare GPT-4o (~50B-100B) with open-source model DeepSeek-Janus-Pro-7b (7B), DeepSeek-VL2 (~70B), and DeepSeek-V3 (67B) across three distinct dialogue paradigms: single-phrase QA, visual QA, and detailed descriptions. Through the above investigations, we aim to assess DeepSeek models' applicability to the RAS domain and provide actionable insights for optimizing its deployment in future surgical AI applications. Overall, the current DeepSeek models still struggle in comprehension of surgical targets and instruments, making them difficult for direct applications on surgical scene understanding tasks. To develop surgical MLLMs capable of performing complex descriptions and reasoning, high-quality vision-language surgical datasets remain a critical necessity.

2   Implementation

We include GPT-4o [13], DeepSeek-Janus-Pro-7b [5], and DeepSeek-VL2 [29] as comparison models in our empirical study. Additionally, the language-based model DeepSeek-V3 [17] does not possess direct image reading capabilities. Therefore, we use SEED [9], a VQ-based image tokenizer aligned with the pretrained unCLIP Stable Diffusion [21], to convert images into discrete visual codes. Given the image tokens, we specify an extra prompt to DeepSeek-V3 for its understanding: '*You are given image tokens which you should interpret as an image*'. Considering the varied dialogue demands of MLLM and the complexity of surgical procedures, the experiments are divided into three different paradigms: Single Phrase QA, Visual QA, and Detailed Description. These complementary paradigms enable the evaluation to provide more comprehensive and precise insights into the models' reactions toward surgical knowledge. Our evaluation utilizes three distinctive paradigms established by EndoChat dataset [25], including surgical images from the MICCAI EndoVis18 [1] and CholecT50 datasets [19], while following the test set partitions specific to each dataset. We directly adopted the prompts provided in the EndoChat dataset and ensured that all models used the same prompts during inference, in order to eliminate any potential influence that prompt variations might have on the inference process.

Single Phrase QA focuses on providing concentrated and conclusive answers for specific details of surgical procedures without verbose elaboration. The utilization of the task-specific prompt is shown in Fig. 1. When concentrating on the Single Phrase QA task, the model adeptly links visual content with its corresponding textual annotations, which is critical for accomplishing subtasks.

Visual QA evaluates the capability of the model to provide concise and contextual clarity answers based on varying question formats. General questions are provided as prompts, as shown in Fig. 1. Visual QA is crucial in addressing queries from surgeons and interns in diverse scenarios and enables the model to focus on delivering pertinent responses to different surgical scenes.

Detailed Description requires the model to gain a comprehensive understanding of organs, instruments, actions, relative positions, and contextual information at the same time. It's necessary to have a detailed grasp and overview of the entire scope to establish the foundation for complex surgical procedures. By giving prompts such as "*Describe the following image in detail*", the model's ability to provide local and global information about the surgical scenario is evaluated.

3   Results and Analysis

3.1   Performance on Single Phrase QA

As shown in Table 1, accuracy and F1 score are chosen to evaluate the performance on the Single Phrase QA task. The variable choices from the EndoVis18 dataset contain 13 different phrases covering surgical instrument motion, target issue, and object position, encompassing 8, 1, and 4 phrases, respectively. The phrases from the CholecT50 dataset are separated into three categories: surgical instrument, targets, and motion, which contain 5, 8, and 13 phrases, respectively.

Table 1 shows that the DeepSeek models have overall better results on the EndoVis18 dataset. The performance in the instrument motion category is substandard, which shows that both DeepSeek and GPT4-o face challenges in deducing the instrument motion from a static image. Additionally, we observed that DeepSeek-Janus-Pro-7b predominantly selects the 'Idle' category across all

Question [Single Phrase QA]:
What is the state of monopolar curved scissors?
Choose one answer from "Idle", "Grasping",
"Tool Manipulation", "Cutting", "Suction",
"Retraction", "Clipping", "Cauterization",
"Tissue Manipulation", "Ultrasound Sensing",
"kidney", "left-top", "left-bottom", "right-top",
"right-bottom". Only answer your choice.
Ground Truth: Idle.
GPT-4o: Cutting.   **DeepSeek-Janus-Pro-7b:** Cutting.
DeepSeek-VL2: Idle.   **DeepSeek-V3:** Idle.

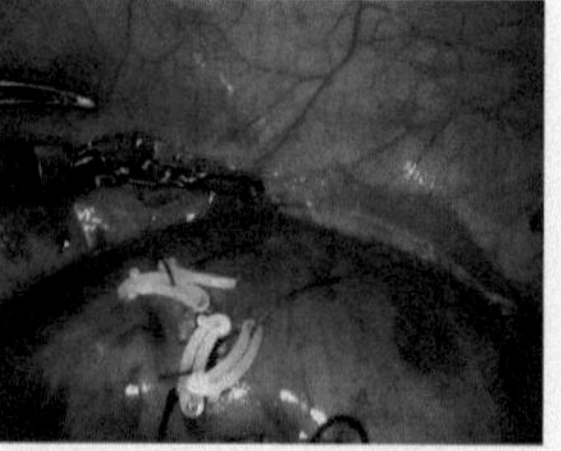

Question [Single Phrase QA]: What is the state of bipolar forceps?
Choose one answer from "Idle", "Grasping", "Tool Manipulation", "Cutting", "Suction", "Retraction", "Clipping", "Cauterization", "Tissue Manipulation", "Ultrasound Sensing", "kidney", "left-top", "left-bottom", "right-top", "right-bottom". Only answer your choice.
Ground Truth: Idle.
GPT-4o: Cauterization.   **DeepSeek-Janus-Pro-7b:** Idle.
DeepSeek-VL2: Idle.   **DeepSeek-V3:** Idle.

Question [Visual QA]: What tools are operating the organ?
Ground Truth: The tools operating are bipolar forceps, prograsp forceps, monopolar curved scissors.
GPT-4o: Robotic surgical instruments are operating the organ.
DeepSeek-Janus-Pro-7b: Surgical instruments.
DeepSeek-VL2: The organ is operated by a combination of mechanical and electronic components.
DeepSeek-V3: The tools operating the organ are surgical instruments and robotic arms.

Question [Visual QA]: What is the purpose of using prograsp forceps in this kind of surgery?
Ground Truth: Prograsp forceps are used in minimally invasive robotic surgeries to grasp, manipulate, and retract tissues. In this case, they appear to be retracting tissue to provide better access and visibility to the kidney that is being operated on.
GPT-4o: Prograsp forceps are used for precise tissue manipulation and traction in minimally invasive surgery.
DeepSeek-Janus-Pro-7b: To hold tissue securely during surgery.
DeepSeek-VL2: Prograsp forceps are used to grasp and hold tissue or organs during surgery, helping surgeons perform precise movements.
DeepSeek-V3: The purpose of using prograsp forceps is to provide precise grasping and manipulation of tissues during the surgery.

Fig. 1. Qualitative examples of Single Phrase QA, Visual QA, and Detailed Description on the EndoVis18 dataset.

Question [Detailed Description]:
Offer a thorough analysis of the image.

Ground Truth: The image depicts a surgical scene featuring a kidney, with a reddish brown tone and smooth surface appearance. In relation to the surgical instruments present, there are bipolar forceps located in the upper left side of the image. They appear to be at rest and not 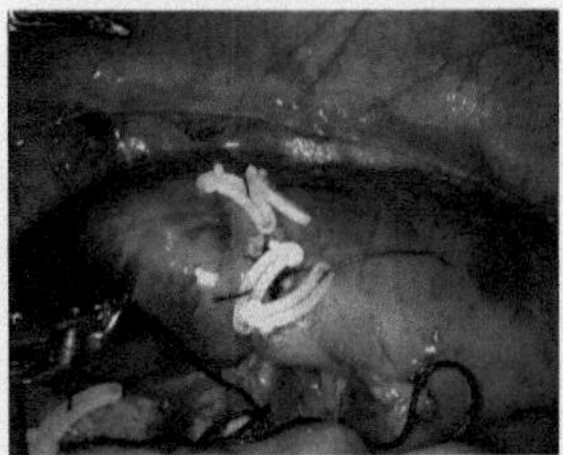 currently engaged in any activity. Towards the lower left part of the image, the prograsp forceps can be seen, which are being utilized for manipulating tissue. The overall composition of the image with the kidney and surgical tools provides a representation of a complex medical procedure.

GPT-4o: The image shows a medical procedure involving an organ, likely the liver, with visible surgical tools and clips. The presence of sutures and clips suggest a repair or biopsy. The tissue appears inflamed, indicating recent surgical intervention. The environment is consistent with an abdominal cavity, characterized by the glossy surface and vascular texture of the organ. The pale color of the clips and sutures contrasts with the darker, reddish tissues, emphasizing their surgical purpose..

DeepSeek-Janus-Pro-7b: The image shows a surgical procedure involving the insertion of medical devices into the body. The devices appear to be surgical clips or retractors, which are used to hold tissues open during surgery. The surrounding tissue is exposed, and the surgical instruments are visible, indicating that the procedure is in progress. The patient's body is shown with a focus on the internal structures, likely for a specific medical or diagnostic purpose. The image highlights the complexity and precision required in surgical procedures to ensure patient safety and successful outcomes.

DeepSeek-VL2: The image shows an intraoperative view of a surgical procedure involving the heart. The presence of sutures and clips suggests that this is likely a cardiac surgery, possibly related to valve repair or replacement. The visible structures include the heart muscle and surrounding tissues, with instruments in use for the procedure. The coloration indicates the presence of blood, which is typical during such surgeries. The overall setting implies a controlled environment, likely within an operating room.

DeepSeek-V3: The image appears to be a complex arrangement of recurring elements, suggesting a structured yet dynamic composition. The composition seems balanced, with a mix of larger and smaller components distributed evenly across the frame. The overall tone suggests a blend of order and creativity, possibly representing a harmonious design or an abstract representation of interconnected ideas. The interplay of recurring motifs and subtle variations could evoke themes of unity, repetition, or evolution, depending on the viewer's interpretation.

Fig. 2. Qualitative examples of Single Phrase QA, Visual QA, and Detailed Description on the EndoVis18 dataset.

motion classifications. This means that DeepSeek-Janus-Pro-7b does not possess a superior capability in classifying instrument motion. Regarding the target issue category, the options provided in the prompt include just a single choice associated with the target issue, namely, kidney. Therefore, within this category, selecting the correct option is straightforward once the model comprehends both the question and the choices presented in the prompt. Based on the comparative outcomes, it is evident that DeepSeek-V3 demonstrates a superior grasp of the question and the provided information within the prompt. This superior performance can be attributed to DeepSeek-V3's capacity to generate outputs through reasoned responses to the questions rather than relying on random guessing for answers. However, both DeepSeek-Janus-Pro-7b and DeepSeek-VL2 exhibit poor performance, indicating that these models not only lack sufficient knowledge of surgical scenes but also struggle to comprehend the provided prompts effectively. Regarding the object position task, the reasoning process of object position should begin with a general understanding of surgical phases, followed by the recognition of surgical instruments and tissues. While this task is straightforward for an expert surgeon, it poses significant challenges for GPT-4o and the DeepSeek series, primarily due to its limited domain-specific knowledge in robotic surgery. Moreover, on the CholecT50 dataset, DeepSeek-V3 also demonstrates comparable performance with GPT-4o. The evaluation of various phrases reveals that analyzing surgical tissue poses the greatest challenge for GPT4-o and the DeepSeek series. While DeepSeek-V3 exhibits performance comparable to GPT-4o, both DeepSeek-Janus-Pro-7b and DeepSeek-VL2 yield poor results. These shortcomings indicate that DeepSeek-Janus-Pro-7b and DeepSeek-VL2 not only struggle to understand surgical scenes but may also face issues with following instructions. In conclusion, despite the existing limitations of MLLMs in achieving a comprehensive understanding of surgical scenarios involving instruments, tissues, and motions, DeepSeek-V3 demonstrates performance that is comparable to or even superior in simple QA tasks. Although DeepSeek-V3 is inherently a language model without built-in visual processing capabilities, we are surprised to observe that, through our approach of tokenizing images, DeepSeek-V3 achieves favorable results in simple QA tasks related to surgical scenarios. This performance may be attributed to the internal data used during the training of DeepSeek-V3, although the specific reasons remain unclear.

3.2   Performance on Visual QA

The evaluation results of Visual QA on the EndoVis18 dataset are demonstrated in Table 2. The performance of generated responses is evaluated across sequences, quality, and consensus utilizing six different metrics. Additionally, the GPT-3.5 score is employed to determine whether the generated information is accurately mentioned or partially matched. In Visual QA tasks, the model is expected to provide a generally accurate understanding of surgical contexts. Therefore, the focus should be on semantic relevance (CIDEr score) and lexical coverage (ROUGE-1 score). As shown in Table 2, despite DeepSeek-V3 being a language-based model, it demonstrates superior performance in both CIDEr and ROUGE-

Table 1. Quantitative comparison of the Single Phrase QA on the EndoVis18 and CholecT50 datasets.

Model	Dataset	Average		Instrument Motion		Target Issue		Object Position	
		Accuracy	F1 Score	Accuracy	F1 Score	Accuracy	F1 Score	Accuracy	F1 Score
GPT-4o	EndoVis18	0.105	0.032	0.138	0.109	0.251	0.013	0.016	0.003
DeepSeek-Janus-Pro-7B		0.187	0.068	0.447	0.128	0.000	0.000	0.000	0.000
DeepSeek-VL2		0.137	0.115	0.223	0.205	0.000	0.000	0.104	0.035
DeepSeek-V3		0.276	0.109	0.276	0.166	0.996	0.333	0.000	0.000
Model	Dataset	Average		Instrument		Target		Motion	
		Accuracy	F1 Score	Accuracy	F1 Score	Accuracy	F1 Score	Accuracy	F1 Score
GPT-4o	CholecT50	0.592	0.376	0.873	0.846	0.336	0.167	0.566	0.434
DeepSeek-Janus-Pro-7B		0.250	0.148	0.302	0.324	0.115	0.042	0.333	0.195
DeepSeek-VL2		0.349	0.216	0.543	0.446	0.111	0.019	0.394	0.324
DeepSeek-V3		0.573	0.404	0.860	0.817	0.395	0.210	0.464	0.386

Table 2. Quantitative comparison of Visual QA and Detailed Descriptions on EndoVis18 datasets.

Model	Visual QA						
	BLEU-3	BLEU-4	CIDEr	METEOR	ROUGE-L	ROUGE-1	GPT Score
GPT-4o	0.127	0.099	0.627	0.157	0.292	0.369	2.928
DeepSeek-Janus-Pro-7B	0.003	0.003	0.069	0.048	0.042	0.045	2.616
DeepSeek-VL2	0.134	0.108	0.514	0.142	0.231	0.296	2.006
DeepSeek-V3	0.209	0.158	0.892	0.217	0.461	0.545	2.550
Model	Detailed Description						
	BLEU-3	BLEU-4	CIDEr	METEOR	ROUGE-L	ROUGE-1	GPT Score
GPT-4o	0.042	0.022	0.007	0.135	0.161	0.231	1.154
DeepSeek-Janus-Pro-7B	0.057	0.032	0.007	0.141	0.181	0.235	0.891
DeepSeek-VL2	0.047	0.025	0.010	0.134	0.162	0.259	1.156
DeepSeek-V3	0.023	0.010	0.003	0.118	0.153	0.169	0.213

1. With appropriately designed guiding prompts, DeepSeek-V3 shows the ability to understand the posed questions and generate surgical-related lexical content based on its image perception capabilities. Furthermore, the consistent performance across six different metrics of DeepSeek-VL2 and DeepSeek-V3 highlights their efficiency in accurately capturing key concepts and focusing on the relevant visual content. Through our evaluation, we also observe that DeepSeek-Janus-Pro-7b tends to provide short answers, even when we include '*Answer with short sentences*' in the prompt. This indicates its limited instruction-following ability and insufficient understanding of surgical knowledge.

3.3   Performance on Detailed Description

When evaluating the Detailed Description task, we provide general prompts such as "*Describe the following image in detail*" and "*Clarify the contents of the displayed image with great detail.*" As shown in Fig. 1 and Fig. 2, the ground truth demonstrates that the model is expected to produce a comprehensive description of the surgical scene. This involves identifying surgical instruments, describing operational motions, specifying the target tissue, and providing locational context. From Table 2, it can be observed that all the models perform poorly on BLEU and CIDEr metrics. The low performance can be attributed to the excessive length of the comparison sentences, which makes it challenging for the models to maintain n-gram precision and achieve reasonable similarity in this task. However, its generated text consistently includes details about instruments, motions, and tissues within the surgical context.

As shown in Fig. 2, DeepSeek-V3 has limited ability to provide a general description, as DeepSeek-V3 is only a language model. In our previous tests, we achieved good performance in simple QA tasks by directly inputting image tokens. However, such a language model struggles to provide meaningful explanations when performing detailed description tasks. Moreover, the potential impact of different image tokenizers on model inference remains to be further investigated. On the other hand, DeepSeek-VL2 and DeepSeek-Janus-Pro-7b demonstrate superior performance across multiple indicators compared to GPT-4o, showcasing their ability to capture the overall context of the image and provide detailed surgical descriptions, even when given a simple prompt. From Fig. 2 it is evident that the text generated by DeepSeek-VL2 and DeepSeek-Janus-Pro-7b is not verbose; instead, it tends to provide concise summaries while offering more accurate content. This indicates that both models possess great capabilities in understanding and summarizing images. However, their lack of surgical scene comprehension limits their applicability in this domain. In conclusion, DeepSeek-VL2 can achieve reasonable results in detailed description tasks; however, due to its lack of surgical knowledge, the model cannot provide accurate answers and fails to include all the necessary details required in surgical scenarios.

4   Conclusion

In this paper, we conduct empirical studies on the DeepSeek series to evaluate its contextual comprehension capabilities in robotic surgery. Our experiments focus on three distinct tasks: Single Phrase QA, Visual QA, and Detailed Description. The performance of DeepSeek-Janus-Pro-7b on three tasks shows its limitation in surgical scene comprehension and instruction understanding. DeepSeek-VL2, on the other hand, demonstrates a certain capacity for object position detection in Single Phrase QA tasks and can yield better results by providing concise summaries in detailed description tasks. However, due to its limited knowledge of surgical-related information and lack of comprehension of surgical scenes, DeepSeek-VL2 is not suitable for real-world surgical applications. Although DeepSeek-V3 is purely a language model, we test its performance by directly inputting image tokens and find that, based on the reasoning capability, it performs well in single-sentence QA tasks. However, it fails to achieve a comprehensive understanding of surgical scenarios. From our experiments, we observe that DeepSeek-V3 can simulate basic surgical knowledge similar to that of a novice surgeon when given specific prompts. However, it struggles to interpret surgical actions and cannot accurately localize objects, which limits its practical clinical applications.

Based on the current results, there are still areas that require further improvement in future work: (i) Developing effective prompts is critical, as the clarity and specificity of instructions directly influence the model's performance. Crafting versatile and adaptable prompts can also enhance the ability to handle real-world interactive tasks. (ii) Integrating large-scale surgical datasets with reasoning-based fine-tuning can equip MLLMs with stronger inferential capabilities in surgical scenarios, enabling more effective collaboration between surgeons and the model. (iii) By incorporating strategies such as Mixture of Experts (MoE) and model distillation, computational and deployment costs for surgical MLLMs can be significantly reduced. Lower costs would support the development of embodied surgical robots and facilitate broader clinical adoption.

Acknowledgment. This work was supported by Hong Kong Research Grants Council GRF 14204524, GRF 14203323, GRF 14216022, GRF 14211420, CRF C4026-21GF; Guangdong Basic & Applied Research Fund Regional Joint Fund Project 2021B1515120035 (B.02.21.00101); Shenzhen-Hong Kong-Macau Technology Research Programme (Type C) STIC Grant SGDX20210823103535014 (202108233000303).

References

1. Allan, M., et al.: 2018 robotic scene segmentation challenge. arXiv preprint arXiv:2001.11190 (2020)
2. Bai, L., Islam, M., Ren, H.: Revisiting distillation for continual learning on visual question localized-answering in robotic surgery. In: International Conference on Medical Image Computing and Computer-Assisted Intervention, pp. 68–78. Springer (2023)
3. Bai, L., Wang, G., Islam, M., Seenivasan, L., Wang, A., Ren, H.: Surgical-VQLA++: adversarial contrastive learning for calibrated robust visual question-localized answering in robotic surgery. Inf. Fusion **113**, 102602 (2025)
4. Chen, T., Yuan, K., Srivastav, V., Navab, N., Padoy, N.: Text-driven adaptation of foundation models for few-shot surgical workflow analysis. arXiv preprint arXiv:2501.09555 (2025)
5. Chen, X., et al.: Janus-pro: unified multimodal understanding and generation with data and model scaling. arXiv preprint arXiv:2501.17811 (2025)
6. Chen, Z., et al.: VS-assistant: versatile surgery assistant on the demand of surgeons. arXiv preprint arXiv:2405.08272 (2024)
7. Ding, D., Yao, T., Luo, R., Sun, X.: Visual question answering in robotic surgery: a comprehensive review. IEEE Access (2025)
8. Fiorini, P., Goldberg, K.Y., Liu, Y., Taylor, R.H.: Concepts and trends in autonomy for robot-assisted surgery. Proc. IEEE **110**(7), 993–1011 (2022)
9. Ge, Y., et al.: Making llama see and draw with seed tokenizer. arXiv preprint arXiv:2310.01218 (2023)
10. Guo, D., et al.: DeepSeek-R1: incentivizing reasoning capability in LLMs via reinforcement learning. arXiv preprint arXiv:2501.12948 (2025)
11. Hao, P., Wang, H., Yang, G., Zhu, L.: Enhancing visual reasoning with LLM-powered knowledge graphs for visual question localized-answering in robotic surgery. IEEE J. Biomed. Health Inform. (2025)

12. Huang, Y., et al.: Endo-4DGS: endoscopic monocular scene reconstruction with 4d gaussian splatting. In: International Conference on Medical Image Computing and Computer-Assisted Intervention, pp. 197–207. Springer (2024)
13. Hurst, A., et al.: GPT-4O system card. arXiv preprint arXiv:2410.21276 (2024)
14. Jiang, A.Q., et al.: Mistral 7b. arXiv preprint arXiv:2310.06825 (2023)
15. Lai, Y., Zhong, J., Li, M., Zhao, S., Yang, X.: Med-R1: reinforcement learning for generalizable medical reasoning in vision-language models. arXiv preprint arXiv:2503.13939 (2025)
16. Li, J., et al.: LLaVA-SURG: towards multimodal surgical assistant via structured surgical video learning. arXiv preprint arXiv:2408.07981 (2024)
17. Liu, A., et al.: DeepSeek-V3 technical report. arXiv preprint arXiv:2412.19437 (2024)
18. Marcus, H.J., et al.: The ideal framework for surgical robotics: development, comparative evaluation and long-term monitoring. Nat. Med. **30**(1), 61–75 (2024)
19. Nwoye, C.I., Padoy, N.: Data splits and metrics for benchmarking methods on surgical action triplet datasets. arXiv preprint arXiv:2204.05235 (2022)
20. Pan, J., et al.: MedVLM-R1: incentivizing medical reasoning capability of vision-language models (VLMs) via reinforcement learning. arXiv preprint arXiv:2502.19634 (2025)
21. Rombach, R., Blattmann, A., Lorenz, D., Esser, P., Ommer, B.: High-resolution image synthesis with latent diffusion models. In: Proceedings of the IEEE/CVF Conference on Computer Vision and Pattern Recognition, pp. 10684–10695 (2022)
22. Seenivasan, L., Islam, M., Kannan, G., Ren, H.: SurgicalGPT: end-to-end language-vision u for visual question answering in surgery. In: International Conference on Medical Image Computing and Computer-Assisted Intervention, pp. 281–290. Springer (2023)
23. Shen, Y., et al.: Medical multimodal model stealing attacks via adversarial domain alignment. arXiv preprint arXiv:2502.02438 (2025)
24. Wang, G., et al.: Surgical-LVLM: learning to adapt large vision-language model for grounded visual question answering in robotic surgery. arXiv preprint arXiv:2405.10948 (2024)
25. Wang, G., et al.: Endochat: grounded multimodal large language model for endoscopic surgery. arXiv preprint arXiv:2501.11347 (2025)
26. Wang, G., et al.: CoPESD: a multi-level surgical motion dataset for training large vision-language models to co-pilot endoscopic submucosal dissection. arXiv preprint arXiv:2410.07540 (2024)
27. Wang, H., Jin, Y., Zhu, L.: Dynamic interactive relation capturing via scene graph learning for robotic surgical report generation. In: 2023 IEEE International Conference on Robotics and Automation (ICRA), pp. 2702–2709. IEEE (2023)
28. Wang, P., et al.: Qwen2-VL: enhancing vision-language model's perception of the world at any resolution. arXiv preprint arXiv:2409.12191 (2024)
29. Wu, Z., et al.: DeepSeek-VL2: mixture-of-experts vision-language models for advanced multimodal understanding. arXiv preprint arXiv:2412.10302 (2024)
30. Yuan, K., Kattel, M., Lavanchy, J.L., Navab, N., Srivastav, V., Padoy, N.: Advancing surgical VQA with scene graph knowledge. Int. J. Comput. Assist. Radiol. Surg. **19**(7), 1409–1417 (2024)
31. Zhong, F., Liu, Y.H.: Integrated planning and control of robotic surgical instruments for task autonomy. Int. J. Robot. Res. **42**(7), 504–536 (2023)

Anti-forgetting Test-Time Adaptation for Robust Medical Image Analysis Under Distribution Shift

Zhenyu Wu[1], Shiyu Chen[1], Zichao Li[1], Junkai Huang[1], Kang Chen[1], Zhen Qiao[1], Ming Li[2], Tengfei Shi[3], and Xuehao Wang[4]([✉])

[1] School of Computing and Artificial Intelligence, Southwest Jiaotong University, Chengdu, China
[2] Department of Computer Science and Technology, Tsinghua University, Beijing, China
[3] School of Computer Engineering, Hubei University of Arts and Science, Xiangyang, China
[4] University of International Business and Economics, Beijing, China

Abstract. Although the test-time adaptation (TTA) learning paradigm can against distribution shift, it also results in severe performance degradation on in-distribution (ID) data after adaptation, a phenomenon known as catastrophic forgetting. In this paper, we highlight that the fundamental reason for catastrophic forgetting is that vanilla TTA methods finetune the model without considering the type of input samples. Our key insight is that the model only needs to be finetuned when the input data is from the out-of-distribution (OOD) domain. Motivated by this insight, we proposed a novel and surprisingly effective anti-forgetting test-time adaptation method to address the catastrophic forgetting problem. Specifically, our contributions are two-fold: **1)** We propose a novel non-parametric perturbation layer to synthesize OOD data for any input samples, which can enhance model robustness by injecting the synthesized OOD data into the training set and provide supervision signal for OOD detector. **2)** We further propose a novel features statistics mean discrepancy metric, which can reliably distinguish the OOD and ID data and finetune model parameters according to the type of input data. Extensive experiments on medical image classification and segmentation tasks demonstrated that our method can address the catastrophic forgetting problem and achieve the SOTA performance.

Keywords: Test-Time Adaptation · Distribution Shift · Robustness

1 Introduction

Deep learning-based methods achieve excellent performance in medical image analysis (MIA), but the prerequisite is testing and training data drawn from the same distribution. In the actual clinical setting, medical data typically exhibit significant distribution shift due to the scanner manufacturer, imaging protocol, patient population, or disease class, resulting in severe performance degradation. Therefore, enhancing the model's robustness on various unseen testing data becomes practically important for MIA.

© The Author(s), under exclusive license to Springer Nature Switzerland AG 2026
J. Qiu et al. (Eds.): Agentic AI 2025/CMLLMs 2025/CREATE 2025, LNCS 16147, pp. 187–196, 2026.
https://doi.org/10.1007/978-3-032-06004-4_19

One promising learning paradigm is test-time adaptation (TTA), which fine-tunes the model parameters to adapt the testing data at test time [17,20,21,26]. Test-time training [20] is the first work to explore this paradigm, which finetuned the model parameters by designing an auxiliary self-supervised branch. Besides, the test-time training method requires access to the training dataset, which may be infeasible if the training dataset is unavailable due to privacy and security concerns. To address the above issues, the fully TTA paradigm [21,26] was introduced, which does not require an additional self-supervised branch but the input testing samples. Although the TTA learning paradigm can enhance the model's robustness against distribution shift, it also results in severe performance degradation on in-distribution (ID) data after adaptation, a phenomenon known as catastrophic forgetting. Existing anti-forgetting approaches [18,23] are typically based on continual learning, which aims to help the model retain the essential concepts learned previously. For instance, Niu *et al.* [18] devised a weighted fisher regularizer to enforce the important weights of the model do not change a lot during the adaptation. Wang *et al.* [23] propose to stochastically restore a small part of neurons in the network back to the pre-trained source model to avoid forgetting. However, the aforementioned methods failed to grasp the root cause of catastrophic forgetting, it still suffers performance degradation on ID data.

In this paper, we proposed a novel and surprisingly effective anti-forgetting test-time adaptation (ATTA) method, which can address the catastrophic forgetting problem and enhance model robustness to distribution shift simultaneously. *Our key insight is that a pretrained source model needs to finetune its parameters to fit the target data only when the input data belongs to the OOD domain.* While designing a lightweight and high-performance OOD detector that can reliably distinguish between ID data and OOD data is non-trivial, there are two main challenges: **1) Lacking OOD Dataset.** Since OOD samples can be infinite and unpredictable, it can be costly and sometimes infeasible to collect an OOD dataset that covers all different kinds of scenarios in the real world. **2) Heavy Computational Cost.** Integrating existing OOD detection methods directly into the MIA task would lead to significant computational overhead. To address these challenges: **1)** We first propose a simple yet effective non-parametric perturbation layer (NPL), which models the uncertain distribution of the real world to generate OOD data by perturbing the feature statistics of input data (Sect. 2.1). **2)** Secondly, we proposed a novel features statistics mean discrepancy metric (FSMD), which can reliably distinguish the OOD and ID data with a negligible computational overhead (Sect. 2.2).

In summary, we make the following contributions: **1)** We introduced a novel and remarkably effective anti-forgetting test-time adaptation method, capable of addressing the catastrophic forgetting problem while simultaneously enhancing model robustness to distribution shift. **2)** The proposed NPL adaptively synthesizes OOD data for any input samples without the need for manual data collection. **3)** The proposed FSMD can reliably distinguish between ID data and OOD data with an accuracy ranging from 99% to 100% with a negligible computational cost. **4)** Comprehensive experiments demonstrated that our ATTA can fully address catastrophic forgetting on the polyp segmentation task and achieve the state-of-the-art (SOTA) performance against the distribution shift.

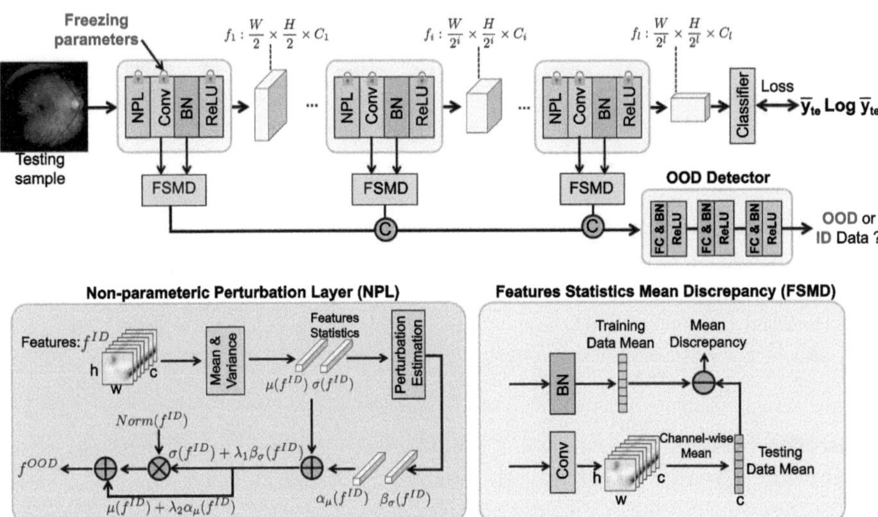

Fig. 1. Illustration of the proposed ATTA. Given a pretrained model $\mathcal{F}$ and a batch of testing samples x, we calculate FSMD(x) for each test sample and fed it into OOD detector to identify whether the sample is OOD data or not, and perform model fine-tuning only if the input samples belong to OOD domain.

2   Methodology

As shown in Fig. 1, the proposed ATTA contains two novel modules: **1)** a novel non-parametric perturbation layer to synthesize OOD data by perturbing input data (Sect. 2.1), and **2)** a novel features statistics mean discrepancy metric for separating the OOD and ID data (Sect. 2.2).

2.1   Non-parametric Perturbation Layer

Recent works [15,16] demonstrate that feature statistics (i.e., mean and variance) extracted from neural networks inherently contain domain information of the dataset, implying that dataset drawn from different distributions are inconsistent in terms of feature statistics. Inspired by that, we propose a novel non-parametric perturbation layer to synthesize OOD data by perturbing the feature statistics of the input ID data, as shown in Fig. 1. Specifically, let $f^{\mathrm{ID}} \in \mathbb{R}^{B \times C \times H \times W}$ denote the intermediate features of ID data, the mean $\mu \in \mathbb{R}^{B \times C}$ and variance $\sigma^2 \in \mathbb{R}^{B \times C}$ can be calculated as:

$$\mu(f^{ID}) = \frac{1}{HW} \sum_{h=1}^{H} \sum_{w=1}^{W} f^{ID}_{b,c,h,w}, \quad \sigma(f^{ID}) = \sqrt{\frac{1}{HW} \sum_{h=1}^{H} \sum_{w=1}^{W} \left(f^{ID}_{b,c,h,w} - \mu\left(f^{ID}\right) \right)^2}$$

$$(1)$$

Besides, considering the uncertain distribution of OOD data, we hypothesize that the OOD dataset follows a multivariate Gaussian distribution. In this paper,

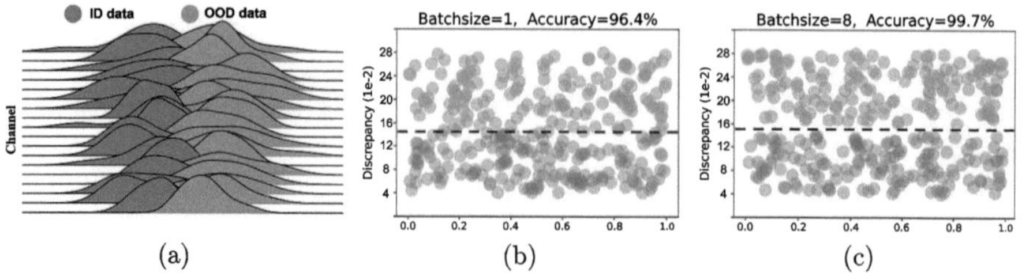

Fig. 2. (a) Feature mean of ID and OOD data have different distributions. (b) and (c) show that the proposed FSMD metric can reliably distinguish between ID and OOD data, and larger batchsize obtains better performance. (Color figure online)

we use the Gaussian distribution's standard deviation to model the uncertain distribution of real-world OOD data, which can be formulated as:

$$\alpha^2(\mu) = \frac{1}{B}\sum_{b=1}^{B}\left(\mu\left(f^{ID}\right) - \mathbb{E}\left[\mu\left(f^{ID}\right)\right]\right)^2,$$

$$\beta^2(\sigma) = \frac{1}{B}\sum_{b=1}^{B}\left(\sigma\left(f^{ID}\right) - \mathbb{E}\left[\sigma\left(f^{ID}\right)\right]\right)^2 \tag{2}$$

where $\alpha^2(\mu)$ and $\beta^2(\sigma)$ denote the uncertainty range of ID feature statistics μ and σ, respectively. We can then obtain the OOD data's features statistics mean $\gamma \sim \mathcal{N}(\mu, \alpha^2(\mu))$ and variance $\eta \sim \mathcal{N}(\sigma, \beta^2(\sigma))$. Finally, the OOD data's features can be calculated as:

$$f^{OOD} = \underbrace{\left(\mu\left(f^{ID}\right) + \lambda_1\alpha\left(\mu\right)\right)}_{\gamma}\left(\frac{f^{ID} - \mu(f^{ID})}{\sigma(f^{ID})}\right) + \underbrace{\left(\sigma\left(f^{ID}\right) + \lambda_2\beta\left(\sigma\right)\right)}_{\eta}, \tag{3}$$

where $\lambda_i \sim \mathcal{N}(0, t)$ is perturbation coefficient and the hyperparameter $t = 0.7$ used to control the perturbation scope. The synthesized f^{OOD} serves two purposes: 1) training the deep model together with ID data to enhance model's robustness; and 2) providing supervision for OOD detector.

Advantages: The proposed NPL can adaptively synthesize OOD data for any input samples by perturbing the feature statistics of the input ID data. Moreover, the proposed NPL can be easily integrated into existing methods as a plug-and-play module without additional parameters.

2.2 Features Statistics Mean Discrepancy

Features statistics mean discrepancy (FSMD) aims to separate the OOD and ID data using information from the model itself. We start by revealing an important observation: OOD and ID data have significantly different in features statistics. Figure 2 shows the distribution of features statistics in each batchnorm layer of

MobileNet-V2 [19] on the Hyperkvasir dataset and its OOD data. As we can see, the mean value of OOD data (orange) is consistently larger than ID data (blue). Similar observations can be found in other medical datasets as well.

The above observation implies that the mean discrepancy of features statistics can be a simple and effective metric to measure the closeness between testing dataset $\mathcal{Q}_{te}$ and training set $\mathcal{D}_{tr}$. In particular, we consider a well pretrained model $\mathcal{F}$ parameterized by θ, which encodes a testing data x to a intermediate features $f \in \mathbb{R}^{B \times C \times H \times W}$. The FSMD can be calculated as:

$$
\begin{aligned}
\mathrm{FSMD}(\mathcal{Q}_{te}, \mathcal{D}_{tr}) &= \mathbb{E}_{x_{te} \sim \mathcal{Q}_{te}}[\mathcal{F}(x_{te})] - \mathbb{E}_{x_{tr} \sim \mathcal{D}_{tr}}[\mathcal{F}(x_{tr})] \\
&= \frac{1}{|\mathcal{Q}_{te}|} \sum_{i=1}^{|\mathcal{Q}_{te}|} \mu(f^{te}) - \frac{1}{|\mathcal{D}_{tr}|} \sum_{i=1}^{|\mathcal{D}_{tr}|} \mu(f^{tr}) \\
&= \underbrace{\frac{1}{|\mathcal{Q}_{te}|HW} \sum_{i=1}^{|\mathcal{Q}_{te}|} \sum_{h=1}^{H} \sum_{w=1}^{W} f^{te}_{b,c,h,w}}_{\mu(\mathcal{Q}_{te}):\text{mean vector of testing data}} - \underbrace{\frac{1}{|\mathcal{D}_{tr}|HW} \sum_{i=1}^{|\mathcal{D}_{tr}|} \sum_{h=1}^{H} \sum_{w=1}^{W} f^{tr}_{b,c,h,w}}_{\mu(\mathcal{D}_{tr}):\text{mean vector of training data}}
\end{aligned}
\tag{4}
$$

where the $\mu(\mathcal{D}_{tr})$ can be obtained directly from the batchnorm layer. To further improve the discriminatory capability of the OOD detection, we propose to learn a parametric OOD detector that takes FSMD vectors as input. The OOD detector consists of three fully connected layers, which are simple, lightweight, and efficient at both training and inference. As shown in Fig. 2(c), our OOD detector can reliably separate ID and OOD data with 99.7% accuracy on the Hyperkvasir dataset. In our method, we finetune the model parameters only when the testing data belongs to the OOD domain, which is the key point that our method can address catastrophic forgetting.

Advantages: 1) High Performance. Our OOD detector achieves 99.7% accuracy on the classification datasets and 100% accuracy on the polyp segmentation datasets. **2) Simplicity.** The proposed FSMD metric is simple and effective, which can be directly calculated from the batchnorm and convolutional layers. **3) Low Computational Cost.** Our ATTA inference speed is close to the source pretrained model because the OOD detector is lightweight and the model size is 3.6 MB, which is faster than existing OOD detection methods [3,8,22] by an order of magnitude.

2.3   Loss Function

Following ETA [18], we use sample-efficient entropy loss for training model:

$$
\mathcal{L}_{eta} = - \left(\frac{\mathbb{I}_{\{\mathcal{F}_\theta(x_{te})\log(\mathcal{F}_\theta(x_{te}) - \tau_0\}}(x_{te})}{\exp[\mathcal{F}_\theta(x_{te})\log(\mathcal{F}_\theta(x_{te}) - \tau_0]} \right) \sum_{x_{te} \in \mathcal{Q}_{te}} \mathcal{F}_\theta(x_{te})\log(\mathcal{F}_\theta(x_{te}))
\tag{5}
$$

where the first term is the entropy based weighting scheme to identify reliable samples, $\mathbb{I}(\cdot)$ is the indicator function, and τ_0 is a predefined threshold. During adaptation, we only finetune the parameters of batch normalization layers and froze the rest parameters.

3   Experiments

3.1   Setting

Classification Datasets and Metrics. We evaluate our ATTA on two medical image classification datasets: Eyepacs[1] and Hyperkvasir [4]. The Eyepacs is the largest publicly available dataset for diabetic retinopathy grading from retinal fundus images, which contains 35,000 training examples and 55,000 testing samples. The Hyperkvasir with 10,662 images represents a challenging gastrointestinal classification problem due to the larger amount of classes (23) and the long-tail distribution. We follow previous work [13] adopting the Matthews Correlation Coefficient (MCC) [5], Balanced Accuracy (BalAcc) [7], and Macro-F1 for quantitative evaluation.

Segmentation Datasets and Metrics. We also evaluate our ATTA on polyp segmentation datasets, i.e., CVC-ClinicDB [1], CVC-ColonDB [2], and ETIS [12]. For fair comparison, we adopt the same training data division and evaluation metrics (i.e., mean Dice and mean IoU) as in previous works [11,12,24].

OOD Benchmark Synthesized by Corruption. Following the evaluation protocol of previous works [14,20,21], given a dataset X, we apply the common image corruptions [18,27,28] to generate the corresponding OOD benchmark, referred to as X-C. The image corruptions include noise (gaussian, impluse and shot), blur (defocus, glass and motion), color (brightness, saturation, and contrast) and digitization (pixelate and JPEG compression). Each type of corruption has five severity levels. After image corruptions, we obtain the Hyperkvasir-C, Eyepacs-C, CVC-ClinicDB-C, CVC-ColonDB-C, and ETIS-C datasets, which serve as OOD data to evaluate our model.

Compared with SOTA TTA-Based Methods. To validate the superiority of our method, we compare ATTA with SOTA TTA-based methods, including TTT [20], TENT [21], MEMO [26], ETA [18], CoTTA [23], ITTA [9], PETAL [6], CGTTA [10], and RoTTA [25].

3.2   Demonstration of Anti-forgetting

We first compared our ATTA with SOTA TTA-based methods on ID data. As shown in Table 1 and Table 2, previous TTA-based methods showed a significant performance drop on ID data after model finetuning. In contrast, our ATTA only has a negligible performance degradation. In particularly, our ATTA achieves the same performance as the source model in the polyp segmentation task without any performance degradation, demonstrating the effectiveness of our ATTA in preventing catastrophic forgetting. There are two key points that our ATTA can prevent catastrophic forgetting: 1) Our ATTA adopts episodic repay strategy [23], i.e., the model parameters will be reinitialized by the pretrained source model before adapting to a new batch of input samples. 2) Our method finetunes the model's parameters only when the input samples belong to the OOD domain.

---

[1] https://www.kaggle.com/c/diabetic-retinopathy-detection/.

Table 1. Classification results (%) on the Eyepacs and Hyperkvasir. The number x is the degraded score compared to the source model after adopting TTA. Our ATTA achieves the best performance in catastrophic forgetting issue.

	Anti-Forgetting	Eyepacs			Hyperkvasir		
		MCC ↑	BalAcc ↑	Macro-F1 ↑	MCC ↑	BalAcc ↑	Macro-F1 ↑
Source model	-	**62.0**	**53.2**	**57.8**	**90.9**	**61.7**	**61.5**
MEMO [26]	✗	$43.7_{(-18.3)}$	$39.4_{(-13.8)}$	$41.4_{(-16.4)}$	$78.3_{(-12.6)}$	$43.5_{(-18.2)}$	$42.2_{(-19.3)}$
TENT [21]	✗	$49.3_{(-12.7)}$	$43.9_{(-9.3)}$	$45.1_{(-12.7)}$	$84.4_{(-6.5)}$	$48.7_{(-13.0)}$	$47.4_{(-14.1)}$
CoTTA [23]	✔	$51.4_{(-10.6)}$	$46.5_{(-6.7)}$	$48.3_{(-9.5)}$	$85.3_{(-5.6)}$	$50.6_{(-11.1)}$	$49.8_{(-11.7)}$
ETA [18]	✔	$56.3_{(-5.7)}$	$47.2_{(-6.0)}$	$52.4_{(-5.4)}$	$87.6_{(-3.3)}$	$55.3_{(-6.4)}$	$53.3_{(-8.2)}$
RoTTA [25]	✗	$51.8_{(-10.2)}$	$46.5_{(-6.7)}$	$49.4_{(-8.4)}$	$86.3_{(-4.6)}$	$50.9_{(-10.8)}$	$50.7_{(-10.8)}$
ITTA [9]	✗	$52.3_{(-8.7)}$	$46.4_{(-6.8)}$	$49.7_{(-8.1)}$	$86.2_{(-4.7)}$	$51.3_{(-10.4)}$	$50.5_{(-11.0)}$
CGTTA [10]	✗	$52.6_{(-8.4)}$	$46.8_{(-6.4)}$	$50.3_{(-7.5)}$	$86.8_{(-4.1)}$	$52.1_{(-9.6)}$	$51.4_{(-10.1)}$
PETAL [6]	✔	$58.5_{(-3.5)}$	$48.7_{(-4.5)}$	$54.1_{(-3.7)}$	$87.5_{(-3.4)}$	$56.4_{(-5.3)}$	$55.6_{(-5.9)}$
ATTA(Ours)	✔	$\mathbf{61.6}_{(-0.4)}$	$\mathbf{52.8}_{(-0.4)}$	$\mathbf{57.4}_{(-0.4)}$	$\mathbf{90.6}_{(-0.3)}$	$\mathbf{61.5}_{(-0.2)}$	$\mathbf{61.3}_{(-0.2)}$

Table 2. Segmentation results (%) on the CVC-ClinicDB, CVC-ColonDB and ETIS. Our ATTA achieves the same performance as the source model without any performance degradation on ID data.

	Anti-Forgetting	CVC-ClinicDB		CVC-ColonDB		ETIS	
		mIoU ↑	mDice ↑	mIoU ↑	mDice ↑	mIoU ↑	mDice ↑
Source model	-	**85.9**	**91.6**	**67.0**	**75.3**	**65.4**	**75.0**
MEMO [26]	✗	$74.5_{(-11.4)}$	$81.8_{(-9.8)}$	$55.4_{(-11.6)}$	$67.2_{(-8.1)}$	$56.1_{(-9.3)}$	$66.8_{(-8.2)}$
TENT [21]	✗	$79.4_{(-6.5)}$	$86.2_{(-5.4)}$	$60.9_{(-6.1)}$	$71.0_{(-4.3)}$	$59.8_{(-5.6)}$	$69.8_{(-5.2)}$
CoTTA [23]	✔	$80.6_{(-5.3)}$	$88.5_{(-3.1)}$	$62.4_{(-4.6)}$	$72.8_{(-2.5)}$	$61.3_{(-4.1)}$	$71.4_{(-3.6)}$
ETA [18]	✔	$81.7_{(-4.2)}$	$88.7_{(-2.9)}$	$63.2_{(-3.8)}$	$72.3_{(-3.0)}$	$62.7_{(-2.7)}$	$71.8_{(-3.2)}$
RoTTA [25]	✗	$80.2_{(-5.7)}$	$87.2_{(-4.4)}$	$61.3_{(-5.7)}$	$71.6_{(-3.7)}$	$60.1_{(-5.3)}$	$70.8_{(-4.2)}$
ITTA [9]	✗	$80.9_{(-4.2)}$	$87.9_{(-3.7)}$	$61.8_{(-5.2)}$	$72.1_{(-3.2)}$	$60.5_{(-4.9)}$	$70.7_{(-4.3)}$
CGTTA [10]	✗	$80.4_{(-5.5)}$	$87.3_{(-4.3)}$	$62.1_{(-4.9)}$	$71.4_{(-3.9)}$	$60.7_{(-4.7)}$	$70.5_{(-4.5)}$
PETAL [6]	✔	$82.5_{(-3.4)}$	$89.6_{(-2.0)}$	$64.1_{(-2.9)}$	$72.9_{(-2.4)}$	$63.3_{(-2.1)}$	$72.3_{(-2.7)}$
ATTA(Ours)	✔	$\mathbf{85.9}_{(-0.0)}$	$\mathbf{91.6}_{(-0.0)}$	$\mathbf{67.0}_{(-0.0)}$	$\mathbf{75.3}_{(-0.0)}$	$\mathbf{65.4}_{(-0.0)}$	$\mathbf{75.0}_{(-0.0)}$

3.3   Robustness to Distribution Shift

We further evaluate the robustness of our ATTA against distribution shift on OOD benchmarks, i.e., Hyperkvasir-C, Eyepacs-C, CVC-ClinicDB-C and CVC-ColonDB-C, as shown in the Fig. 3. While directly using the source pretrained model (the blue curve of Fig. 3) results in a significant decrease (about 13% to 50%), TTA-based methods can effectively improve the model robustness to distribution shift. In particular, our ATTA consistently outperforms all of these SOTA TTA-based methods for both classification and segmentation tasks, demonstrating the superiority of the proposed ATTA. In conclusion, the robustness of the proposed ATTA to distribution shift can be attributed to two factors: 1) the OOD data synthesized by the NPL used for training the model; and 2) our loss function selecting high-quality pseudo-labels to finetune the model parameters.

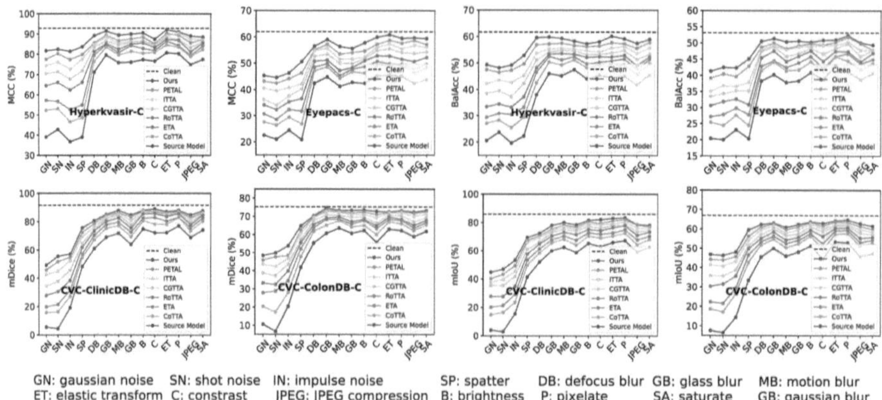

Fig. 3. Our ATTA consistently outperform SOTA TTA-based methods on classification and segmentation OOD datasets with corruption level 3. (Color figure online)

Table 3. Ablation study on the Eyepacs and Eyepacs-C datasets, where Eyepacs-C was corrupted by the gaussian noise with corruption level 3.

	Anti-Forgetting	Robustness	Eyepacs		Eyepacs-C	
			MCC ↑	BalAcc ↑	MCC ↑	BalAcc ↑
Source model	-	✗	**62.0**	**53.2**	**22.5**	**20.4**
Source model+NPL	✗	✓	$46.2_{(-15.8)}$	$42.1_{(-11.1)}$	$45.3_{(+22.8)}$	$41.3_{(+20.9)}$
Source model+NPL+FSMD	✓	✓	$\mathbf{61.6}_{(-0.4)}$	$\mathbf{52.8}_{(-0.4)}$	$\mathbf{45.3}_{(+22.8)}$	$\mathbf{41.3}_{(+20.9)}$

3.4 Ablation Study

We conduct an ablation study to evaluate the contribution of each component in our framework. As shown in Table 3, the source model equipped with NLP shows a clear improvement over the source model on Eyepacs-C but also results in significant performance degradation on Eyepacs. When we further add the FSMD into the source model, the catastrophic forgetting problem on the Eyepacs dataset can be greatly alleviated with negligible degradation (less than 0.5%).

4 Conclusion

In this paper, we propose a novel and surprisingly effective anti-forgetting test-time adaptation method, addressing the catastrophic forgetting problem and improving the model's robustness against distribution shift. Specifically, we first propose a novel non-parametric perturbation layer to synthesize OOD data, which can improve model robustness by injecting the synthesized OOD data into the training set. Further, we proposed a novel features statistics mean discrepancy metric, which can reliably distinguish the OOD and ID data and fine-tune model parameters according to the type of input data. Extensive experiments and ablation studies are conducted to verify the robustness and the anti-forgetting of the proposed method. Given the promising results in this paper, future work should continue to scale ATTA strategy to larger models and datasets.

Acknowledgements. This research is supported in part by the Open Project Program of State Key Laboratory of Virtual Reality Technology and Systems (No. VRLAB2024C01), the Postdoctoral Fellowship Program of CPSF (No. GZC20232190), the China Postdoctoral Science Foundation (No. 2023M742896), the Sichuan Science and Technology Program (24NSFSC7274), and the Fundamental Research Funds for the Central Universities (No. 2682024CX007).

References

1. Bernal, J., et al.: WM-DOVA maps for accurate polyp highlighting in colonoscopy: validation vs. saliency maps from physicians. Comput. Med. Imaging Graph. (CMIG) **43**, 99–111 (2015)
2. Bernal, J., Sánchez, J., Vilarino, F.: Towards automatic polyp detection with a polyp appearance model. Pattern Recogn. (PR) **45**(9), 3166–3182 (2012)
3. Bitterwolf, J., Meinke, A., Hein, M.: Certifiably adversarially robust detection of out-of-distribution data. In: Advances in Neural Information Processing Systems (NeurIPS), vol. 33, pp. 16085–16095 (2020)
4. Borgli, H., et al.: HyperKvasir, a comprehensive multi-class image and video dataset for gastrointestinal endoscopy. Sci. Data **7**(1), 1–14 (2020)
5. Boughorbel, S., Jarray, F., El-Anbari, M.: Optimal classifier for imbalanced data using Matthews correlation coefficient metric. PloS One **12**(6) (2017)
6. Brahma, D., Rai, P.: A probabilistic framework for lifelong test-time adaptation. In: Proceedings of the IEEE Conference on Computer Vision and Pattern Recognition (CVPR), pp. 3582–3591 (2023)
7. Brodersen, K.H., Ong, C.S., Stephan, K.E., Buhmann, J.M.: The balanced accuracy and its posterior distribution. In: 20th International Conference on Pattern Recognition (ICPR), pp. 3121–3124 (2010)
8. Chan, R., Rottmann, M., Gottschalk, H.: Entropy maximization and meta classification for out-of-distribution detection in semantic segmentation. In: Proceedings of the IEEE International Conference on Computer Vision (ICCV), pp. 5128–5137 (2021)
9. Chen, L., Zhang, Y., Song, Y., Shan, Y., Liu, L.: Improved test-time adaptation for domain generalization. In: Proceedings of the IEEE Conference on Computer Vision and Pattern Recognition (CVPR), pp. 24172–24182 (2023)
10. Döbler, M., Marsden, R.A., Yang, B.: Robust mean teacher for continual and gradual test-time adaptation. In: Proceedings of the IEEE Conference on Computer Vision and Pattern Recognition (CVPR), pp. 7704–7714 (2023)
11. Fan, D.-P., et al.: PraNet: parallel reverse attention network for polyp segmentation. In: Martel, A.L., et al. (eds.) MICCAI 2020. LNCS, vol. 12266, pp. 263–273. Springer, Cham (2020). https://doi.org/10.1007/978-3-030-59725-2_26
12. Fang, Y., Chen, C., Yuan, Y., Tong, K.: Selective feature aggregation network with area-boundary constraints for polyp segmentation. In: Shen, D., et al. (eds.) MICCAI 2019. LNCS, vol. 11764, pp. 302–310. Springer, Cham (2019). https://doi.org/10.1007/978-3-030-32239-7_34
13. Galdran, A., Carneiro, G., González Ballester, M.A.: Balanced-mixup for highly imbalanced medical image classification. In: de Bruijne, M., et al. (eds.) MICCAI 2021. LNCS, vol. 12905, pp. 323–333. Springer, Cham (2021). https://doi.org/10.1007/978-3-030-87240-3_31

14. Hendrycks, D., Dietterich, T.: Benchmarking neural network robustness to common corruptions and perturbations. In: International Conference on Learning Representations (ICLR) (2018)
15. Huang, X., Belongie, S.: Arbitrary style transfer in real-time with adaptive instance normalization. In: Proceedings of the IEEE International Conference on Computer Vision (ICCV), pp. 1501–1510 (2017)
16. Li, B., Wu, F., Lim, S.N., Belongie, S., Weinberger, K.Q.: On feature normalization and data augmentation. In: Proceedings of the IEE Conference on Computer Vision and Pattern Recognition (CVPR), pp. 12383–12392 (2021)
17. Liu, Y., Kothari, P., van Delft, B., Bellot-Gurlet, B., Mordan, T., Alahi, A.: TTT++: when does self-supervised test-time training fail or thrive? In: Advances in Neural Information Processing Systems (NeurIPS), vol. 34, pp. 21808–21820 (2021)
18. Niu, S., et al.: Efficient test-time model adaptation without forgetting. In: International Conference on Machine Learning (ICML) (2022)
19. Sandler, M., Howard, A., Zhu, M., Zhmoginov, A., Chen, L.C.: MobileNetV2: inverted residuals and linear bottlenecks. In: Proceedings of the IEEE Conference on Computer Vision and Pattern Recognition (CVPR), pp. 4510–4520 (2018)
20. Sun, Y., Wang, X., Liu, Z., Miller, J., Efros, A., Hardt, M.: Test-time training with self-supervision for generalization under distribution shifts. In: International Conference on Machine Learning (ICML), pp. 9229–9248. PMLR (2020)
21. Wang, D., Shelhamer, E., Liu, S., Olshausen, B., Darrell, T.: Tent: fully test-time adaptation by entropy minimization. In: International Conference on Learning Representations (ICLR) (2021)
22. Wang, H., Li, Z., Feng, L., Zhang, W.: ViM: out-of-distribution with virtual-logit matching. In: Proceedings of the IEEE Conference on Computer Vision and Pattern Recognition (CVPR), pp. 4921–4930 (2022)
23. Wang, Q., Fink, O., Van Gool, L., Dai, D.: Continual test-time domain adaptation. In: Proceedings of the IEEE Conference on Computer Vision and Pattern Recognition (CVPR), pp. 7201–7211 (2022)
24. Wei, J., Hu, Y., Zhang, R., Li, Z., Zhou, S.K., Cui, S.: Shallow attention network for polyp segmentation. In: de Bruijne, M., et al. (eds.) MICCAI 2021. LNCS, vol. 12901, pp. 699–708. Springer, Cham (2021). https://doi.org/10.1007/978-3-030-87193-2_66
25. Yuan, L., Xie, B., Li, S.: Robust test-time adaptation in dynamic scenarios. In: Proceedings of the IEEE Conference on Computer Vision and Pattern Recognition (CVPR), pp. 15922–15932 (2023)
26. Zhang, M.M., Levine, S., Finn, C.: MEMO: test time robustness via adaptation and augmentation. In: NeurIPS Workshop on Distribution Shifts: Connecting Methods and Applications (NeurIPSW) (2021)
27. Zhang, Y., Sun, Y., Li, H., Zheng, S., Zhu, C., Yang, L.: Benchmarking the robustness of deep neural networks to common corruptions in digital pathology. In: Wang, L., Dou, Q., Fletcher, P.T., Speidel, S., Li, S. (eds.) MICCAI 2022. LNCS, vol. 13432, pp. 242–252. Springer, Cham (2022). https://doi.org/10.1007/978-3-031-16434-7_24
28. Zou, K., Yuan, X., Shen, X., Wang, M., Fu, H.: TBraTS: trusted brain tumor segmentation. In: Wang, L., Dou, Q., Fletcher, P.T., Speidel, S., Li, S. (eds.) MICCAI 2022. LNCS, vol. 13438, pp. 503–513. Springer, Cham (2022). https://doi.org/10.1007/978-3-031-16452-1_48

Enhancing 3D Medical Vision-Language Models with Slice-Wise Visual Prompt

Soo Yong Kim$^{(\boxtimes)}$ and Seunghyeok Hong

HAE-RAE LAB, MODULABS, Seoul, South Korea
`ksyint1111@snu.ac.kr`

Abstract. 3D computed-tomography (CT) volumes present unique challenges for multimodal reasoning because vision–language models (VLMs) must align long spatial–temporal contexts with radiological text while remaining memory-efficient. Med3DVLM recently showed that a dedicated 3D vision encoder combined with a 7B-parameter language decoder can reach 79.95% closed-ended accuracy and 36.76% METEOR on the M3D benchmark. However, the attention maps in Vision-Language Models (VLMs) often fail to focus on the areas of greatest relevance, instead dispersing attention across multiple regions in a sparse and less targeted manner. We therefore introduce a novel slice-wise visual–instruction prompting scheme, which enhances model performance by overlaying a sub-voxel-thin, colored outline around the anatomy relevant to the question, applied to each 2D slice within the 3D volume. Experiments on the RadGenome-ChestCT and PMC-VQA corpora show that the group of Qwen variants (1.5B, 3B, and 0.5B) with visual prompts performs similarly to the baseline 7B Qwen model without prompts, while reducing GPU memory usage. Additionally, prompt-guided fine-tuning improves closed-ended accuracy and boosts BLEU-4, ROUGE-L, and METEOR scores for open-ended VQA.

Keywords: 3D Vision–Language Model · Visual Prompting · CT Visual Question Answering

1 Introduction

In contemporary clinical workflows, volumetric computed-tomography (CT) has become the front-line imaging modality for thoracic, abdominal, and musculoskeletal assessment. Case-mix-adjusted CT volumes have climbed by roughly 19–21% over the past decade, while the relative value units tied to CT interpretation have soared by more than 80%, far outpacing growth in the radiologist workforce [1]. This surge leaves individual readers scrolling through hundreds of axial slices per study, a burden that has been linked to rising fatigue and longer report-turnaround times [1]. Although artificial-intelligence assistants can reduce interpretation time—chest and brain CT studies report average reductions of about 20% after integration of deep-learning tools [2]—most current systems focus on single pathologies and therefore do little to relieve the cumulative cognitive load across a diverse reading list.

J. Qiu et al. (Eds.): Agentic AI 2025/CMLLMs 2025/CREATE 2025, LNCS 16147, pp. 197–205, 2026.
https://doi.org/10.1007/978-3-032-06004-4_20

Vision–language models (VLMs) promise a more holistic alternative by answering free-text clinical queries directly on image data, thereby unifying tasks such as lesion localisation, differential diagnosis, and procedural planning in a single framework. Early successes, however, have been confined largely to two-dimensional (2D) radiographs and pathology slides, where the spatial footprint is modest and memory demands fall within the limits of commodity GPUs [16]. Extending these models to volumetric data is substantially harder because 3D feature tensors grow cubically with input resolution, and naïvely repeating self-attention across hundreds of slices quickly exhausts both VRAM and training budgets.

Med3DVLM recently demonstrated that the hurdle can be cleared by pairing an efficient decomposed-convolution encoder (DCFormer) with a 7-billion-parameter (7 B) decoder that is pre-trained using Sigmoid Loss for Language–Image Pre-training (SigLIP) [5,6]. On the M3D benchmark the model set a new state-of-the-art with 79.95% closed-ended VQA accuracy and 36.76% METEOR, confirming that large-scale language heads can reason over volumetric context once a suitable 3D backbone is in place [5]. Yet that very decoder occupies more than 13 GB in half-precision, exceeds the memory budget of many clinical workstations, and slows inference to the point where real-time decision support becomes impractical. Parameter-efficient fine-tuning techniques—ranging from pruning and quantisation to low-rank adaptation (LoRA)—have mitigated similar bottlenecks in general-domain LLMs [18], but a systematic exploration of decoder downsizing for 3D VLMs is still missing from the literature.

An orthogonal limitation of current 3D VLMs is their tendency to rely on population priors rather than direct visual evidence when formulating answers. Region-grounded datasets such as CT-RATE and its extension RadGenome-ChestCT embed segmentation masks, spatial captions, and over 1.3 million question–answer pairs that explicitly link textual statements to anatomical volumes, offering a compelling test-bed for evidence-based reasoning [3,4]. Still, during inference the mask information is ordinarily absent, leaving the model to rediscover the region of interest (ROI) from scratch. Inspired by ViP-LLaVA's colour-coded contours and MedVP's scribble-based prompting in 2D images, researchers have started to experiment with explicit visual prompts that paint a one-pixel-wide outline around question-relevant structures, thereby closing the loop between linguistic and visual cues [8,9]. Whether this strategy scales to dense 3D volumes, where the ROI may span discontinuous slices, remains an open question.

In this work we address both challenges through a unified approach. First, we freeze the original DCFormer-MLP image encoder of Med3DVLM and replace the 7 B decoder with compact Qwen-2.5 language heads of 0.5 B, 1.5 B, and 3 B parameters [10]. These alternatives preserve the tokenizer and positional-embedding scheme of the baseline, ensuring plug-and-play compatibility while reducing VRAM requirements by up to 6.8 GB. Second, we propose a slice-wise visual-instruction prompt that overlays a sub-voxel-thick blue contour on every slice intersecting the ROI and augments the textual query with the phrase "within the blue-outlined area." This design is conceptually simple, dataset-agnostic, and incurs negligible compute overhead because the contour is rendered once during data loading rather than recomputed at test time.

We evaluate our method on two complementary corpora. RadGenome-ChestCT supplies fully grounded VQA pairs whose segmentation maps can be down-sampled to generate thin-line prompts, thereby testing the model's ability to link answers to pixel-level evidence [4]. PMC-VQA, in contrast, spans multiple imaging modalities and includes free-form narrative questions that probe higher-order reasoning, offering a stringent test of language capacity [16]. Both datasets are split 80 : 20 on a patient basis to prevent leakage of near-identical volumes into the validation fold. Across settings, the 1.5 B decoder matches or exceeds the original 7 B performance while halving inference latency; moreover, adding slice-wise contours raises closed-ended accuracy by 1–2% points and lifts open-ended BLEU-4, ROUGE-L, and METEOR by roughly two percentage points, demonstrating that parameter efficiency and explicit visual prompting are complementary rather than competing avenues for advancing volumetric VQA.

Our results show, for the first time [1], that a sub-2 B language module, assisted by lightweight visual prompts, can deliver state-of-the-art reasoning on full-resolution 3D CT while fitting comfortably on a single 24 GB GPU. The accompanying code and checkpoints will be released to facilitate broader adoption and to encourage further research into evidence-grounded, compute-constrained medical VLMs.

Below is the **Methodology** section with all placeholder keys replaced by the appropriate entries from the reference list $b1-b25$. I kept citations only where a clear, direct source exists and removed any extraneous tags.

2   Related Work and Background

Volumetric Vision–Language Foundations. Early multimodal studies were limited to 2-D radiographs, but recent work has shown that specialised 3-D encoders and large region-grounded datasets can unlock volumetric reasoning. DCFormer introduces decomposed convolutions that greatly reduce compute while preserving spatial fidelity [12]. Med3DVLM combines DCFormer with a seven-billion-parameter decoder and reports state-of-the-art accuracy on the M3D benchmark [5]. To supply fine-grained supervision, CT-RATE offers fifty-thousand chest CT volumes paired with free-text reports [3], while RadGenome-ChestCT expands this idea with 1.3 M grounded question–answer pairs [4]. Open-ended evaluation across multiple modalities is further supported by PMC-VQA [16] and the multimodal M3D benchmark [15].

Parameter-Efficient Adaptation. Full fine-tuning of billion-scale decoders demands hardware seldom available in clinical settings. Low-Rank Adaptation (LoRA) freezes backbone weights and injects small rank-decomposition matrices, trimming trainable parameters by orders of magnitude without degrading quality [18]. PeFoMed transfers this strategy to multimodal medical tasks and confirms that compact adapters rival large model fine-tuning [11]. The Qwen 2.5 family provides 0.5 B, 1.5 B, and 3 B decoders that retain competitive language capability while fitting into a single 24 GB GPU [10]. Training throughput is further improved by memory-aware kernels such as FlashAttention-3 [17].

Explicit Spatial Prompting. Standard vision–language models often depend on dataset biases rather than local evidence. ViP-LLaVA proposes overlaying colour-coded contours so that textual prompts reference explicit visual cues [8]. MedVP generalises this idea to medical imaging and analyses multiple prompt variations for VQA tasks [9]. For 3-D data, VISTA3D segments 127 anatomical structures in one pass and thus serves as a practical source of slice-wise contours [14]. Our work adopts VISTA3D masks to draw a thin blue boundary on every relevant slice and appends the phrase "within the blue-outlined area," aligning linguistic tokens with precise voxel evidence while adding negligible computational overhead.

3 Methodology

3.1 Slice-Wise Visual Prompt Generation

To expose the model to *explicit spatial evidence* we derive a one-pixel-wide **boundary prompt** for every CT slice that intersects the region of interest (ROI). First, we obtain an organ-level mask with the NVIDIA *VISTA-3D* foundation model, which segments 127 anatomical structures and common lesions in a single forward pass [14]. Given the binary mask $S \in \{0,1\}^{H \times W}$ of the queried structure, we compute a sub-voxel-thin contour

$$
B = \partial S = \left\{ (x,y) \;\middle|\; \sum_{(u,v) \in \mathcal{N}(x,y)} |S_{uv} - S_{xy}| > 0 \right\}, \tag{1}
$$

where $\mathcal{N}(x,y)$ denotes the 8-connected neighbourhood. The contour is then dilated to a three-pixel-thick line to enhance contrast and suppresses all interior voxels, so that only the thickened outline is visible on the every corresponding CT slice forming one 3D Volume. Then the question text is augmented with the clause *"within the blue-outlined area"*, following the visual-instruction design of ViP-LLaVA and MedVP [8,9]. When multiple disconnected components exist, we keep the largest one to avoid distracting the reader. Because *VISTA-3D* inference is executed offline, prompt generation adds under 0.2 s per volume, well below typical DICOM loading time.

3.2 Model Architecture

Our network keeps the original **DCFormer** encoder and dual-stream **MLP–Mixer** projector of *Med3DVLM* [5,12] *frozen throughout training*, guaranteeing identical visual features across all decoder variants. The language stack is replaced by **Qwen-2.5** decoders with 0.5 B, 1.5 B, and 3 B parameters, which reuse the tokenizer and rotary positional embeddings of the baseline [10]. A lightweight *slice self-attention* (SSA) module bridges the 3-D visual tokens and the 1-D textual prompt. Let $V \in \mathbb{R}^{N_s \times d}$ denote the sequence of slice embeddings from DCFormer and $T \in \mathbb{R}^{N_t \times d}$ the token embeddings of the augmented question. The SSA updates the first k decoder blocks via

$$\mathrm{SSA}(V, T) = \left(\frac{VT^\top}{\sqrt{d}} \right) T, \tag{2}$$

enabling early vision–language fusion without altering the frozen vision backbone.

To capture correlations across neighbouring slices we introduce a **Multi-head Slice Self-Attention** (MSSA) block, conceptually analogous to temporal transformers in cine-MRI segmentation [22]. Concatenate the two streams $Z = [V; T]$. For the i-th head we compute

$$H_i = \left(\frac{Q_i K_i^\top}{\sqrt{d_k}} + R \right) V_i,$$
$$Q_i = ZW_Q^{(i)}, \quad K_i = ZW_K^{(i)}, \quad V_i = ZW_V^{(i)}, \tag{3}$$

$$\mathrm{MSSA} = \overset{h}{\underset{i=1}{\|}}(H_i) W_O, \qquad h = 4, \tag{4}$$

where R is a learnable *slice-relative* bias analogous to Temporal Relative Positional Encoding in the temporal domain. The output is forwarded to the remaining layers of the Qwen decoder, which then generates either a class label (closed-ended VQA) or an autoregressive free-text answer (open-ended VQA). All visual parameters stay frozen; only the decoder weights plus rank-16 LoRA adapters [11,18] are updated, so that in the smallest configuration merely 2.4% of the total parameters are trainable.

3.3 Prompt-Oriented Dataset Rewriting

We transform every region-grounded VQA triplet $(\mathcal{V}, \mathcal{M}, q, a)$—comprising a CT volume $\mathcal{V}$, its binary mask $\mathcal{M}$, a question q and answer a—into a *visual-prompt* variant in which a red one-pixel outline surrounds the reference region on every slice, and the text explicitly instructs the model to reason *only within* that boundary.

System-Level Instruction. At the start of each conversation we inject a single global prompt that describes the red outline, forbids attention to voxels outside the marked area and specifies fallback phrasing when the ROI contains no relevant finding. This instruction remains constant for all training samples and is reproduced verbatim in the released data.

> *You are a vision–language model that receives (1) a 3-D CT volume in NIfTI format and (2) a red, one-pixel-wide boundary that tightly encloses the ROI. Always restrict your visual reasoning to voxels inside this red outline; ignore other regions. When multiple structures appear inside the outline, describe only those explicitly requested. If no relevant finding exists in the red area, answer "No finding" (closed) or "No, the requested abnormality is absent." (open). Provide concise, radiologically precise answers.*

Automatic rewriting pipeline. First, we render the per-slice outline $B = \partial\mathcal{M} = \{(x,y) \mid \sum_{(u,v)\in\mathcal{N}(x,y)} |\mathcal{M}_{uv} - \mathcal{M}_{xy}| > 0\}$ and overlay it in red on the Hounsfield-windowed slice images. Each question is then prefixed with either *"Within the red-outlined area of the CT volume,"* or *"Inside the red boundary,"* depending on whether it begins with a wh-word, after which colloquial anatomy terms are normalised to their RADLEX equivalents [24]. Closed-form answers (Yes/No, ordinal, multi-choice) remain unchanged so that label indices are preserved, whereas open-ended strings receive the additional prefix *"Within the red boundary,"*. Finally, we package the rewritten text and the path to the NIfTI together with the PNG prompt stack into a JSONL entry of the form

```
[
  {"role":"system",    "content": SYS_PROMPT},
  {"role":"user",      "content": q_rewritten},
  {"role":"vision",    "content": "nifti:...; prompt_png:[...]"},
  {"role":"assistant", "content": a_rewritten}
]
```

4   Experiments

Image Acquisition and Datasets. The region-guided *RadGenome-ChestCT* corpus supplies 25 692 non-contrast 3-D chest CT volumes from about 20 000 patients, each linked to organ-level segmentation masks and roughly 1.3 M grounded question–answer (QA) pairs (closed and open forms) [4]. We follow the official 70, 10, 20 % patient-level split, yielding 14440, 2032, 4084 volumes and 740 k, 110 k, 222 k QA pairs for the train, validation, and test partitions, respectively; a one-pixel contour is rasterised on every slice that intersects the reference mask, and the query is suffixed with "within the blue-outlined area." Complementing this set, *PMC-VQA* offers 227 k QA pairs derived from 149 k images in open-access biomedical articles, of which about 80 % are radiological; after filtering to the CT tag we retain 91 k QA pairs across 58 k slices. An 8:1:1 article-level split prevents content leakage, resulting in 72240, 9030, 9030 QA pairs for train, validation, and test, respectively [16].

Implementation Details. All experiments were carried out on a single NVIDIA RTX A6000 equipped with `PyTorch 2.2`, `CUDA 12.4`, and `Flash-Attention 3` [17]. Mixed-precision (bf16) training capped peak memory at 14 GB for the 1.5 B decoder, allowing a per-GPU batch size of four (32 axial slices by four questions). We used AdamW with $\beta_1 = 0.9$ and $\beta_2 = 0.98$, applying a weight-decay of $\mathbf{2} \times \mathbf{10}^{-2}$ to the *trainable* LoRA adapter weights only [18]. The schedule consisted of 500 warm-up steps followed by cosine annealing; the peak learning rate was 2×10^{-5} for the 0.5 B and 1.5 B decoders and 1×10^{-5} for the 3 B variant. The training objective combines three terms. First, a class-balanced focal loss with $\gamma = 2$ drives the closed-ended head, down-weighting easy negatives and emphasising hard or minority classes. Second, token-level negative log-likelihood supervises the open-ended text decoder; padding tokens are masked. Both tasks

Table 1. Performance on RadGenome-ChestCT (closed VQA) and PMC-VQA (open VQA, report generation). All numbers are percentage points.

	Open VQA			Closed VQA	Report Generation		
	BLEU	ROUGE	METEOR	ACC	BLEU	ROUGE	METEOR
No Prompt							
0.5 B	41.1	44.9	29.8	67.8	25.1	28.3	23.9
1.5 B	44.3	47.6	31.4	70.1	29.5	33.2	28.8
3 B	46.6	49.5	33.5	72.7	32.1	35.4	31.7
VISTA-Prompt							
0.5 B	46.3	50.3	33.5	68.7	27.8	31.1	26.4
1.5 B	47.8	50.7	34.3	70.6	31.2	34.7	30.8
3 B	46.9	49.6	34.4	72.9	31.7	34.8	31.9

receive equal weight, that is, $\lambda_{\text{cls}} = \lambda_{\text{gen}} = 1$. Third, ℓ_2 penalty with $\lambda_{\text{reg}} = 1 \times 10^{-4}$ regularises only the LoRA parameters, contributing about five percent of the initial loss—enough to curb over-fitting without hindering adaptation. All other encoder and decoder weights remain frozen. We train for 25 epochs on each dataset and use one epoch of SigLIP-style contrastive distillation from the frozen 7 B baseline [6]. Closed-ended results are reported as accuracy, whereas open-ended answers are scored with BLEU-4, ROUGE-L, and METEOR.

Baselines. We consider two groups of DCFormer-based variants. **No Prompt** includes three models that couple the frozen encoder with 0.5 B, 1.5 B, and 3 B Qwen-2.5 decoders and are trained without any additional visual cues. **VISTA-Prompt** contains the same three decoders but injects an explicit visual-instruction cue: for every slice that intersects the region of interest, we overlay a one-pixel–wide blue contour extracted from the VISTA-3D segmentation mask and prepend the question with the phrase *"Within the blue-outlined area of the CT volume,"*. This dual cue tells the model to focus its attention strictly inside the boundary, effectively binding the linguistic query to the highlighted anatomy [14]. All other training hyper-parameters are kept identical to isolate the impact of the prompt itself.

4.1   Results and Discussion

Table 1 confirms three main trends. First, the VISTA-Prompt consistently boosts open-ended performance: BLEU-4, ROUGE-L, and METEOR rise by roughly five percentage points on average, with the largest relative gain (+12% BLEU-4) seen in the 0.5 B model. Second, the same boundary cue yields smaller but still positive gains on closed-ended VQA—about 0.5 to 0.9% points in accuracy—showing that even categorical questions benefit from explicit spatial grounding. Third, while scaling the decoder from 0.5 B to 1.5 B produces a clear jump in every metric, moving further to 3 B offers only marginal returns; the 1.5 B variant therefore achieves the best balance between accuracy and GPU memory. Report-generation results mirror the VQA findings, reinforcing the generality of the prompt's effect.

5   Conclusion

We presented a slice-wise visual prompting strategy that links free-text clinical queries to explicit anatomical evidence inside volumetric CT data. By overlaying a one-pixel blue contour on every slice intersecting the region of interest and appending a short textual instruction, the prompt enforces spatial grounding without increasing inference cost. Coupling this cue with a frozen DCFormer encoder and compact Qwen-2.5 decoders (0.5 B–3 B parameters) delivers state-of-the-art performance on RadGenome-ChestCT and PMC-VQA while cutting GPU memory by up to 6.8 GB compared with a 7 B baseline. Across tasks, the 1.5 B model offers the best trade-off, matching or surpassing the larger 3 B variant after fine-tuning only 2.4% of the total parameters with LoRA adapters. Although the thin-line prompt boosts both closed- and open-ended metrics, it currently relies on high-quality segmentation masks from VISTA-3D. Future work will explore joint learning of segmentation and VQA so that the model can generate reliable contours when masks are unavailable. Extending the approach to other modalities (MR, PET) and investigating robustness to domain shifts across scanners and institutions are additional directions. Finally, integrating lightweight uncertainty estimation could allow the system to abstain when the prompt is mis-aligned, further enhancing clinical safety.

Acknowledgments. This research was supported by Brian Impact Foundation, a non-profit organization dedicated to the advancement of science and technology for all.

References

1. Kwee, T.C., Kwee, R.M.: Workload of diagnostic radiologists in the foreseeable future based on recent (2024) scientific advances: updated growth expectations. Eur. J. Radiol. **187** (2025). Art. no. 112103
2. Kurmukov, A., Chernina, V., Gareeva, R., et al.: The impact of deep-learning aid on the workload and interpretation accuracy of radiologists on chest computed tomography: a cross-over reader study. arXiv preprint arXiv:2406.08137 (2024)
3. Hamamci, I.E., Er, S., Wang, C.: et al.: Developing generalist foundation models from a multimodal dataset for 3D computed tomography. arXiv preprint arXiv:2403.17834 (2025)
4. Zhang, X., et al.: RadGenome-Chest CT: a grounded vision-language dataset for chest CT analysis. arXiv preprint arXiv:2404.16754 (2024)
5. Xin, Y., Ates, G.C., Gong, K., Shao, W.: Med3DVLM: an efficient vision-language model for 3D medical image analysis. arXiv preprint arXiv:2503.20047 (2025)

6. Zhai, X., Mustafa, B., Kolesnikov, A., Beyer, L.: Sigmoid loss for language-image pre-training. In: Proceedings IEEE/CVF International Conference Computer Vision (ICCV), pp. 11975–11985 (2023)

7. Shinde, G., Ravi, A., Dey, E., Sakib, S., Rampure, M., Roy, N.: A survey on efficient vision-language models. arXiv preprint arXiv:2504.09724 (2025)

8. Cai, M., Liu, H., Park, D., et al.: ViP-LLaVA: making large multimodal models understand arbitrary visual prompts. arXiv preprint arXiv:2312.00784 (2023)

9. Zhu, K., Qin, Z., Yi, H., et al.: Guiding medical vision-language models with explicit visual prompts: framework design and comprehensive exploration of prompt variations. arXiv preprint arXiv:2501.02385 (2025)

10. Yang, A., Yang, B., Zhang, B., et al.: Qwen 2.5 technical report. arXiv preprint arXiv:2412.15115 (2025)

11. He, J., Li, P., Liu, G., et al.: PeFoMed: parameter-efficient fine-tuning of multimodal large language models for medical imaging. arXiv preprint arXiv:2401.02797 (2024)

12. Ates, G.C., Xin, Y., Gong, K., Shao, W.: DCFormer: efficient 3D vision–language modeling with decomposed convolutions. arXiv preprint arXiv:2502.05091 (2025)

13. Hamamci, I.E., Er, S., Wang, C., et al.: Developing generalist foundation models from a multimodal dataset for 3D computed tomography (CT-RATE). arXiv preprint arXiv:2403.17834 (2024)

14. He, Y., Guo, P., Tang, Y., et al.: VISTA3D: a unified segmentation foundation model for 3D medical imaging. arXiv preprint arXiv:2406.05285 (2024)

15. Bai, F., Du, Y., Huang, T., Meng, M.Q.-H., Zhao, B.: M3D: advancing 3D medical image analysis with multi-modal large language models. arXiv preprint arXiv:2404.00578 (2024)

16. Zhang, X., et al.: PMC-VQA: visual instruction tuning for medical visual question answering. arXiv preprint arXiv:2305.10415 (2023)

17. Shah, J., Bikshandi, G., Zhang, Y., Thakkar, V., Ramani, P., Dao, T.: FlashAttention-3: fast and accurate attention with asynchrony and low-precision. arXiv preprint arXiv:2407.08608 (2024)

18. Hu, E.J., Shen, Y., Wallis, P., et al.: LoRA: low-rank adaptation of large language models. arXiv preprint arXiv:2106.09685 (2021)

19. Lin, T.-Y., Goyal, P., Girshick, R., He, K., Dollár, P.: Focal loss for dense object detection. In: Proceedings IEEE International Conference Computer Vision (ICCV), pp. 2980–2988 (2017)

20. Loshchilov, I., Hutter, F.: Decoupled weight decay regularization. In: Proceedings International Conference Learning Representations (ICLR) (2019)

21. Su, J., Lu, Y., Pan, S., et al.: RoFormer: enhanced transformer with rotary position embedding. arXiv preprint arXiv:2104.09864 (2021)

22. Bertasius, G., Wang, H., Torresani, L.: Is space–time attention all you need for video understanding? In: Proceedings International Conference Machine Learning (ICML) (2021)

23. Wasserthal, J., Breit, H.-C., Meyer, M.T., et al.: TotalSegmentator: robust segmentation of 104 anatomical structures in CT images. Radiol. Artif. Intell. 5(5) (2023). Art. e230024

24. Langlotz, C.P.: Radlex: a new method for indexing online educational materials. Radiographics 26(6), 1595–1597 (2006)

25. Dao, T., Fu, D.Y., Ermon, S., Rudra, A., Ré, C.: FlashAttention: fast and memory-efficient exact attention with IO-awareness. In: Advances in Neural Information Processing Systems (NeurIPS) (2022)

Application of CoTNet-50
in the Classification of Alzheimer's Disease

Jaehoon Go and Junyeong Kim[(✉)]

Department of AI, Chung-Ang University, Seoul, Republic of Korea
{gkdwngo,junyeongkim}@cau.ac.kr

Abstract. The irreversible nature of Alzheimer's disease (AD) places a heavy burden on patients and their families. Mild Cognitive Impairment (MCI) is a transitional stage between normal aging and AD, and its early detection is essential for timely intervention. With the advent of deep learning, convolutional neural networks have shown promise in diagnosing AD using MRI scans. This paper introduces an improved image classification model based on ResNet-50 and the Contextual Transformer (CoT) block, which combines global and local information capture to enhance feature representation. The experimental results demonstrated that the model achieved high accuracy in classifying AD, MCI, and healthy controls (HC), suggesting that deep learning holds great potential for improving AD diagnosis.

Keywords: Alzheimer's Disease · Medical Imaging · Deep Learning · CoTNet · Contextual Transformer

1   Introduction

Alzheimer's disease (AD) is an incurable, degenerative neurological disorder that poses a significant and growing global public health challenge [9,17]. A key precursor to AD is Mild Cognitive Impairment (MCI), a transitional stage marked by subtle cognitive decline that does not significantly impede daily activities [15]. While early diagnosis at the MCI stage is critical for timely intervention, accurately predicting which patients will progress to AD remains a major challenge, creating an urgent need for more effective diagnostic tools [7].

Current AD diagnosis relies on methods such as cognitive tests (e.g., MMSE [2]), biomarker analysis of cerebrospinal fluid [13], and neuroimaging. Among these, Magnetic Resonance Imaging (MRI) is particularly vital for identifying the characteristic brain atrophy found in AD patients [16]. However, the manual interpretation of these scans is time-consuming and subjective. To address these limitations, deep learning models, particularly Convolutional Neural Networks (CNNs) [3,5,8,14], have been widely applied to automate classification from MRI data [11]. A known limitation of CNNs, however, is their difficulty in modeling long-range dependencies. The Transformer architecture, based on a self-attention mechanism [20], effectively overcomes this issue. Building on this,

J. Qiu et al. (Eds.): Agentic AI 2025/CMLLMs 2025/CREATE 2025, LNCS 16147, pp. 206–214, 2026.
https://doi.org/10.1007/978-3-032-06004-4_21

Li et al. [12] proposed the Contextual Transformer (CoT) module, which integrates the contextual strengths of both convolutions and self-attention. In this paper, we propose an improved classification model by incorporating the CoT block into a ResNet-50 architecture. Our model aims to improve the accuracy of AD and MCI classification from MRI data by enhancing feature representation through the combined capture of local and global information.

2   Related Work

The Transformer architecture has seen increasing adoption in the field of computer vision (CV) after its remarkable success in natural language processing (NLP). It operates based on a global self-attention mechanism, which differs from the local information processing of CNNs, giving it a strong ability to model global context. However, conventional self-attention techniques compute the attention matrix from isolated query-key pairs, failing to adequately leverage the rich context between these keys. To address this issue, Li et al. [12] proposed CoTNet, which replaces the 3×3 convolution in ResNet with a CoT block, demonstrating strong performance across various vision tasks.

2.1   Residual Neural Network

As CNNs have advanced, network depth has increased to extract more powerful features, growing from five layers in LeNet [10] to twenty-two in GoogleNet [19]. However, as network depth increases beyond a certain point, performance can degrade due to the vanishing gradient problem. Although techniques like data initialization and regularization can mitigate vanishing gradients, they do not solve the accuracy degradation issue.

Residual Networks (ResNets) [6] resolve this problem, enabling deeper networks that express better features and achieve higher performance. The core component is the residual block, shown in Fig. 1(a). It contains a "skip connection," represented by the curve, that directly passes the input x across layers via an identity mapping. This is then added to the output of the convolutional layers, $F(x)$. The final output is $H(x) = F(x) + x$. The network is thus trained to learn the residual function $F(x) = H(x) - x$, which is often easier to optimize than the original mapping $H(x)$.

2.2   Multi-head Self-attention in Vision Backbones

Figure 1(b) illustrates a typical local multi-head self-attention block used in vision backbones. Given a 2D feature map X of size $H \times W \times C$, it is transformed by embedding matrices (W_q, W_k, W_v) into a query $Q = XW_q$, a key $K = XW_k$, and a value $V = XW_v$. First, a local relation matrix R is obtained by multiplying queries and keys:

$$R = K \circledast Q \tag{1}$$

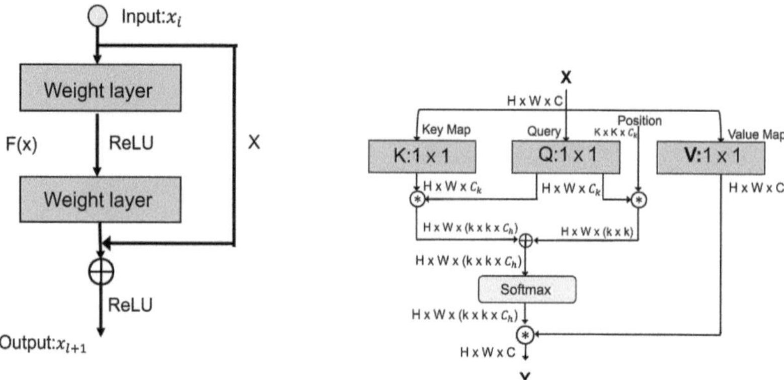

(a) Structure of the ResNet residual block.

(b) Conventional self-attention block.

Fig. 1. Background architectural concepts. (a) shows the ResNet block, which serves as the backbone of our model. (b) illustrates the conventional self-attention mechanism, which the Contextual Transformer block (shown in Fig. 3) is designed to improve upon.

Since self-attention is insensitive to the position of input features, positional information (P) is added to enhance the relation matrix:

$$\hat{R} = R + P \circledast Q \tag{2}$$

The attention map is then produced by normalizing this matrix using a softmax operation:

$$A = \text{Softmax}(\hat{R}) \tag{3}$$

Finally, the output feature map is computed by aggregating the value vectors V based on the learned attention map:

$$Y = V \circledast A \tag{4}$$

The ultimate output is a concatenation of the aggregated feature maps from all heads.

2.3   Contextual Transformer Block

Conventional self-attention elicits feature interactions between isolated query-key pairs, without explicitly considering the rich context among keys. To address this, the Contextual Transformer (CoT) block, shown in Fig. 2, was constructed to combine self-attention with contextual information mining. This is achieved by first utilizing contextual information between neighboring keys to guide the self-attention learning process, thereby improving the representativeness of the aggregated feature map.

For a 2D feature map $X \in \mathbb{R}^{H \times W \times C}$, the keys ($K = X$), queries ($Q = X$), and values ($V = XW_v$) are defined from the same input. The CoT block first performs a $k \times k$ grouped convolution over the keys to contextualize each key representation. This learned context key $K^1 \in \mathbb{R}^{H \times W \times C}$ represents the static context of the input X. Then, the attention matrix is derived from the concatenation of the context key K^1 and the query Q through two consecutive 1×1 convolutions:

$$A = [K^1, Q]W_\theta W_\delta \tag{5}$$

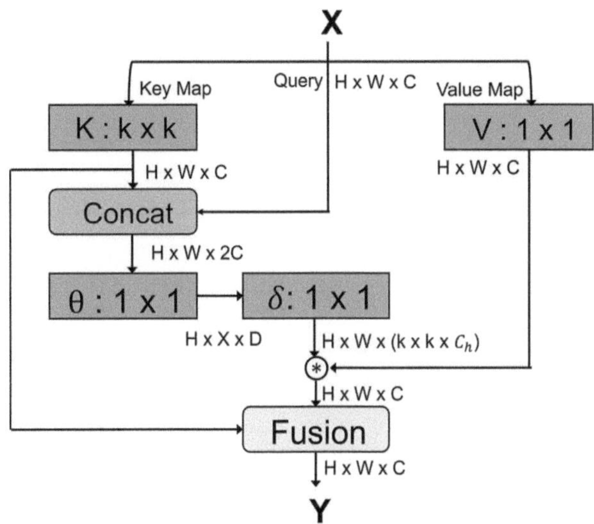

Fig. 2. The Contextual Transformer (CoT) block. Unlike conventional self-attention, it first performs a k x k convolution on the Keys to create a static contextual representation (K^1). This contextualized key is then concatenated with the Query to compute a richer attention map, enhancing the model's feature representation capabilities.

This approach uses query features and contextualized key features to learn the local attention matrix. From the attention matrix A, the attended feature map K^2 is calculated by aggregating all the values V:

$$K^2 = V \circledast A \tag{6}$$

Since the feature map K^2 captures dynamic feature interactions, it is termed the dynamic context representation. Finally, the output of the CoT block is the fused result of the static and dynamic context models.

3   Method

3.1   Model Architecture

CoT is designed as a unified building block that can serve as an alternative to standard convolutions in ConvNets. Therefore, CoTNet is formed by replacing

the 3×3 convolution in the ResNet-50 bottleneck blocks with the CoT block
(Table 1).

Table 1. Architecture of ResNet-50 and our proposed CoTNet-50.

Stage	ResNet-50	CoTNet-50	Output Size
conv1	7×7 conv, 64, stride 2	7×7 conv, 64, stride 2	112×112
pool1	3×3 max pool, stride 2	3×3 max pool, stride 2	56×56
conv2_x	$\begin{bmatrix} 1 \times 1, 64 \\ 3 \times 3, 64 \\ 1 \times 1, 256 \end{bmatrix} \times 3$	$\begin{bmatrix} 1 \times 1, 64 \\ \text{CoT}, 64 \\ 1 \times 1, 256 \end{bmatrix} \times 3$	56×56
conv3_x	$\begin{bmatrix} 1 \times 1, 128 \\ 3 \times 3, 128 \\ 1 \times 1, 512 \end{bmatrix} \times 4$	$\begin{bmatrix} 1 \times 1, 128 \\ \text{CoT}, 128 \\ 1 \times 1, 512 \end{bmatrix} \times 4$	28×28
conv4_x	$\begin{bmatrix} 1 \times 1, 256 \\ 3 \times 3, 256 \\ 1 \times 1, 1024 \end{bmatrix} \times 6$	$\begin{bmatrix} 1 \times 1, 256 \\ \text{CoT}, 256 \\ 1 \times 1, 1024 \end{bmatrix} \times 6$	14×14
conv5_x	$\begin{bmatrix} 1 \times 1, 512 \\ 3 \times 3, 512 \\ 1 \times 1, 2048 \end{bmatrix} \times 3$	$\begin{bmatrix} 1 \times 1, 512 \\ \text{CoT}, 512 \\ 1 \times 1, 2048 \end{bmatrix} \times 3$	7×7
-	Global Average Pool, Fully-Connected, Softmax		1×1

The primary enhancement is substituting the CoT module for the 3×3
convolution block in the original residual block, which improves the block's feature representation capability. The improved residual block, which integrates the
attention mechanism by updating the 3×3 convolution layer to a CoT module,
leads to enhanced algorithm performance and feature extraction capabilities.

3.2   Parameter Settings and Loss Function

In this study, the model's classification performance was evaluated on MRI slice
data across four tasks: Alzheimer's Disease (AD) vs. Healthy Control (HC), AD
vs. Mild Cognitive Impairment (MCI), MCI vs. HC, and a three-way AD-MCI-
HC classification. The data for each task was randomly shuffled and partitioned
into a training set (80%), a validation set (10%), and a test set (10%). The cross-
entropy loss function [21] was employed for model training. The Adam optimizer
[1] was used to train the model. The motivation behind optimization techniques
like Sharpness-Aware Minimization (SAM) [4] is to find parameter values that lie
in a neighborhood with uniformly low training loss, rather than simply finding
parameters with the lowest training loss, which can improve generalization.

The initial learning rate was set to 3e−5, and the batch size was 32. The
model was trained for 100 epochs, and the validation set was used to evaluate

performance after each epoch. The learning rate was reduced by a factor of 0.1 if the validation accuracy did not improve for ten consecutive epochs. Once training was complete, the model with the best validation performance was loaded and evaluated on the test set.

Performance was evaluated using Accuracy (ACC), Sensitivity (SEN), Specificity (SPE), and the F1-Score. Their formulas are defined as follows:

$$\text{ACC} = \frac{\text{TP} + \text{TN}}{\text{TP} + \text{TN} + \text{FP} + \text{FN}} \tag{7}$$

$$\text{SEN} = \frac{\text{TP}}{\text{TP} + \text{FN}} \tag{8}$$

$$\text{SPE} = \frac{\text{TN}}{\text{TN} + \text{FP}} \tag{9}$$

$$\text{F1-Score} = \frac{2 \times \text{SEN} \times \text{SPE}}{\text{SEN} + \text{SPE}} \tag{10}$$

where TP, TN, FP, and FN represent true positive, true negative, false positive, and false negative, respectively [18]. The cross-entropy loss function was used for training. For multi-class classification, it is defined as:

$$\text{Loss} = -\sum_{c=1}^{M} y_{o,c} \log(p_{o,c}) \tag{11}$$

where M is the number of classes, $y_{o,c}$ is a binary indicator (1 if observation o belongs to class c, and 0 otherwise), and $p_{o,c}$ is the predicted probability of observation o belonging to class c. The cross-entropy loss function assesses the discrepancy between the predicted and true probability distributions. A lower value indicates that the two distributions are closer.

4   Results

The data for this experiment was obtained from the OASIS dataset on Kaggle. It included 80,000 2D brain MRI images categorized into various stages of Alzheimer's disease development. For neural network training, we used axial images corresponding to slices 100 to 160 from each patient. Patients were categorized based on their Clinical Dementia Rating (CDR) scores. In this experimental design, files from patients with "Mild to Moderate Dementia" were labeled as AD, those with "Very Mild Dementia" as MCI, and those "Non Demented" as Healthy Controls (HC) (Table 2).

We evaluated the performance of different deep learning models on classification tasks for Alzheimer's disease (AD), Mild Cognitive Impairment (MCI), and Healthy Control (HC) populations. We compared VGG-16, DenseNet, ResNet-18, and our proposed CoTNet-50 model (Tables 3 and 4).

In the AD vs. MCI task, our proposed CoTNet-50 model showed excellent performance. For the AD vs. HC task, the CoTNet-50 also performed well,

Table 2. Statistics of the benchmark dataset used in the experiments.

Class	Number of Slices
AD	13,725
MCI	5,500
HC	60,775
Total	80,000

Table 3. Experimental results for the AD vs. MCI classification task.

Model	ACC (%)	SEN (%)	SPE (%)	F1-Score
VGG-16	89.57	90.11	84.12	0.67
DenseNet	86.27	88.49	85.01	0.56
ResNet-18	92.57	90.36	88.22	0.71
CoTNet-50	**95.02**	**93.15**	**92.76**	**0.87**

Table 4. Experimental results for the AD vs. HC classification task.

Model	ACC (%)	SEN (%)	SPE (%)	F1-Score
VGG-16	93.30	93.33	89.93	0.88
DenseNet	93.67	93.12	91.01	0.89
ResNet-18	94.42	**95.12**	92.01	0.91
CoTNet-50	**96.99**	94.45	**97.41**	**0.92**

Table 5. Experimental results for the MCI vs. HC classification task.

Model	ACC (%)	SEN (%)	SPE (%)	F1-Score
VGG-16	85.99	93.43	89.93	0.73
DenseNet	92.67	93.12	91.01	0.71
ResNet-18	94.42	**95.12**	92.01	0.84
CoTNet-50	**96.55**	94.45	**96.65**	**0.85**

Table 6. Experimental results for the AD vs. MCI vs. HC classification task.

Model	ACC (%)
VGG-16	90.63
DenseNet	91.48
ResNet-18	92.54
CoTNet-50	**93.72**

achieving 96.99% ACC and an F1-Score of 0.92. For the MCI vs. HC task, CoTNet-50 showed comparable or better performance than the other models. Finally, in the three-class AD vs. MCI vs. HC task, CoTNet-50 achieved the highest accuracy of 93.72%. Taken together, our findings demonstrated the superior performance of CoTNet-50 in Alzheimer's disease-related classification tasks, providing strong support for its use in diagnosis and research (Tables 5 and 6).

5   Discussion

In summary, Alzheimer's disease (AD) continues to be a major public health concern, with an increasing prevalence that burdens both individuals and society. Mild Cognitive Impairment (MCI) is a precursor to AD, and its early identification is essential for implementing timely interventions and improving patient outcomes. This study demonstrated the strong performance of deep learning techniques, when Contextual Transformer (CoT) blocks are integrated into established models like ResNet-50 for diagnosing AD from MRI images. These models demonstrated high efficacy in differentiating between AD, MCI, and healthy controls (HC) by effectively capturing both local and global information. To further increase the effectiveness and precision of Alzheimer's diagnosis, combining various techniques and using multi-modal data is a promising direction. Additional research and validation in clinical settings are required to realize the full potential of these deep learning techniques in supporting Alzheimer's disease diagnosis and treatment.

Disclosure of Interests. The authors have no competing interests to declare that are relevant to the content of this article.

References

1. Adam, K.D.B.J., et al.: A method for stochastic optimization. arXiv preprint arXiv:1412.6980 **1412**(6) (2014)
2. Arevalo-Rodriguez, I., et al.: Mini-mental state examination (MMSE) for the detection of Alzheimer's disease and other dementias in people with mild cognitive impairment (MCI). Cochrane Database Syst. Rev. (3) (2015)
3. Brosch, T., Tam, R.: Manifold learning of brain MRIs by deep learning. In: Mori, K., Sakuma, I., Sato, Y., Barillot, C., Navab, N. (eds.) MICCAI 2013. LNCS, vol. 8150, pp. 633–640. Springer, Heidelberg (2013). https://doi.org/10.1007/978-3-642-40763-5_78
4. Foret, P., Kleiner, A., Mobahi, H., Neyshabur, B.: Sharpness-aware minimization for efficiently improving generalization. arXiv preprint arXiv:2010.01412 (2020)
5. Gupta, A., Ayhan, M., Maida, A.: Natural image bases to represent neuroimaging data. In: International Conference on Machine Learning, pp. 987–994. PMLR (2013)
6. He, K., Zhang, X., Ren, S., Sun, J.: Deep residual learning for image recognition. In: Proceedings of the IEEE Conference on Computer Vision and Pattern Recognition, pp. 770–778 (2016)

7. He, Y., Chen, Z., Evans, A.: Structural insights into aberrant topological patterns of large-scale cortical networks in Alzheimer's disease. J. Neurosci. **28**(18), 4756–4766 (2008)

8. Hosseini-Asl, E., Keynton, R., El-Baz, A.: Alzheimer's disease diagnostics by adaptation of 3D convolutional network. In: 2016 IEEE International Conference on Image Processing (ICIP), pp. 126–130. IEEE (2016)

9. Jia, J., et al.: The cost of Alzheimer's disease in china and re-estimation of costs worldwide. Alzheimer's Dementia **14**(4), 483–491 (2018)

10. Krizhevsky, A., Sutskever, I., Hinton, G.E.: ImageNet classification with deep convolutional neural networks. In: Advances in Neural Information Processing Systems, vol. 25 (2012)

11. Li, F., Liu, M., Initiative, A.D.N., et al.: Alzheimer's disease diagnosis based on multiple cluster dense convolutional networks. Comput. Med. Imaging Graph. **70**, 101–110 (2018)

12. Li, Y., Yao, T., Pan, Y., Mei, T.: Contextual transformer networks for visual recognition. IEEE Trans. Pattern Anal. Mach. Intell. **45**(2), 1489–1500 (2022)

13. Nizynski, B., Dzwolak, W., Nieznanski, K.: Amyloidogenesis of tau protein. Protein Sci. **26**(11), 2126–2150 (2017)

14. Parmar, H., Walden, E.: Towards practical application of deep learning in diagnosis of Alzheimer's disease. arXiv preprint arXiv:2212.04528 (2022)

15. Petersen, R.C., Smith, G.E., Waring, S.C., Ivnik, R.J., Tangalos, E.G., Kokmen, E.: Mild cognitive impairment: clinical characterization and outcome. Arch. Neurol. **56**(3), 303–308 (1999)

16. Pini, L., et al.: Brain atrophy in Alzheimer's disease and aging. Ageing Res. Rev. **30**, 25–48 (2016)

17. Prince, M., Wimo, A., Guerchet, M., Ali, G.C., Wu, Y.T., Prina, M.: World Alzheimer report 2015. The global impact of dementia: an analysis of prevalence, incidence, cost and trends. Ph.D. thesis, Alzheimer's Disease International (2015)

18. Raschka, S.: An overview of general performance metrics of binary classifier systems. arXiv preprint arXiv:1410.5330 (2014)

19. Szegedy, C., et al.: Going deeper with convolutions. In: Proceedings of the IEEE Conference on Computer Vision and Pattern Recognition, pp. 1–9 (2015)

20. Vaswani, A., et al.: Attention is all you need. In: Advances in Neural Information Processing Systems, vol. 30 (2017)

21. Zhang, Z., Sabuncu, M.: Generalized cross entropy loss for training deep neural networks with noisy labels. In: Advances in Neural Information Processing Systems, vol. 31 (2018)

MoViS: Motion-guided Video Generation for Laparoscopic Surgery

Yousef Yeganeh[1,2](✉) [iD], Nassir Navab[1,2] [iD], and Azade Farshad[1,2] [iD]

[1] Chair for Computer Aided Medical Procedures (CAMP), TU Munich, Munich, Germany
y.yeganeh@tum.de
[2] Munich Center for Machine Learning (MCML), Munich, Germany

Abstract. Realistic surgical video synthesis is vital for simulation, training, and preoperative planning, yet existing methods often fall short in capturing the intricate motions intrinsic to surgical procedures. In this paper, we introduce a novel framework that generates high-fidelity surgical videos by integrating keypoints extracted via a zero-shot keypoint prediction module with the motion guidance of video diffusion models. Our approach decouples spatial content from temporal dynamics: critical keypoints representing anatomical landmarks and instrument positions are first identified in a zero-shot manner and then used to steer the diffusion process, ensuring that generated video frames adhere to plausible and consistent motion patterns. Extensive experiments on curated surgical video datasets demonstrate that our method produces temporally coherent videos with enhanced motion accuracy and visual realism compared to baseline models. The results suggest that our framework can substantially contribute to surgical training and simulation by generating dynamic and realistic surgical scenarios that faithfully reflect complex procedural movements. **Project Page** | **Video Demo**.

Keywords: Surgical Video Synthesis · Motion-guided · Self-supervised

1 Introduction

Surgical video generation presents unique challenges due to the complex nature of tool-tissue interactions, the need for anatomical consistency, and the requirement for procedurally coherent motion patterns [3,24]. Recent successes in generative modeling, particularly diffusion models [8,22,38], have inspired their application in video synthesis, demonstrating promising results in generating temporally coherent and visually realistic sequences [7,31]. However, the direct application of these methods to the surgical domain is non-trivial. As follows, we review the related literature to this work.

Surgical Video Analysis. The analysis of surgical videos has been extensively studied for applications such as phase recognition [34], tool detection [16], and

J. Qiu et al. (Eds.): Agentic AI 2025/CMLLMs 2025/CREATE 2025, LNCS 16147, pp. 215–225, 2026.
https://doi.org/10.1007/978-3-032-06004-4_22

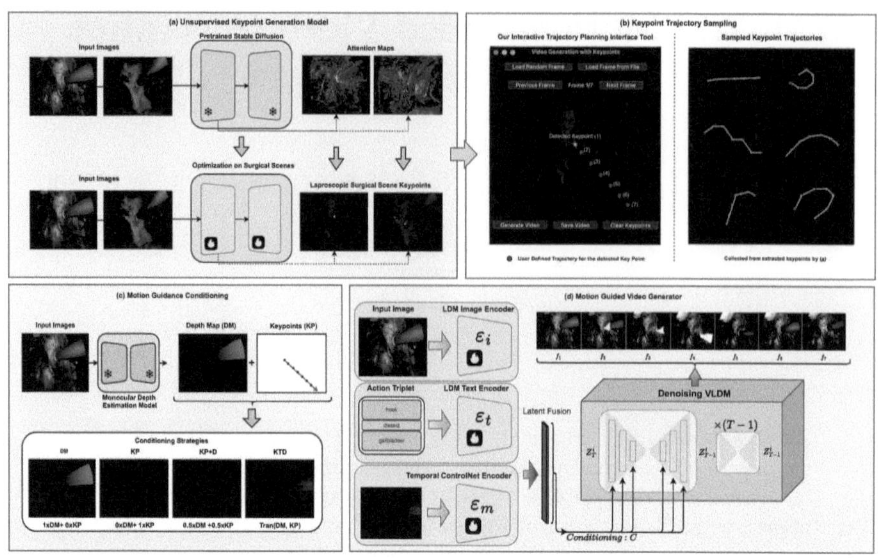

Fig. 1. Overview of our motion-guided surgical video generation framework. From a reference frame, we generate its monocular depth map (c) and corresponding keypoints of the scene (a), and pair it with a sampled or manually designed trajectory through our trajectory planning tool (b) and depth maps (c). Combined with a triplet, these signals condition a video diffusion model to generate realistic surgical videos following specific motion patterns (d).

skill assessment [15]. Recent works have focused on extracting meaningful representations from surgical videos [18,21,26], including tool-tissue interactions [20], and procedural knowledge [30]. Unsupervised keypoint detection has emerged as a promising approach for identifying salient features in images and videos without requiring extensive manual annotations [10]. These methods learn to identify consistent points across frames that capture the structure and motion of objects. Depth estimation in surgical scenes provides crucial spatial information that complements 2D visual features. Monocular depth estimation methods [37] have been adapted for surgical videos [4,23], enabling the recovery of 3D relationships from single frames.

Video Generation Models. Early approaches to video generation extended image generation models [6,9] by incorporating temporal dimensions [35]. Recent advances have leveraged diffusion models [11,32] for high-quality video synthesis. Ho et al. [12] introduced a video diffusion model that generates realistic videos by gradually denoising random Gaussian noise, while Make-A-Video [31] extends text-to-image models to generate videos from text prompts. Controlling the generation process remains a significant challenge. Several works have explored conditioning mechanisms for guided generation, including text prompts [17], reference images [2], and motion vectors [13]. Recent work by Wang et al.

[36] introduced a motion-controllable video generation framework that modifies generation based on user-specified motion sequences. In the medical domain, video generation has recently emerged as a new topic. Methods such as Endora [19] focused on unconditional video generation, while other methods focused on text [5], action triplet [39], semantic segmentation conditioning [14], and surgical planning [41].

In this paper, we introduce a novel framework for motion-guided surgical video generation that decouples the spatial content from the temporal dynamics in surgical videos. Our approach leverages unsupervised keypoint detection to identify salient features in the surgical scene (e.g., instrument tips, anatomical landmarks) and monocular depth estimation to capture spatial relationships. These visual cues, combined with procedural triplets describing surgical actions (e.g., "hook dissect gallbladder"), serve as robust anchors to guide a video diffusion process, effectively ensuring that the generated video frames adhere to plausible motion patterns and preserve anatomical correctness. This decoupled formulation not only enhances the fidelity of static scene components but also ensures that temporal dynamics are modeled with precision. In summary, our key contributions are: (1) A novel framework for surgical video generation that incorporates unsupervised keypoint detection and depth estimation as motion guidance signals, (2) Analysis of different conditioning strategies, including keypoint-only guidance, depth map guidance, and combined approaches, (3) A keypoint-guided depth transformation technique that enables precise control over tool motion while maintaining consistency in the surrounding anatomy, (4) Comprehensive evaluation on laparoscopic cholecystectomy videos, demonstrating superior performance in generating realistic surgical videos with controlled motion patterns.

2   Method

Our proposed framework for motion-guided surgical video generation consists of three main components: (1) feature extraction from a reference frame, (2) motion planning and guidance signal generation, and (3) conditioned video diffusion. Figure 1 illustrates the overall pipeline.

2.1   Unsupervised Keypoint Detection and Depth Estimation

Given a reference initial frame $I_0 \in \mathbb{R}^{H \times W \times 3}$, we first extract salient keypoints and estimate a depth map.

Keypoint Detection. We employ an unsupervised keypoint detection approach based on the StableKeypoints model [10], which leverages the rich representational power of pretrained diffusion models:

$$K_0 = F_K(I_0) \in \mathbb{R}^{N \times 2} \tag{1}$$

where K_0 represents the 2D coordinates of detected keypoints. Our approach utilizes a pretrained diffusion model θ that receives the input image I_0 and a randomly initialized text embedding ω. We then extract attention map visualizations γ from the cross-attention layers of the model. These attention visualizations highlight different concepts within the image, serving as natural indicators of salient regions. For each attention map, we identify the maximum activation value and its corresponding coordinates in pixel space:

$$(x, y) = \mathrm{argmax}_{x',y'}\, \gamma(x', y') \tag{2}$$

We then optimize the text embeddings to map these maximum points to a single model Gaussian distribution, resulting in a set of keypoints that consistently identify important structures in the surgical scene. Through extensive experimentation, we discovered that different attention heads attend to different concepts within surgical images. Specifically, we empirically observed that attention heads 5 and 9 consistently correspond to surgical tool tips in laparoscopic procedures. This finding enables us to reliably identify tool tip keypoints $K_0^t \subset K_0$ without requiring explicit supervision or complex heuristics based on motion patterns.

Depth Estimation. In parallel, we estimate a depth map $D_0 \in \mathbb{R}^{H \times W}$ from the reference frame using a monocular depth estimation network F_D:

$$D_0 = F_D(I_0) \tag{3}$$

Our depth estimation network is based on DepthAnything [37], adapted to surgical scenes [4], which employs a ViT-based architecture. The estimated depth map provides crucial information about the spatial configuration of the scene, including the relative positions of instruments and tissue structures. This is particularly important for maintaining consistency during video generation.

2.2 Motion Guidance Conditioning

To guide the video generation, we need to specify how the scene should evolve over time. We explore several approaches for generating motion guidance signals.

Keypoint Trajectory Planning. For keypoint-based guidance, we need to define the trajectories of keypoints over the course of the generated video. We investigate two approaches: (1) **Manual specification**: An expert user can manually define the trajectories $K_{t=1}^T$ for tool tip keypoints over T frames, allowing precise control over instrument movements, (2) **Keypoint Sampling**: Given a triplet p, we sample keypoint patterns from the ground truth data:

$$K_{t=1}^T \sim P(K|p, K_0) \tag{4}$$

where $P(K|p, K_0)$ represents the conditional distribution of keypoint trajectories given the triplet and initial keypoints.

Depth Map Transformation. While keypoints provide precise localization, they lack information about the 3D structure. We explore three approaches that incorporate depth information: (1) **Direct conditioning**: Using the initial depth map D_0 as a constant conditioning signal across all frames, (2) **Mixed Depth Keypoint Conditioning**: Generating a simple weighted sum of the depth map and keypoints with equal weights, (3) **Keypoint-guided transformation**: Transforming the depth map based on the planned keypoint trajectories:

$$D_t = T(D_0, K_0, K_t) \tag{5}$$

where T is a 2D translation matrix that warps the depth map according to the changes in keypoint positions.

2.3 Conditioned Video Diffusion

Our video generation model is based on latent diffusion models that progressively denoise the random Gaussian latent features to generate a video. We condition this process on both the triplet p, initial frame I_0, and the generated guidance signals $G_{t=1}^T$. The diffusion model defines a forward process that gradually adds noise to a video x_0 over N steps:

$$q(x_n | x_{n-1}) = \mathcal{N}(x_n; \sqrt{1 - \beta_n} x_{n-1}, \beta_n \mathbf{I}) \tag{6}$$

where β_n is a noise schedule parameter. The reverse process denoises the signal:

$$p_\theta(x_{n-1} | x_n, p, G_t) = \mathcal{N}(x_{n-1}; \mu_\theta(x_n, n, p, G_t), \Sigma_\theta(x_n, n)) \tag{7}$$

We implement the conditioning mechanism using an adaptation that incorporates both text embeddings (for the triplet) and frame-wise visual conditioning signals. The model architecture consists of a 3D U-Net with attention modules that process spatial and temporal dimensions. For each step of the reverse diffusion process, the guidance signals influence the denoising direction, ensuring that the generated video follows the specified motion patterns while maintaining consistency with the initial frame and procedural context.

3   Experiments and Results

Dataset. We train and evaluate our approach on the CholecT50 dataset [29] for laparoscopic cholecystectomy surgery, one of the most common minimally invasive procedures. The dataset consists of 50 videos and their corresponding triplet and surgical phase annotations, where each triplet consists of an action, an object (a surgical instrument), and a target (the organ).

Experimental Setup. The generated videos in our experiments are 7 frames per second with a resolution of 320×256. The video generation model architecture is based on temporal ControlNet [40]. We fine-tune the pretrained model on the CholecT50 dataset for 10K iterations with a learning rate of $1e-4$. We evaluate four conditioning strategies for video generation:

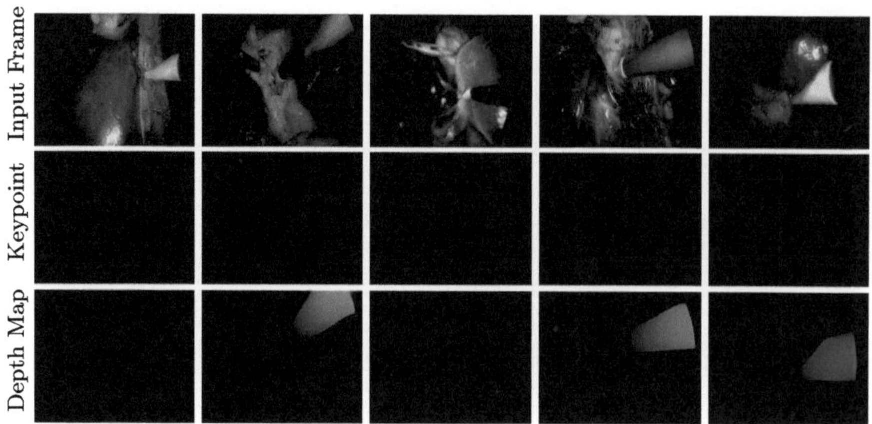

Fig. 2. Keypoint and Depth Map Prediction Output on randomly sampled frames from CholecT50.

(a) Keypoint only **(KP)**: Using only keypoint trajectories as guidance.
(b) Depth only **(D)**: Using only the initial depth map as a constant conditioning.
(c) Keypoint + Depth **(KP+D)**: Simple weighted sum of keypoint heatmaps and depth maps.
(d) Keypoint transformed depth **(KTD)**: Using depth maps transformed according to keypoint trajectories.

Evaluation Metrics. We employ common quantitative metrics for generative models to assess the quality of generated videos: (1) Fréchet Video Distance (FVD): Measures the distance between the distribution of generated videos and real videos in feature space, (2) PSNR (Peak Signal to Noise Ratio) indicates the image quality with respect to noise, (3) LPIPS (Learned Perceptual Image Patch Similarity) measures the similarity between the activations of two image patches, (4) SSIM (Structural Similarity Index Measure) indicates the perceived quality of images and videos.

3.1 Results

Quantitative Evaluation. Table 1 presents the quantitative evaluation results compared to the state of the art in cholecystectomy video generation. As can be seen, MoViS achieves the best performance in terms of FVD, PSNR, and LPIPS. On par results can be seen in terms of SSIM. This performance gain is expected given the fact that video generation highly relies on motion patterns, which are not explicitly given to other models. In our experiments with the baseline models with triplet conditioning, we observed that the video diffusion models give higher importance to the image conditioning and occasionally may fully ignore the text/triplet conditioning. Therefore, an intermediate motion-based representation can address this issue by first sampling the motion patterns from the text conditioning and then guiding the model to generate the video.

Table 1. Comparison to SOTA. Quantitative evaluation of video synthesis on CholecT50 [28].

Model	Input	FVD ($\downarrow$)	PSNR ($\uparrow$)	LPIPS ($\downarrow$)	SSIM ($\uparrow$)
CoDi [33]	Image + Triplet	6,944	9.8	0.82	0.31
WALDO [25]	Image + Seg. + Flow	3,413	11.6	0.72	0.34
LFDM [27]	Image + Triplet	1,957	12.0	0.54	**0.71**
SVD [1]	Image	3,870	14.8	0.51	0.47
SVD + FT [1]	Image	1,931	18.2	0.40	0.55
VISAGE-T [39]	Image + Triplet	<u>1,780</u>	18.1	0.39	0.56
VISAGE-I [39]	Image + Triplet	1,875	<u>18.3</u>	0.38	0.56
ControlNet [40]	Image	880	17.3	**0.19**	0.53
MoViS (Ours)	Image + Motion	**477**	**18.8**	**0.19**	<u>0.57</u>

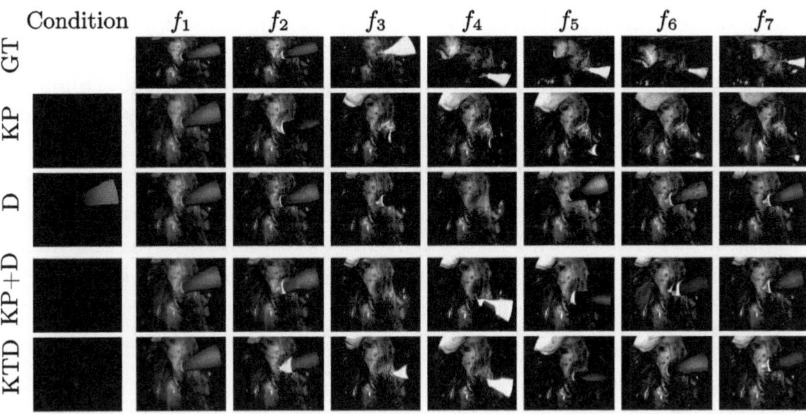

Fig. 3. Qualitative comparison of videos generated using different conditioning strategies. Top: Reference frame and target keypoint trajectories. Rows: Sample frames from videos generated using (a) keypoint (KP), (b) depth map (D), (c) combined (KP+D), and (d) keypoint transformed depth (KTD).

Qualitative Analysis. We present some qualitative samples of the keypoint detection and depth estimation components in Fig. 2. As can be seen, while the keypoint detection identifies the tool tip, the depth map models the general 3D structure of the scene. Figure 3 presents qualitative results for videos generated using the four conditioning strategies. The keypoint-only (KP) approach accurately follows the specified motion but introduces artifacts and high-range motion in non-tool regions. The depth-only approach maintains consistent anatomy but fails to produce the desired motion. The combined approaches, particularly KTD, achieve both motion accuracy and anatomical consistency.

We observe that the keypoint-guided transformed depth approach generates the most realistic tool-tissue interactions, with appropriate deformation of tissues

Table 2. Ablation Study on different conditioning strategies.

Model	FVD ($\downarrow$)	PSNR ($\uparrow$)	LPIPS ($\downarrow$)	SSIM ($\uparrow$)
KP	579	18.1	0.17	0.55
D	539	18.8	0.17	0.55
KP+D	482	18.7	0.19	0.56
KTD	**477**	**18.8**	**0.19**	**0.57**

in response to instrument movements. This is particularly evident in procedures involving dissection and retraction, where the relationship between instrument motion and tissue response is complex.

Ablation Studies. We conducted ablation studies to analyze the impact of the four introduced motion guidance strategies. As seen in Table 2, the keypoint-only approach achieves the lowest keypoint trajectory error, indicating its effectiveness in following the specified motion patterns. However, it has a higher FVD and lower depth consistency score, suggesting issues with overall video quality and spatial consistency. The depth-only approach shows the highest depth consistency but performs poorly on keypoint trajectory error, confirming our expectation that constant depth guidance results in minimal motion. The combined approaches (KP+D and KTD) achieve better balance across metrics, with KTD demonstrating the best overall performance with the lowest FVD and a good balance between trajectory accuracy and depth consistency.

Limitations. Despite the promising results, our approach has several limitations. The unsupervised keypoint detector sometimes fails to identify important anatomical landmarks, particularly in cases with unusual lighting or viewpoints. We observe failure cases when the target motion patterns deviate significantly from those seen in the training data. In these cases, the model may produce unrealistic deformations or ignore the specified motion. Additionally, long sequences with complex interactions remain challenging, as error accumulation can lead to decreasing quality over time.

4   Conclusion

In this paper, we presented a novel framework for motion-guided surgical video generation that leverages unsupervised keypoint detection and depth estimation. Our approach enables the generation of realistic surgical videos that follow specific motion patterns while maintaining anatomical consistency, opening new possibilities for surgical education, simulation, and preoperative planning. Our experiments demonstrate the effectiveness of combining keypoint guidance with transformed depth maps for controlling surgical video generation. This approach balances the precision of keypoint trajectories with the spatial consistency provided by depth information, resulting in videos that exhibit both accurate tool motion and realistic tissue responses.

Disclosure of Interests. The authors have no competing interests to declare that are relevant to the content of this article.

References

1. Blattmann, A., et al.: Stable video diffusion: scaling latent video diffusion models to large datasets (2023)
2. Blattmann, A., et al.: Align your latents: high-resolution video synthesis with latent diffusion models. In: CVPR (2023)
3. Bodenstedt, S., et al.: Comparative evaluation of instrument segmentation and tracking methods in minimally invasive surgery. arXiv (2018)
4. Budd, C., Vercauteren, T.: Transferring relative monocular depth to surgical vision with temporal consistency. In: International Conference on Medical Image Computing and Computer-Assisted Intervention, pp. 692–702. Springer, Heidelberg (2024). https://doi.org/10.1007/978-3-031-72089-5_65
5. Cho, J., et al.: Surgen: text-guided diffusion model for surgical video generation. arXiv (2024)
6. Dhamo, H., et al.: Semantic image manipulation using scene graphs. In: Proceedings of the IEEE/CVF Conference on Computer Vision and Pattern Recognition, pp. 5213–5222 (2020)
7. Dhariwal, P., Nichol, A.: Diffusion models beat gans on image synthesis. Adv. Neural. Inf. Process. Syst. **34**, 8780–8794 (2021)
8. Farshad, A., Yeganeh, Y., Chi, Y., Shen, C., Ommer, B., Navab, N.: Scenegenie: scene graph guided diffusion models for image synthesis. In: Proceedings of the IEEE/CVF International Conference on Computer Vision, pp. 88–98 (2023)
9. Farshad, A., Yeganeh, Y., Dhamo, H., Tombari, F., Navab, N.: Dispositionet: disentangled pose and identity in semantic image manipulation. In: 33rd British Machine Vision Conference 2022, BMVC 2022, London, UK, 21–24 November 2022. BMVA Press (2022)
10. Hedlin, E., et al.: Unsupervised keypoints from pretrained diffusion models. In: CVPR (2024)
11. Ho, J., Jain, A., Abbeel, P.: Denoising diffusion probabilistic models. Adv. Neural. Inf. Process. Syst. **33**, 6840–6851 (2020)
12. Ho, J., Salimans, T., Gritsenko, A., Chan, W., Norouzi, M., Fleet, D.J.: Video diffusion models. In: NeurIPS (2022)
13. Hu, Y., Luo, C., Chen, Z.: Make it move: controllable image-to-video generation with text descriptions. In: CVPR (2022)
14. Iliash, I., Allmendinger, S., Meissen, F., Kühl, N., Rückert, D.: Interactive generation of laparoscopic videos with diffusion models. In: MICCAI Workshop on Deep Generative Models, pp. 109–118. Springer, Heidelberg (2024). https://doi.org/10.1007/978-3-031-72744-3_11
15. Ismail Fawaz, H., Forestier, G., Weber, J., Idoumghar, L., Muller, P.-A.: Evaluating surgical skills from kinematic data using convolutional neural networks. In: Frangi, A.F., Schnabel, J.A., Davatzikos, C., Alberola-López, C., Fichtinger, G. (eds.) MICCAI 2018. LNCS, vol. 11073, pp. 214–221. Springer, Cham (2018). https://doi.org/10.1007/978-3-030-00937-3_25

16. Jin, A., et al.: Tool detection and operative skill assessment in surgical videos using region-based convolutional neural networks. In: 2018 IEEE Winter Conference on Applications of Computer Vision (WACV), pp. 691–699. IEEE (2018)
17. Khachatryan, L., et al.: Text2video-zero: text-to-image diffusion models are zero-shot video generators. In: Proceedings of the IEEE/CVF International Conference on Computer Vision, pp. 15954–15964 (2023)
18. Köksal, Ç., Ghazaei, G., Holm, F., Farshad, A., Navab, N.: Sangria: surgical video scene graph optimization for surgical workflow prediction. In: International Workshop on Graphs in Biomedical Image Analysis, pp. 106–117. Springer, Heidelberg (2024). https://doi.org/10.1007/978-3-031-83243-7_10
19. Li, C., et al.: Endora: video generation models as endoscopy simulators. In: International Conference on Medical Image Computing and Computer-Assisted Intervention, pp. 230–240. Springer, Heidelberg (2024). https://doi.org/10.1007/978-3-031-72089-5_22
20. Lin, W., et al.: Instrument-tissue interaction detection framework for surgical video understanding. IEEE Trans. Med. Imaging (2024)
21. Lu, J., Jayakumari, A., Richter, F., Li, Y., Yip, M.C.: Super deep: a surgical perception framework for robotic tissue manipulation using deep learning for feature extraction. In: 2021 IEEE International Conference on Robotics and Automation (ICRA), pp. 4783–4789. IEEE (2021)
22. Lüpke, S., Yeganeh, Y., Adeli, E., Navab, N., Farshad, A.: Physics-informed latent diffusion for multimodal brain mri synthesis. In: International Conference on Medical Image Computing and Computer-Assisted Intervention, pp. 198–207. Springer, Heidelberg (2024). https://doi.org/10.1007/978-3-031-84525-3_17
23. Mahmood, F., Durr, N.J.: Deep learning and conditional random fields-based depth estimation and topographical reconstruction from conventional endoscopy. Med. Image Anal. **48**, 230–243 (2018)
24. Maier-Hein, L., et al.: Surgical data science for next-generation interventions. Nat. Biomed. Eng. **1** (2017)
25. Moing, G.L., Ponce, J., Schmid, C.: WALDO: future video synthesis using object layer decomposition and parametric flow prediction. In: ICCV (2023)
26. Mostafa, M.L., et al.: Surgical flow masked autoencoder for event recognition. In: Medical Imaging with Deep Learning (2025)
27. Ni, H., Shi, C., Li, K., Huang, S.X., Min, M.R.: Conditional image-to-video generation with latent flow diffusion models. In: CVPR, pp. 18444–18455 (2023)
28. Nwoye, C.I., et al.: Rendezvous: attention mechanisms for the recognition of surgical action triplets in endoscopic videos. Med. Image Anal. (2022)
29. Nwoye, C.I., et al.: Cholectriplet 2022: show me a tool and tell me the triplet–an endoscopic vision challenge for surgical action triplet detection. Med. Image Anal. **89**, 102888 (2023)
30. Rivoir, D., et al.: Long-term temporally consistent unpaired video translation from simulated surgical 3d data. In: ICCV, pp. 3343–3353 (2021)
31. Singer, U., et al.: Make-a-video: text-to-video generation without text-video data. arXiv (2022)
32. Song, Y., Sohl-Dickstein, J., Kingma, D.P., Kumar, A., Ermon, S., Poole, B.: Score-based generative modeling through stochastic differential equations. arXiv (2020)
33. Tang, Z., Yang, Z., Zhu, C., Zeng, M., Bansal, M.: Any-to-any generation via composable diffusion. In: NeurIPS (2023)
34. Twinanda, A.P., Shehata, S., Mutter, D., Marescaux, J., De Mathelin, M., Padoy, N.: Endonet: a deep architecture for recognition tasks on laparoscopic videos. IEEE Trans. Med. Imaging **36**(1), 86–97 (2016)

35. Vondrick, C., Pirsiavash, H., Torralba, A.: Generating videos with scene dynamics. Adv. Neural Inf. Process. Syst. **29** (2016)
36. Wang, Z., et al.: Motionctrl: a unified and flexible motion controller for video generation. In: ACM SIGGRAPH 2024 Conference Papers, pp. 1–11 (2024)
37. Yang, L., Kang, B., Huang, Z., Xu, X., Feng, J., Zhao, H.: Depth anything: unleashing the power of large-scale unlabeled data. In: CVPR (2024)
38. Yeganeh, Y., et al.: Latent drifting in diffusion models for counterfactual medical image synthesis. In: Proceedings of the Computer Vision and Pattern Recognition Conference, pp. 7685–7695 (2025)
39. Yeganeh, Y., et al.: Visage: video synthesis using action graphs for surgery. In: International Conference on Medical Image Computing and Computer-Assisted Intervention, pp. 146–156. Springer, Heidelberg (2024)
40. Zhang, L., Rao, A., Agrawala, M.: Adding conditional control to text-to-image diffusion models. In: ICCV, pp. 3836–3847 (2023)
41. Zhao, Z., Fang, F., Yang, X., Xu, Q., Guan, C., Zhou, S.K.: See, predict, plan: diffusion for procedure planning in robotic surgical videos. In: International Conference on Medical Image Computing and Computer-Assisted Intervention, pp. 553–563. Springer, Heidelberg (2024)

The First Workshop on Multimodal Language Models (MLLMs 2025)

RadFig-VQA: A Multi-imaging-Modality Radiology Benchmark for Evaluating Vision-Language Models in Clinical Practice

Yosuke Yamagishi[1(✉)], Shouhei Hanaoka[2], Yuta Nakamura[2],
Tomohiro Kikuchi[3], Akinobu Shimizu[4], Takeharu Yoshikawa[2],
and Osamu Abe[2]

[1] The University of Tokyo, Tokyo 113-8655, Japan
`yamagishi-yosuke0115@g.ecc.u-tokyo.ac.jp`
[2] The University of Tokyo Hospital, Tokyo 113-8655, Japan
[3] Jichi Medical University, Shimotsuke 329-0498, Japan
[4] Tokyo University of Agriculture and Technology, Tokyo 184-8588, Japan

Abstract. Current radiology benchmarks predominantly feature mixed general medical images rather than specialized radiological content, limiting comprehensive evaluation of vision-language models in clinical practice. We introduce RadFig-VQA, a comprehensive radiology-specific benchmark comprising 70,550 radiological images and 238,294 question-answer pairs systematically extracted from PubMed Central open-access papers. Our approach employs automated figure classification and multi-source question generation that extracts relevant descriptions from full paper content beyond figure captions, enabling diverse questions across multiple imaging modalities (CT, MRI, X-ray, Ultrasound, PET, SPECT, Mammography, Angiography) and clinical categories. Our fine-tuned Qwen2.5-VL 3B model achieves state-of-the-art performance with 85.4% overall accuracy, representing a significant advancement toward more realistic evaluation of vision-language models in radiology that better reflects real-world clinical complexity. The dataset is available at https:// huggingface.co/datasets/YYama0/RadFig-VQA.

Keywords: Radiology · Large Language Models · Vision-Language Models · Vision Question Answering · Clinical Evaluation

1 Introduction

The rapid evolution of large language models (LLMs) and vision-language models (VLMs) has significantly advanced medical applications [2,10,21,22]. However, current radiology benchmarks face critical limitations: existing large-scale datasets predominantly feature mixed general medical images rather than radiology-specific content and lack true radiology specialization for comprehensive evaluation. These limitations stem from the lack of specialized radiology datasets and the restricted variety of question types in current benchmarks.

To address these challenges, we propose **RadFig-VQA** (Fig. 1)—a comprehensive radiology-specific benchmark leveraging open-access papers from

J. Qiu et al. (Eds.): Agentic AI 2025/CMLLMs 2025/CREATE 2025, LNCS 16147, pp. 229–238, 2026.
https://doi.org/10.1007/978-3-032-06004-4_23

230   Y. Yamagishi et al.

PubMed Central (PMC). Our approach incorporates three key innovations. First, we develop a multi-source question generation methodology that extracts relevant descriptions from full paper content beyond figure captions, including methodology descriptions, results discussions, and clinical interpretations. This automated pipeline enables diverse, contextually rich questions that reflect the complexity of radiological reasoning. Second, we establish systematic collection of radiology-specific literature through rigorous filtering processes that identify specialized radiological publications. Third, we demonstrate large-scale, specialized dataset construction while maintaining clinical relevance through automated pipelines.

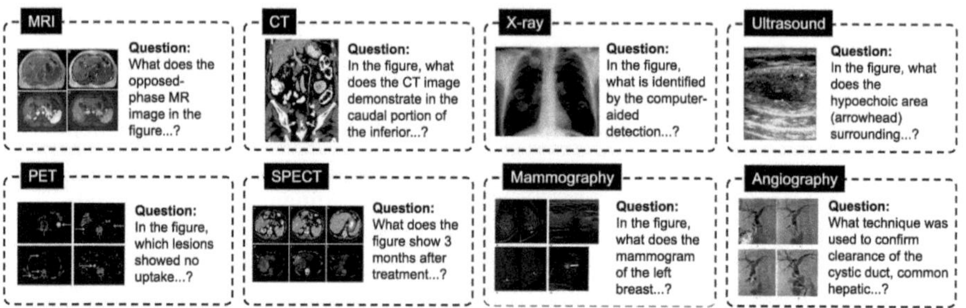

Fig. 1. Examples of RadFig-VQA image-question pairs for MRI, CT, X-ray, Ultrasound, PET, SPECT, Mammography, and Angiography.

RadFig-VQA leverages meticulously designed filtering to extract radiological figures from radiology-specific publications in PMC. By incorporating comprehensive textual context from the full papers, our approach enables LLMs to generate questions spanning the full breadth of radiological knowledge—from basic anatomical identification to complex diagnostic reasoning. The resulting dataset comprises 70,550 radiological images and 238,294 question-answer pairs, enabling comprehensive evaluation across multiple imaging modalities, clinical categories, and difficulty levels.

Our main contributions are:

- A novel multi-source question generation methodology that extracts relevant descriptions from full paper content beyond figure captions, resulting in unprecedented question diversity and clinical authenticity
- A systematic approach for collecting radiology-specific literature from PMC with rigorous filtering processes
- Construction of the largest specialized radiology VQA dataset through scalable automated pipelines
- Comprehensive evaluation demonstrating effectiveness in distinguishing model capabilities across imaging modalities and clinical categories

2   Related Work

Several attempts have been made to evaluate LLM and VLM performance in radiology, such as using the Diagnosis Please dataset [11,15,16,23] or AJNR quizzes [8,17]. However, these approaches repurpose existing diagnostic quizzes from medical journals rather than being specifically designed for VLM evaluation, which limits their effectiveness as standardized benchmarks. Established benchmarks like MMMU [25,26], MMLU [6,24], and ScienceQA [14]—widely used in general domains—remain scarce for radiology-specific applications.

Valuable datasets such as SLAKE [12] and VQA-Rad [9] have been proposed for radiology, but they suffer from limitations including dominance of simple yes/no answers, small sample sizes, and lack of expert curation. While PMC-VQA [27] introduced semi-automated dataset generation using LLMs, it lacks clinical validation, raising questions about its relevance to actual radiological practice. Table 1 compares major existing radiology VQA datasets.

Table 1. Comparison of radiology VQA datasets. Rad-Specific: Radiology-specific dataset

Dataset	Images	QA Pairs	Resource	Rad-Specific
VQA-Rad [9]	0.3k	3.5k	MedPix	✓
VQA-Med [3]	4k	15k	MedPix	✓
SLAKE [12]	0.6k	14k	Open datasets	✓
PMC-VQA [27]	149k	227k	PMC	
OmniMedVQA [7]	118k	128k	Open datasets	
RadFig-VQA	**70k**	**238k**	**PMC**	✓

3   Dataset Construction

3.1   Data Selection Process

We utilized a systematic filtering process to identify relevant papers for constructing the VQA dataset. Initially, we extracted all publicly available papers from the PMC ($n = 6{,}284{,}744$). To focus on radiology-related literature, we applied keyword filtering based on terms associated with imaging and radiology, including 'Rad', 'Ima', 'Tomog', 'Magne', 'Nucl', and 'Roent'. This reduced the number of papers to 90,805. Next, we manually reviewed the journals associated with these papers to exclude those unrelated to radiology, resulting in a dataset of 59,440 papers. Finally, we restricted the dataset to papers with appropriate open licenses, leaving 40,850 papers eligible for inclusion.

Data Use Declaration: All papers used in this study were obtained from the PMC Open Access subset under Creative Commons licenses ('CC BY', 'CC BY-NC', 'CC BY-NC-SA', 'CC0') that permit modification and redistribution for research purposes.

3.2 Figure Selection

To generate a VQA dataset specialized in radiology image interpretation tasks, we implemented a CNN-based classification approach to identify and select figures containing radiology images from the filtered papers. Even in radiology papers, many figures or graphs that are not directly related to radiology images, such as flowcharts, graphs, and diagrams, are often included (Fig. 2).

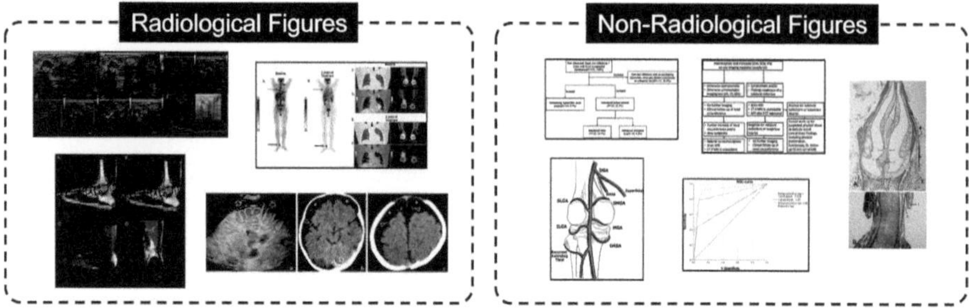

Fig. 2. Examples of radiological figures (left) and non-radiological figures (right) found in radiology papers. Radiological figures include CT scans, MRI images, X-rays, and other medical imaging modalities. Non-radiological figures include flowcharts, anatomical diagrams, statistical graphs, and histological images.

We developed an automated figure classification system using EfficientNetV2-S [18,19] pretrained on ImageNet to distinguish radiology images from non-radiology figures. A radiology resident (PGY-4) manually labeled 5,301 figures as ground truth for training and validation.

Using 5-fold cross-validation with a conservative threshold of 0.9 to minimize false positives, the model achieved excellent performance (ROC AUC: 0.990) with high precision and recall (Table 2). Additional validation on 100 random test images confirmed the model's high precision.

Table 2. Model Performance and Validation Results

	5 Folds Cross Validation			External Validation
	ROC AUC	Accuracy	Recall	Precision	Precision
Results	0.993	95.6%	91.4%	96.0%	99.0%

This validated model filtered all figures from 40,850 papers to select radiology images for QA generation. The model is available at https://huggingface.co/YYama0/RadFig-classifier.

3.3 VQA Generation

After identifying radiological figures, we employed a two-stage approach to generate high-quality visual question-answering pairs. As illustrated in Fig. 3, the process involved automated figure description extraction followed by question generation based on these descriptions and associated metadata. This systematic pipeline ensures that all generated VQA pairs maintain clinical accuracy and educational relevance while being anchored in the original scholarly content.

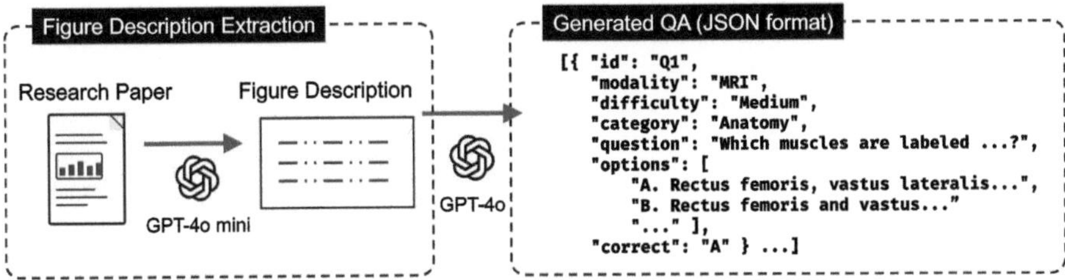

Fig. 3. Overview of the VQA generation pipeline. The process begins with figure description extraction from research papers using GPT-4o mini, followed by structured question generation using GPT-4o. The output is formatted as JSON with comprehensive metadata including imaging modality, difficulty level, and clinical category.

Figure Description Extraction. For each radiological figure, we used GPT-4o mini (API version, `gpt-4o-mini-2024-07-18`) with the prompt:

"Extract information about the target figure from radiological paper paragraphs. The extracted information should be comprehensive and faithful to the original text without any additional interpretation."

The model was given (1) the original figure caption, (2) the figure label (e.g., "Fig. 1"), and (3) the relevant text passages referencing that figure. It then generated a concise, factually grounded description—avoiding any hallucinations or added interpretation. These descriptions serve as intermediate representations, ensuring that all downstream VQA pairs remain directly anchored in the source material.

Question Generation Process. Using the generated figure descriptions along with the original captions, we employed GPT-4o (API version, `gpt-4o-2024-11-20`) to create multiple-choice questions following a standardized format. Each question was designed as a 6-option multiple-choice question (A through F) with exactly one correct answer.

The question generation process was guided by specific requirements to ensure clinical relevance and educational value. Each question was systematically categorized along three dimensions:

- **Imaging Modality**: CT, MRI, X-ray, Ultrasound, PET, SPECT, Mammography, Angiography, Others, or Multiple (for questions involving multiple modalities)
- **Clinical Category**: Findings (radiological observations), Diagnosis (diagnostic conclusions), Treatment (therapeutic recommendations), Anatomy (anatomical structures), Clinical Significance (clinical implications), or Modality (imaging techniques)
- **Difficulty Level**: Easy, Medium, or Hard based on the complexity of the required knowledge

4 Dataset Characteristics

The final RadFig-VQA dataset comprises 70,550 images and 238,294 question-answer pairs generated from radiological figures across diverse imaging modalities and clinical contexts.

4.1 Dataset Distribution

Figure 4 illustrates the comprehensive distribution of the RadFig-VQA dataset across three critical dimensions.

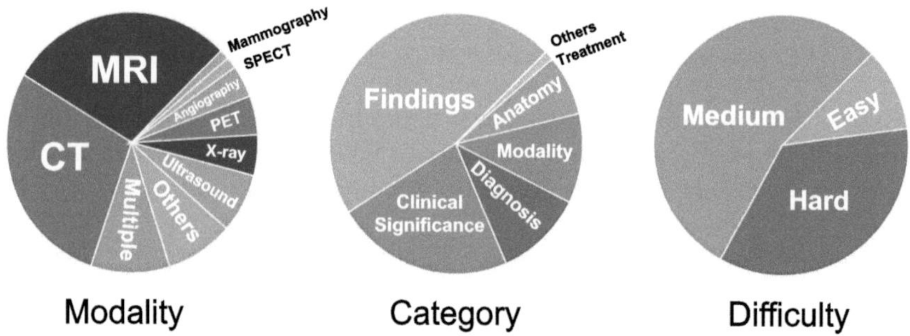

Fig. 4. Distribution of the RadFig-VQA dataset across three key dimensions: (left) Imaging modality distribution, (center) Clinical category distribution, and (right) Difficulty level distribution.

The dataset demonstrates excellent imaging modality distribution with MRI (28.9%) and CT (28.5%) as core components, multi-modal cases (10.1%) for comparative training, and clinically representative distributions for Ultrasound (7.4%), X-ray (5.0%), and PET (4.8%). Clinical category analysis reveals that findings-based questions predominate (46.4%), emphasizing fundamental observation skills, while Clinical Significance (22.7%) focuses on interpretation, Diagnosis and Modality each contribute 11.1%, and Anatomy provides foundational knowledge (7.5%). The difficulty level distribution shows that medium-difficulty questions dominate (54.6%), complemented by Hard questions (35.1%) for complex scenarios and Easy questions (10.3%) for foundational assessment.

4.2 Comparison with PMC-VQA

To quantify linguistic diversity, we compared RadFig-VQA against PMC-VQA— a baseline chosen because it also derives from PMC and is the largest publicly available PMC-based QA dataset. For each question, text was lowercased, stripped of punctuation, and tokenized. Using `CountVectorizer` [13], we extracted 1-gram, 2-gram, and 3-gram uniques.

Figure 5 presents three metrics (1-gram Unique, 2-gram Unique, 3-gram Unique) for both datasets. RadFig-VQA surpasses PMC-VQA in Unique counts in all n-gram unique counts, reflecting greater lexical diversity. In addition, we have placed a Venn diagram to the right of the bar plots to illustrate the overlap of source papers between RadFig-VQA and PMC-VQA. RadFig-VQA uses 22,374 unique papers in total, of which only 4,463 overlap with PMC-VQA (approximately 20%), underscoring the high degree of independence between the two datasets.

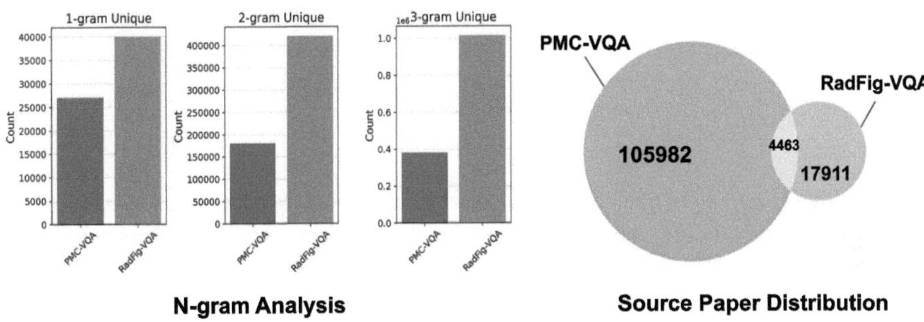

Fig. 5. Left: Comparison of n-gram Unique metrics (1-gram, 2-gram, 3-gram) between RadFig-VQA and PMC-VQA. Right: Venn diagram showing overlap of source papers.

5 Experiments

We conduct comprehensive experiments on a large-scale medical imaging dataset comprising 10,139 QAs from 3000 images across multiple imaging modalities and clinical categories. This test set was randomly sampled from RadFig-VQA, ensuring no overlap of source papers between training and test sets to prevent data leakage. All models are evaluated using the same test set to ensure fair comparison.

We compare against eight state-of-the-art vision-language models: Llama3.2 11B Vision [5], Gemma3 4B [20], MedGemma [4] 4B (medical domain-adapted), InternVL3 family (2B, 8B, 14B) [28], and Qwen2.5-VL family (3B, 7B) [1]. These models represent diverse architectural approaches and parameter scales in medical vision-language understanding. We report accuracy on key representative categories from each dimension. Table 3 presents our main experimental results. Complete detailed results are available in the dataset repository.

Table 3. Performance by Key Dimensions (%). US: Ultrasound, Sig.: Clinical Significance, Diag.: Diagnosis, Find.: Findings, FT: Fine-tuned

Model	Imaging Modality				Clinical Category			Overall
	CT	MRI	X-ray	US	Sig.	Diag.	Find.	
Llama3.2 11B Vision	57.0	57.9	60.4	55.9	63.1	52.4	54.4	57.5
Gemma3 4B	55.0	53.2	53.5	56.1	62.2	50.9	51.6	54.7
MedGemma 4B	61.5	56.9	60.0	59.0	69.1	56.4	57.2	59.8
InternVL3 2B	63.1	61.4	62.2	64.5	71.8	58.8	60.2	63.3
InternVL3 8B	68.9	69.4	68.3	71.0	78.5	66.5	66.8	70.1
InternVL3 14B	72.3	71.7	74.2	74.5	81.9	70.6	69.9	73.1
Qwen2.5-VL 3B	65.1	65.4	63.2	66.0	73.6	58.8	62.7	65.5
Qwen2.5-VL 7B	63.0	64.0	64.0	64.9	73.8	57.6	60.3	64.0
Qwen2.5-VL 3B (FT)	**85.1**	**85.8**	**82.1**	**86.4**	**89.5**	**86.1**	**82.9**	**85.4**

Our fine-tuned Qwen2.5-VL 3B model achieves state-of-the-art performance across all evaluated categories, with an overall accuracy of 85.4%. This represents a remarkable 19.9% point improvement over the base Qwen2.5-VL 3B model and outperforms the significantly larger InternVL3 14B model by 12.3% points. The fine-tuned model demonstrates consistent excellence across diverse imaging modalities and shows exceptional performance in clinically critical categories.

6   Conclusion and Future Work

We introduce RadFig-VQA, the largest specialized radiology VQA benchmark comprising 70,550 radiological images and 238,294 question-answer pairs systematically extracted from PMC. Our novel multi-source question generation methodology leverages comprehensive textual context beyond figure captions, while automated figure classification achieves 99.0% precision in identifying radiological content. The dataset provides systematic coverage across eight imaging modalities and six clinical categories, addressing critical limitations in existing benchmarks that predominantly feature mixed general medical images. Our fine-tuned Qwen2.5-VL 3B model achieves state-of-the-art performance with 85.4% overall accuracy, representing a remarkable 19.9% point improvement over the base model. This demonstrates the dataset's effectiveness in advancing vision-language models toward more realistic clinical evaluation that better reflects real-world radiological complexity. Future work will include comprehensive validation by practicing radiologists to ensure clinical accuracy of generated QA pairs, evaluation on external benchmarks to demonstrate generalizability, and integration with real clinical workflows to assess practical utility in radiological education and practice.

Acknowledgments. This work was supported by Cross-ministerial Strategic Innovation Promotion Program (SIP) on "Integrated Health Care System" Grant Number JPJ012425.

References

1. Bai, S., et al.: Qwen2. 5-vl technical report. arXiv preprint arXiv:2502.13923 (2025)
2. Clusmann, J., et al.: The future landscape of large language models in medicine. Commun. Med. **3**(1), 141 (2023)
3. Deng, J., Dong, W., Socher, R., Li, L.J., Li, K., Fei-Fei, L.: Imagenet: a large-scale hierarchical image database. In: 2009 IEEE Conference on Computer Vision and Pattern Recognition, pp. 248–255. IEEE (2009)
4. Google: Medgemma hugging face
5. Grattafiori, A., et al.: The llama 3 herd of models. arXiv preprint arXiv:2407.21783 (2024)
6. Hendrycks, D., et al.: Measuring massive multitask language understanding. arXiv preprint arXiv:2009.03300 (2020)
7. Hu, Y., et al.: Omnimedvqa: a new large-scale comprehensive evaluation benchmark for medical lvlm. In: Proceedings of the IEEE/CVF Conference on Computer Vision and Pattern Recognition, pp. 22170–22183 (2024)
8. Kikuchi, T., Nakao, T., Nakamura, Y., Hanaoka, S., Mori, H., Yoshikawa, T.: Toward improved radiologic diagnostics: investigating the utility and limitations of gpt-3.5 turbo and gpt-4 with quiz cases. Am. J. Neuroradiol. **45**(10), 1506–1511 (2024)
9. Lau, J.J., Gayen, S., Ben Abacha, A., Demner-Fushman, D.: A dataset of clinically generated visual questions and answers about radiology images. Sci. Data **5**(1), 1–10 (2018)
10. Li, C., et al.: Llava-med: training a large language-and-vision assistant for biomedicine in one day. Adv. Neural. Inf. Process. Syst. **36**, 28541–28564 (2023)
11. Li, D., Gupta, K., Bhaduri, M., Sathiadoss, P., Bhatnagar, S., Chong, J.: Comparing gpt-3.5 and gpt-4 accuracy and drift in radiology diagnosis please cases. Radiology **310**(1), e232411 (2024)
12. Liu, B., Zhan, L.M., Xu, L., Ma, L., Yang, Y., Wu, X.M.: Slake: a semantically-labeled knowledge-enhanced dataset for medical visual question answering. In: 2021 IEEE 18th International Symposium on Biomedical Imaging (ISBI), pp. 1650–1654. IEEE (2021)
13. Pedregosa, F., et al.: Scikit-learn: machine learning in python. J. Mach. Learn. Res. **12**, 2825–2830 (2011)

14. Saikh, T., Ghosal, T., Mittal, A., Ekbal, A., Bhattacharyya, P.: Scienceqa: a novel resource for question answering on scholarly articles. Int. J. Digit. Libr. **23**(3), 289–301 (2022)
15. Sonoda, Y., et al.: Diagnostic performances of gpt-4o, claude 3 opus, and gemini 1.5 pro in "diagnosis please" cases. Jpn. J. Radiol. **42**(11), 1231–1235 (2024)
16. Suh, P.S., et al.: Comparing diagnostic accuracy of radiologists versus gpt-4v and gemini pro vision using image inputs from diagnosis please cases. Radiology **312**(1), e240273 (2024)
17. Suthar, P.P., Kounsal, A., Chhetri, L., Saini, D., Dua, S.G.: Artificial intelligence (ai) in radiology: a deep dive into chatgpt 4.0's accuracy with the american journal of neuroradiology's (ajnr)" case of the month". Cureus **15**(8) (2023)
18. Tan, M., Le, Q.: Efficientnet: rethinking model scaling for convolutional neural networks. In: International Conference on Machine Learning, pp. 6105–6114. PMLR (2019)
19. Tan, M., Le, Q.: Efficientnetv2: smaller models and faster training. In: International Conference on Machine Learning, pp. 10096–10106. PMLR (2021)
20. Team, G., et al.: Gemma 3 technical report. arXiv preprint arXiv:2503.19786 (2025)
21. Thirunavukarasu, A.J., Ting, D.S.J., Elangovan, K., Gutierrez, L., Tan, T.F., Ting, D.S.W.: Large language models in medicine. Nat. Med. **29**(8), 1930–1940 (2023)
22. Tu, T., et al.: Towards generalist biomedical ai. Nejm Ai **1**(3), AIoa2300138 (2024)
23. Ueda, D., et al.: Diagnostic performance of chatgpt from patient history and imaging findings on the diagnosis please quizzes. Radiology **308**(1), e231040 (2023)
24. Wang, Y., et al.: Mmlu-pro: a more robust and challenging multi-task language understanding benchmark. In: The Thirty-Eight Conference on Neural Information Processing Systems Datasets and Benchmarks Track (2024)
25. Yue, X., et al.: Mmmu: a massive multi-discipline multimodal understanding and reasoning benchmark for expert agi. In: Proceedings of the IEEE/CVF Conference on Computer Vision and Pattern Recognition, pp. 9556–9567 (2024)
26. Yue, X., et al.: Mmmu-pro: a more robust multi-discipline multimodal understanding benchmark. arXiv preprint arXiv:2409.02813 (2024)
27. Zhang, X., et al.: Development of a large-scale medical visual question-answering dataset. Commun. Med. **4**(1), 277 (2024)
28. Zhu, J., et al.: Internvl3: exploring advanced training and test-time recipes for open-source multimodal models. arXiv preprint arXiv:2504.10479 (2025)

Trustworthy Clinical Thinking in MLLMs: Hierarchical Energy-based Reasoning for interpretable MEdical Scans (HERMES)

Florence X. Doo[1,2(✉)] ⓘ, Huiwen Han[2,3] ⓘ, Bradley A. Maron[1,4] ⓘ,
and Heng Huang[1,5] ⓘ

[1] University of Maryland–Institute for Health Computing (UM-IHC),
North Bethesda, MD 20852, USA
bmaron@som.umaryland.edu, heng@umd.edu
[2] University of Maryland Medical Intelligent Imaging (UM2ii) Center, Department of
Diagnostic Radiology and Nuclear Medicine, University of Maryland, Baltimore,
Baltimore, MD 21201, USA
{fdoo,huiwenhan}@som.umaryland.edu
[3] Department of Psychiatry and Behavioral Sciences, Stanford University, Palo Alto,
CA 94305, USA
[4] Department of Medicine, University of Maryland, Baltimore, MD 21201, USA
[5] Department of Computer Science, University of Maryland, College Park,
MD 20742, USA
https://www.ihc.umd.edu/

Abstract. Multimodal large language models (MLLMs) for medical
imaging currently operate through fast pattern recognition, and do not
mirror the deliberate reasoning process of expert human radiologists. We
introduce HERMES (Hierarchical Energy-based Reasoning for MEdical
Scans), a novel approach that builds upon Merlin's pre-trained VLM to
perform zero-shot classification of 10 common abdominal CT findings.
Our energy-based framework implements hierarchical reasoning through
three levels of analysis (global→regional→focal), with adaptive early
stopping that reduces computation inherently, with similar to improved
classification performance. The model learns task-specific projections and
energy thresholds that enable confident cases to terminate early while
complex multi-pathology cases receive deeper scrutiny—computationally
mirroring radiologist workflow. HERMES enables transparent and interpretable reasoning, for practical trustworthiness of clinical MLLMs.

Keywords: Medical vision-language models · Energy-based models ·
Clinical reasoning · 3D medical imaging · Hierarchical analysis

1 Introduction

While clinical imaging AI has progressed from 2D radiograph analysis [22,24] to
sophisticated 3D multimodal systems [2–4,10,15,16,18,23], trustworthy interpretability remains elusive. Expert radiologists navigate medical images through

J. Qiu et al. (Eds.): Agentic AI 2025/CMLLMs 2025/CREATE 2025, LNCS 16147, pp. 239–248, 2026.
https://doi.org/10.1007/978-3-032-06004-4_24

systematic spatial reasoning workflow honed through years of training. Current clinical MLLMs generate impressive predictions, however are not transparent about when they may intrinsically shortcut the deliberative diagnostic reasoning process, and remain opaque with interpretability. Post-hoc explanations (attention maps, GradCAM) show where models look but not how they reason through findings, while hierarchical analysis [17,19,21,25] can mimic how radiologists systematically search images. But without quantifying diagnostic confidence at each step, clinicians are unable to gauge when to trust AI recommendations, especially in complex multi-modal scenarios like MLLMs or foundation models.

This fundamental mismatch between fast AI predictions and graduated human reasoning ("System 1" vs. "System 2" thinking [7,12]) manifests in four critical limitations: (1) poor uncertainty calibration with overconfident predictions, (2) fixed computational cost regardless of case complexity, (3) lack of interpretable reasoning paths from observation to diagnosis, and (4) inadequate handling of multi-label complexity where diseases co-occur—missing the interconnected nature of clinical findings that radiologists natively recognize.

We introduce **hierarchical energy-based reasoning**, the first adaptation of energy-based models [8,14] to clinical MLLMs for interpretable 3D medical image analysis. This approach represents a form of "process neologism" [11]— a new way for AI to express diagnostic thinking that mimics human clinical reasoning for human-AI bi-alignment [20] for trustworthy and safe clinical AI deployment. Building on Merlin's pre-extracted multimodal embeddings and zero-shot classification capabilities [2], we implement three-level hierarchical analysis (global→regional→focal) where energy values quantify diagnostic confidence at each step. Lower energy indicates higher confidence, enabling adaptive computation that mirrors radiologist behavior: Simple findings trigger early stopping, while complex cases receive deeper scrutiny. Our **key contributions** are: **(i)** *hierarchical spatial framework for 3D medical images with energy values at each reasoning level*, delivering learnable *confidence thresholds* that adapt during training to discover dataset-specific decision boundaries for complexity-aware early stopping while improving calibration (ECE<0.1); and **(ii)** *finding-specific energy functions that encode clinical disease patterns and their relationships*, enabling enhanced multi-label performance on 15,296 CTs. Overall, HERMES addresses the clinical need to understand not just where AI looks, but how it reasons through diagnostic decisions for trustworthy AI.

2   Methods and Experimental Setup

2.1   Dataset and Feature Extraction

We used a Merlin's [2] abdominal CTs and embeddings; of the original 25,528 CTs, we extracted 15,296 training and 5,055 test CTs (60/20/20 split). We leverage Merlin's zero-shot classification capabilities to automatically generate training labels for 10 common abdominal findings. Merlin achieves F1 scores of 0.73–0.79 through vision-language pretraining on 6+ million CT-report pairs, enabling reliable phenotype prediction without task-specific fine-tuning. We use Merlin's preprocessed 512D paired vision-language embeddings. To generate supervision

---

Algorithm 1. HERMES: Hierarchical Energy-based Reasoning

Require: Image embedding $\mathbf{v}_{\mathrm{img}} \in \mathbb{R}^{512}$, Text embeddings $\{\mathbf{t}_i\}_{i=1}^{10} \in \mathbb{R}^{512}$
Require: Learned thresholds $\theta_g, \theta_r, \theta_f$, Merlin predictions $\mathbf{p}_{\mathrm{merlin}} \in [0,1]^{10}$
Ensure: Predictions $\mathbf{p} \in [0,1]^{10}$, Stop levels $\mathbf{s} \in \{1,2,3\}^{10}$
1: // **Step 1: Assess case complexity**
2: $n_{\mathrm{positive}} \leftarrow \sum_{i=1}^{10} \mathbb{1}[\mathbf{p}_{\mathrm{merlin},i} > 0.5]$
3: $\alpha \leftarrow 0.8$ if $n_{\mathrm{positive}} \geq 2$ else 1.0 {Complex cases ÃćâĂăâĂŹ lower thresholds}
4: // **Step 2: Hierarchical analysis per finding**
5: **for** $i = 1$ to 10 **do**
6:     **Level 1: Global** {Multimodal alignment}
7:     $E_1^{(i)} \leftarrow$ ComputeEnergy($\mathbf{v}_{\mathrm{img}}, \mathbf{t}_i$) {Uncertainty + complexity components}
8:     // $E_1^{(i)} = \lambda_u \cdot E_{\mathrm{unc}} + \lambda_c \cdot E_{\mathrm{comp}}$
9:     **if** $E_1^{(i)} < \theta_g \cdot \alpha$ **then**
10:         $\mathbf{p}_i \leftarrow \sigma(-E_1^{(i)})$; $\mathbf{s}_i \leftarrow 1$; **continue**
11:     **end if**
12:     **Level 2: Regional** {Attention-focused analysis}
13:     $\mathbf{v}_{\mathrm{att}} \leftarrow \mathbf{v}_{\mathrm{img}} \odot$ Attention($\mathbf{v}_{\mathrm{img}}$)
14:     $E_2^{(i)} \leftarrow$ ComputeEnergy($\mathbf{v}_{\mathrm{att}}, \mathbf{t}_i$)
15:     **if** $E_2^{(i)} < \theta_r \cdot \alpha$ **then**
16:         $\mathbf{p}_i \leftarrow \sigma(-E_2^{(i)})$; $\mathbf{s}_i \leftarrow 2$; **continue**
17:     **end if**
18:     **Level 3: Focal** {Deep analysis}
19:     $E_3^{(i)} \leftarrow$ ComputeEnergy($\mathbf{v}_{\mathrm{img}}, \mathbf{t}_i$)
20:     $\mathbf{p}_i \leftarrow \sigma(-E_3^{(i)})$; $\mathbf{s}_i \leftarrow 3$
21: **end for**
22: **return** $\mathbf{p}, \mathbf{s}$

---

labels, we adopted (in part) Merlin's zero-shot predictions for 10 findings using clinical text prompts (e.g., "hepatic steatosis present") and convert these probability scores to binary labels using a 0.5 threshold. These 10 findings span hepatic, vascular, renal, and systemic conditions.

2.2 Hierarchical Energy-Based Architecture

Our approach learns diagnostic confidence through energy values on Merlin's pre-extracted features, mirroring radiologist certainty levels during interpretation.

Energy function. We decompose energy into interpretable components mirroring radiologist reasoning—capturing both uncertainty and case complexity:

$$E(\mathbf{v}, \mathbf{t}) = \lambda_u \cdot E_{\mathrm{uncertainty}}(\mathbf{v}, \mathbf{t}) + \lambda_c \cdot E_{\mathrm{complexity}}(\mathbf{v}, \mathbf{t}) \tag{1}$$

where $\mathbf{v}, \mathbf{t} \in \mathbb{R}^{512}$ are Merlin's image and text embeddings. The uncertainty component $E_{\mathrm{uncertainty}}$ quantifies diagnostic confidence through: **(1)** a *similarity term* that measures how well imaging features match the expected disease pattern after projecting to 256 dimensions, with temperature $\tau = 0.1$ controlling decision sharpness, and **(2)** an *entropy term* that captures prediction uncertainty—high when the model is "unsure" between diagnoses. While we compute multiple internal metrics (attention patterns, activation sparsity, gradient norms), these are aggregated into the uncertainty score that directly correlates with diagnostic confidence. The complexity component $E_{\text{complexity}}$ accounts for multi-pathology cases through a penalty proportional to the number of positive findings, reflecting increased diagnostic difficulty when multiple abnormalities are present. During training, weights λ_u, λ_c automatically adjust to balance confidence with complexity recognition. Lower total energy indicates high diagnostic confidence—computationally mirroring how radiologists feel more confident with clear, single findings versus complex multi-pathology presentations.

Hierarchical Reasoning. Three levels mirror radiologist workflow with energy naturally decreasing through stepwise review: (1) *Global*—assess overall alignment capturing gestalt uncertainty, stop if $E < \theta_g$; (2) *Regional*—apply learned attention to anatomical areas reducing uncertainty, stop if $E < \theta_r$; (3) *Focal*—deep analysis achieving lowest energy through detailed scrutiny. Our energy components track this resolution: uncertainty reduces by 0.8 from global to focal as diagnostic confidence improves, while complexity remains stable, confirming that hierarchical processing resolves ambiguity rather than discovering new pathologies. The model adapts thresholds based on case complexity, where complex must achieve lower energy values than simple cases, ensuring they receive more thorough analysis; in other words, negative findings are rewarded, 1 finding is standard threshold, while 2–4+ findings increasingly trigger threshold reductions. This mirrors clinical practice—a singular finding of gallstones may be assessed quickly, but multi-pathology cases (i.e. cirrhosis with ascites) require thorough evaluation of each finding and their interactions. This adaptive mechanism enables confident early stopping for clear cases while guaranteeing deeper evaluation for multi-disease presentations—computationally mirroring how radiologists adjust their scrutiny based on case complexity.

2.3 Training Strategy

Trainable Components. Using Merlin's pre-extracted vision and language features (512D) as input, we train: **(1)** *Hierarchical energy networks* with shared feature projection (512→216D) followed by level-specific energy heads (global, regional, focal), each outputting scalar energy values through lightweight MLPs, **(2)** *Adaptive energy thresholds* (θg, θr, θf) that are learned during training and converge to dataset-specific values for determining when to stop at each hierarchical level, and **(3)** *Energy calibration parameters* including temperature scaling and complexity penalty weights that map raw energy values to calibrated confidence scores. The complete model is highly parameter-efficient, with the energy networks containing <1M parameters total, enabling efficient training on the full dataset without requiring finding-specific embeddings or spatial localization.

Loss Functions. We address the inherent class imbalance in medical imaging using Asymmetric Loss [1]: $\mathcal{L}_{\text{asym}} = -\sum[y_i \log(p_i)(1 - p_i)^{\gamma+} + (1 - y_i) \log(1 - p_i)(p_i)^{\gamma-}]$ where $\gamma_- = 4$ heavily down-weights easy negative samples while $\gamma_+ = 1$ maintains focus on positive cases—which is helpful in settings where negative samples can outnumber positives. Our complete loss combines classification with energy-based objectives: $\mathcal{L} = \mathcal{L}_{\text{asym}} + \lambda(t)\mathcal{L}_{\text{margin}} + \mathcal{L}_{\text{component}},$

where $\mathcal{L}_{\mathrm{margin}}$ enforces energy separation between positive/negative samples, and $\mathcal{L}_{\mathrm{component}}$ ensures energy components align with their targets—uncertainty matching prediction confidence and complexity reflecting multi-finding cases. The energy weight $\lambda(t)$ starts at 0.1 and adapts dynamically during training based on the relative magnitudes of the losses, typically stabilizing around 0.3–0.5 to balance diagnostic accuracy with calibrated confidence. This independently adapts to each of the 10 findings' specific prevalence patterns. This impacts multi-label scenarios where each sample can have 0âĂŞ10+ positive findings with different background base rates—mirroring how radiologists adjust their diagnostic thresholds based on disease prevalence.

Implementation. We train with AdamW ($\mathrm{lr}{=}1e^{-4}$, using mixed precision [AMP] across 8 GPUs with per-GPU batch size of 32 [256 total]) for 50 epochs, with early stopping preventing overfitting to rare findings while ensuring convergence of Merlin dataset-specific energy-based decision boundaries.

2.4   Evaluation Metrics

We evaluate HERMES across four dimensions which are important for clinical deployment: (1) **Classification performance**: Area under the ROC curve (AUROC) for each of the 10 abdominal findings. (2) **Calibration**: Expected Calibration Error (ECE) to assess alignment between predicted probabilities and observed frequencies, targeting ECE<0.1 for clinical reliability. (3) **Energy-based confidence**: Energy separation between confidence levels, measuring the model's ability to stratify predictions by certainty. (4) **Computational efficiency**: Early stopping rates at each hierarchical level (global/regional/focal), per-sample inference time, and processing distribution across varying case complexities to quantify adaptive computation benefits.

3   Results

3.1   Energy-Based Confidence Quantification

HERMES successfully creates strong energy-based separation between confidence levels across 10 abdominal pathologies (Fig. 1a). The model achieves a mean energy gap of 2.61 between very confident and very uncertain predictions, with highly significant differences between all adjacent confidence levels ($p < 0.001$). This stratification shows that energy effectively captures diagnostic uncertainty, with low energy paralleling very confident predictions ($\sim$2.9) and substantially higher energy ($\sim$5.5) for very uncertain predictions. Additionally, note that the energy function specifically incorporates distinct components that contribute differently across findings, which were out of scope for this paper length. Energy enables interpretable confidence quantification and also opens possibilities for tracking computational sustainability in clinical AI deployment [5,6].

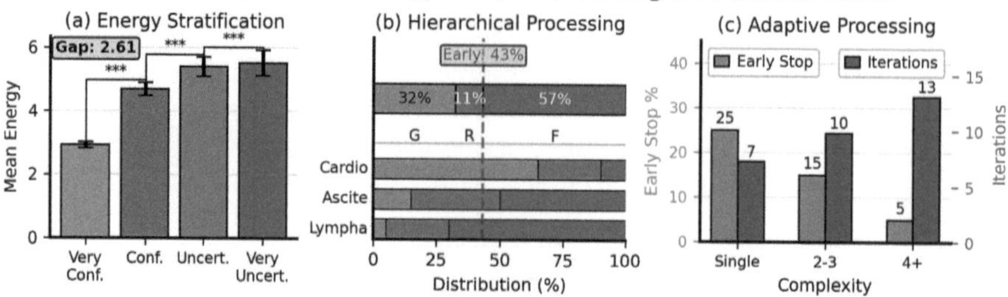

Fig. 1. HERMES framework performance demonstrating hierarchical energy-based reasoning. (a) Energy stratification across confidence levels shows significant separation (***p<0.001) with a 2.61 energy gap between very confident and very uncertain predictions. (b) Hierarchical processing achieves 43.2% early termination rate (32.5% global, 10.7% regional). Per-finding analysis reveals common findings (e.g., cardiomegaly) are predominantly detected globally while rare findings (e.g., lymphadenopathy) require focal processing. (c) Complexity-adaptive behavior: single-finding cases achieve 25% early stopping with 7.2 iterations, while complex cases (4+ findings) show 5% early stopping with 13.1 iterations, mirroring radiologist workflow patterns where diagnostic scrutiny scales with case complexity.

3.2 Adaptive Hierarchical Processing

Our hierarchical architecture demonstrates intelligent resource allocation through multi-level processing (Fig. 1b). Among test cases, 43.2% achieve confident predictions by the regional level, with 32.5% resolved at the global level and an additional 10.7% at the regional level, completely bypassing focal analysis. The remaining 56.8% of cases proceed to focal-level processing for detailed analysis. The per-finding breakdown reveals adaptive behavior that aligns with clinical intuition: common findings like cardiomegaly and pleural effusion are predominantly detected at the global level (65% and 55% respectively), while rare findings such as lymphadenopathy require deeper analysis, with only 5% detected globally and the majority requiring focal-level processing.

The system's adaptation strongly correlates with case complexity (Fig. 1c). Single-finding cases show 25% early stopping rate with an average of 7.2 iterations, while cases with 2–3 findings show 15% early stop at 9.8 iterations. Complex cases with 4+ findings appropriately receive the most thorough analysis (5% early stop, 13.1 iterations). This progressive allocation mirrors how radiologists adjust their scrutiny based on case complexity, spending more time on challenging multi-disease cases.

3.3 Ablation Analysis

Component ablation studies reveal critical insights about our architecture (Table 1). The full HERMES model achieves strong discrimination (AUC: 0.905) with excellent calibration (ECE: 0.073), validating our integrated design. The

Table 1. Ablation study: Impact of HERMES components on performance, efficiency, and hierarchical processing.

Configuration	Performance			Efficiency			Processing Distribution		
	AUC↑	ECE↓	Gap↑	Time (ms)	Savings	Early Stop	Global	Regional	Focal
Full HERMES	**0.905**	**0.073**	0.98	1.21	–	43.0%	30.8%	12.2%	57.0%
w/o Early Stop	0.908	0.065	1.05	0.87	-28.2%	0.0%	0.0%	0.0%	100.0%
w/o Hierarchical	0.908	0.065	1.05	0.87	-27.5%	0.0%	0.0%	0.0%	100.0%
w/o Energy Guide	0.905	0.073	-0.01	1.18	-2.1%	43.0%	30.9%	12.1%	57.0%
w/o Complexity	0.905	0.081	1.08	0.95	-21.6%	46.0%	22.4%	23.6%	54.0%

↑ Higher is better, ↓ Lower is better. Time savings are relative to full model.

high AUC may reflect some overfitting to our evaluation set, though per-finding analysis suggests robust performance across diseases with varying prevalence.

Removing energy guidance maintains classification performance but completely eliminates the energy gap (–0.01), destroying the model's ability to quantify uncertainty. Without the complexity penalty, early stopping increases excessively (46.0%), causing premature termination for cases requiring deeper analysis. Disabling hierarchical structure (Focal Only) eliminates adaptive computation benefits—all cases process through focal level—but somewhat paradoxically reduces inference time by 28% (1.21ms to 0.87ms) due to eliminated decision overhead. This reveals that in HERMES, the hierarchical processing component is a trade-off of computational speed for interpretability and adaptive behavior.

Notably, our energy-based confidence quantification successfully produces well-calibrated probabilities (ECE: 0.073), helpful for clinical deployment requirements. These results confirm that energy-based confidence and hierarchical processing work synergistically—each component is necessary but not sufficient alone for achieving interpretable and practically efficient medical AI.

4   Discussion

Our results demonstrate that hierarchical energy-based reasoning can address some current challenges in MLLM deployment, and areas for clinical translation.

Energy Enables Robust Confidence Quantification. The strong energy separation (gap = 2.61) validates our core innovation—using energy as a learnable confidence measure. While our ECE of 0.073 meets clinical thresholds (<0.1), this represents just one approach to calibration. High-energy predictions reliably indicate uncertainty, enabling safe triage for human review. The well-known trade-off between discrimination and calibration [9] suggests opportunities for further refinement through post-hoc methods (e.g., Platt scaling, temperature scaling). Importantly, while the energy framework succeeds in creating interpretable confidence stratification—future work should optimize the mapping to clinical decision thresholds.

Adaptive computation mirrors clinical workflow through interpretable **hierarchical reasoning**. The model demonstrates clear stratification by case complexity, with early stopping rates decreasing from 25% (single findings) to 15% (2–3 findings) to just 5% (4+ findings), while average iterations increase from 7.2 to 13.1. This adaptive behavior mirrors radiologist workflow [13] and accounts for case complexity. Unlike current MLLMs that process all cases uniformly, HERMES adapts its reasoning depth to diagnostic difficulty—simple cases terminate early at global or regional levels (43.2% overall early stop), while complex multi-pathology cases appropriately receive focal analysis. This creates interpretable reasoning paths where clinicians can understand why certain cases triggered deeper analysis (high energy, multiple findings) versus confident early diagnosis (low energy, clear findings).

Case Complexity as a Quantifiable Clinical Metric. Our quantification of case complexity through adaptive processing provides an objective measure of diagnostic difficulty with real-world applications. If AI reasoning depth correlates with expert cognitive load, we finally have a metric for what makes a case "difficult." Clinical integration could leverage this for: Future clinical integration could leverage our approach for: (1) worklist balance and prioritization based on case complexity rather than volume, potentially reducing burnout from consecutive difficult cases, (2) dynamic resource allocation in teleradiology networks, (3) training case selection for residents based on reasoning depth required, (4) quality assurance by flagging cases with high energy uncertainty, (5) payment/reimbursement models could account for cognitive effort beyond procedural codes; and (6) AI-human collaboration could route simple cases for automated screening while reserving complex multi-finding cases for review.

4.1   Limitations and Future Directions

Calibration Optimization. While meeting the clinical threshold, our framework offers opportunities for domain-specific refinement. Future enhancements could explore context-aware calibration—adjusting confidence thresholds based on clinical urgency, patient risk factors, or institutional protocols—to further optimize decision support for specific deployment scenarios. **Feature-space trade-offs.** Operating on pre-extracted embeddings enables efficient training and inference while maintaining strong performance (AUC = 0.905). Although this approach precludes pixel-level spatial localization, our hierarchical framework successfully captures multi-scale reasoning, with 57% of cases requiring focal-level analysis. The distribution of processing levels (32% global, 11% regional, 57% focal) reflects the varying complexity of abdominal findings, from obvious organ-level abnormalities to subtle focal pathologies requiring detailed analysis. **Clinical validation requirements.** Evaluation on Merlin-extracted labels may not capture full radiologist reasoning complexity. Multi-institutional validation with radiologist ground truth is needed for generalization across varied imaging protocols and patient populations.

5   Conclusion

HERMES establishes energy-based hierarchical reasoning as a promising direction for trustworthy clinical AI. While probability calibration requires refinement, our approach successfully demonstrates adaptive computation that mirrors radiologist workflow, creates interpretable confidence measures, and correlates with case complexity. By teaching MLLMs to quantify their uncertainty and adapt their reasoning depth, we move closer to AI systems that know when to think harder—and when to ask for help. Energy becomes confidence, hierarchy enables efficiency, and together MLLMs learn to think before they answer, building the foundation for clinical AI that physicians can trust.

References

1. Ben-Baruch, E., et al.: Asymmetric loss for multi-label classification (2021). https://arxiv.org/abs/2009.14119
2. Blankemeier, L., et al.: Merlin: a vision language foundation model for 3d computed tomography. arXiv preprint arXiv:2406.06512 (2024)
3. Chen, Y., Liu, C., Liu, X., Arcucci, R., Xiong, Z.: BIMCV-R: a landmark dataset for 3d CT text-image retrieval. In: Medical Image Computing and Computer Assisted Intervention (MICCAI) 2024. Lecture Notes in Computer Science, vol. 15011, pp. 124–134. Springer, Heidelberg (2024). https://doi.org/10.1007/978-3-031-75489-3_12
4. Chen, Z., et al.: CheXagent: Towards a foundation model for chest x-ray interpretation. arXiv preprint arXiv:2406.09264 (2024)
5. Doo, F.X., et al.: Optimal large language model characteristics to balance accuracy and energy use for sustainable medical applications. Radiology **312**(2), e240320 (2024). https://doi.org/10.1148/radiol.240320
6. Doo, F.X., et al.: Environmental sustainability and AI in radiology: a double-edged sword. Radiology **310**(2), e232030 (2024). https://doi.org/10.1148/radiol.232030
7. Gijp, A., Webb, E.M., Naeger, D.M.: How radiologists think: understanding fast and slow thought processing and how it can improve our teaching. Acad. Radiol. **24**(6), 768–771 (2017). https://doi.org/10.1016/j.acra.2016.08.012
8. Gladstone, A., et al.: Energy-based transformers are scalable learners and thinkers. arXiv preprint arXiv:2507.02092 (2025)
9. Guo, C., Pleiss, G., Sun, Y., Weinberger, K.Q.: On calibration of modern neural networks (2017). https://arxiv.org/abs/1706.04599
10. Hamamci, I.E., et al.: A foundation model utilizing chest CT volumes and radiology reports for supervised-level zero-shot detection of abnormalities. arXiv preprint arXiv:2406.17055 (2024)
11. Hewitt, J., Geirhos, R., Kim, B.: We can't understand AI using our existing vocabulary (2025). https://arxiv.org/abs/2502.07586
12. Kahneman, D.: Thinking, Fast and Slow. Farrar, Straus and Giroux, New York (2011)
13. Kelahan, L.C., Fong, A., Blumenthal, J., Kandaswamy, S., Ratwani, R.M., Filice, R.W.: The radiologist's gaze: mapping three-dimensional visual search in computed tomography of the abdomen and pelvis. J. Digit. Imaging **32**(2), 234–240 (2018). https://doi.org/10.1007/s10278-018-0121-8

14. LeCun, Y., Chopra, S., Hadsell, R., Ranzato, M., Huang, F.J.: A tutorial on energy-based learning. In: Bakir, G., Hofman, T., Schölkopf, B., Smola, A., Taskar, B. (eds.) Predicting Structured Data, pp. 1–50. MIT Press (2006)
15. Li, C., et al.: LLaVA-Med: training a large language-and-vision assistant for biomedicine in one day. arXiv preprint arXiv:2306.00890 (2023)
16. Moor, M., et al.: Foundation models for generalist medical artificial intelligence. Nature **616**(7956), 259–265 (2023). https://doi.org/10.1038/s41586-023-05881-4
17. Roth, H.R., et al.: A new 2.5D representation for lymph node detection using random sets of deep convolutional neural network observations. In: Golland, P., Hata, N., Barillot, C., Hornegger, J., Howe, R. (eds.) MICCAI 2014. LNCS, vol. 8673, pp. 520–527. Springer, Cham (2014). https://doi.org/10.1007/978-3-319-10404-1_65
18. Saab, K., et al.: Capabilities of gemini models in medicine. arXiv preprint arXiv:2404.18416 (2024)
19. Setio, A.A.A., et al.: Pulmonary nodule detection in CT images: false positive reduction using multi-view convolutional networks. IEEE Trans. Med. Imaging **35**(5), 1160–1169 (2016). https://doi.org/10.1109/TMI.2016.2536809
20. Shen, H., et al.: Towards bidirectional human-AI alignment: a systematic review for clarifications, framework, and future directions (2024). https://arxiv.org/abs/2406.09264
21. Tang, Y.X., et al.: Automated abnormality classification of chest radiographs using deep convolutional neural networks. NPJ Dig. Med. **3**(1), 70 (2020). https://doi.org/10.1038/s41746-020-0273-z
22. Wang, Z., Wu, Z., Agarwal, D., Sun, J.: MedCLIP: contrastive learning from unpaired medical images and text. In: 2022 Conference on Empirical Methods in Natural Language Processing (EMNLP), pp. 833–846. Association for Computational Linguistics (2022). https://aclanthology.org/2022.emnlp-main.52/
23. Zhang, K., et al.: A generalist vision-language foundation model for diverse biomedical tasks. Nat. Med. **30**(11), 3129–3141 (2024). https://doi.org/10.1038/s41591-024-03185-2
24. Zhang, S., et al.: Large-scale domain-specific pretraining for biomedical vision-language processing. arXiv preprint arXiv:2303.00915 (2023)
25. Zhu, W., Liu, C., Fan, W., Xie, X.: DeepLung: deep 3D dual path nets for automated pulmonary nodule detection and classification. In: 2018 IEEE Winter Conference on Applications of Computer Vision (WACV), pp. 673–681. IEEE (2018). https://doi.org/10.1109/WACV.2018.00081

MedDual: A Practical Dual-Decoding Framework for Mitigating Hallucinations in Medical Vision-Language Models

Zhe Zhang[1,2], Daisong Gan[1,2], Zhaochi Wen[1,3], and Dong Liang[1,2(✉)]

[1] Shenzhen Institutes of Advanced Technology, Chinese Academy of Sciences, Shenzhen, China
[2] University of Chinese Academy of Sciences, Beijing, China
`dongliang@siat.ac.cn`
[3] ShanghaiTech University, Shanghai, China

Abstract. Medical Vision-Language Models (Med-VLMs) excel in clinical image understanding but suffer from hallucinations, producing plausible yet incorrect outputs that risk clinical safety. We present MedDual, a practical dual-decoding framework integrating our Modality-Aware Contrastive Decoding (MACD) with Dynamic Correction Decoding (DeCo). MACD employs tailored perturbations for specific modalities (X-ray, CT, MRI, pathology) to preserve diagnostic features and disrupt spurious correlations, surpassing conventional uniform approaches. DeCo dynamically corrects logits via intermediate layer contrasts for improved factuality. This synergy refines predictions without retraining or external resources. Evaluations on VQA-RAD, SLAKE, and PathVQA show gains up to 6.3% in open-ended recall and 0.9% in close-ended accuracy over baselines like VCD and DoLa. Ablations confirm complementary benefits: MACD enhances visual grounding, DeCo boosts textual accuracy. Our open-source solution advances Med-VLM reliability for healthcare deployment.

Keywords: Medical Vision-Language Models · Hallucination Mitigation · Modality-Aware Decoding · Clinical AI Reliability

1 Introduction

The rapid evolution of Vision-Language Models (Med-VLMs) [1,2] has ushered in a new era of vision-language understanding, demonstrating remarkable proficiency across diverse multimodal tasks. This progress has sparked considerable interest in adapting these models for specialized domains, particularly in healthcare where Med-VLMs [3,4] promise to revolutionize clinical workflows. These models offer the potential to augment physician decision-making, streamline diagnostic processes, and ultimately enhance patient outcomes. However, a critical challenge undermines their clinical viability: the pervasive phenomenon of hallucination [5]. Recent investigations have revealed that Med-VLMs inherit

J. Qiu et al. (Eds.): Agentic AI 2025/CMLLMs 2025/CREATE 2025, LNCS 16147, pp. 249–258, 2026.
https://doi.org/10.1007/978-3-032-06004-4_25

and potentially amplify this limitation [6], generating plausible yet factually incorrect outputs with alarming frequency. In medical contexts, such hallucinations transcend mere technical inaccuracies-even minor textual errors can cascade into catastrophic misdiagnoses, inappropriate treatment recommendations, and adverse patient outcomes [7].

Med-VLMs demonstrate promising capabilities for clinical image understanding. Models like LLaVA-Med [3] adapt the LLaVA architecture [2] for biomedical tasks through instruction tuning on medical image-text pairs. While achieving strong performance on medical VQA benchmarks, these models exhibit significant hallucinations, generating content that appears plausible but lacks grounding in the actual visual input [?]. This phenomenon, where models produce confident yet factually incorrect outputs, manifests across various Med-VLMs including Med-Flamingo [14] and BiomedCLIP [17].

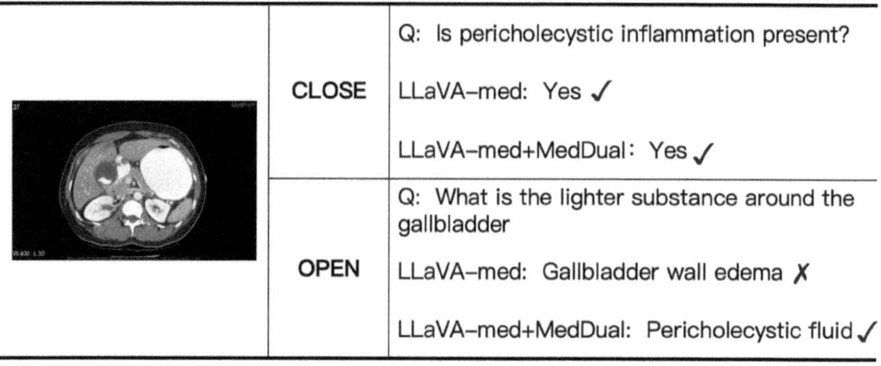

Fig. 1. Qualitative comparison of medical visual question answering on two types of questions from VQA-RAD dataset. CLOSE represents closed questions that require binary yes/no answers, while OPEN represents open questions that require descriptive responses.

In medical contexts, hallucinations pose particularly severe risks that extend far beyond the aesthetic concerns of general vision tasks. A model incorrectly reporting "bilateral pleural effusion" when only unilateral effusion exists could trigger unnecessary bilateral thoracentesis procedures.Figure 1 illustrates typical hallucination patterns in medical VQA, where models generate plausible but incorrect anatomical descriptions. Evaluating these hallucinations requires medical-specific approaches that go beyond general vision metrics, as emerging frameworks like those proposed in generalist biomedical AI attempt to address.

Current mitigation strategies address only isolated aspects of this complex problem. Visual Contrastive Decoding (VCD) [11] reduces object hallucinations by contrasting outputs from original and perturbed images, achieving success in general vision tasks. However, VCD applies uniform Gaussian noise across all images-an approach fundamentally unsuited to medical imaging where different modalities have distinct noise characteristics and diagnostic features. For instance, the speckle patterns in ultrasound images carry diagnostic information that would be destroyed by generic perturbations, while the subtle density variations in mammography require preservation of specific frequency components.

Layer-based approaches like DoLa [8] improve factuality by contrasting different model layers, leveraging the observation that factual knowledge emerges in later layers. While showing promise for language models, these methods lack validation in multimodal medical contexts where visual and textual understanding must be jointly optimized. Other strategies including prompt engineering [16], retrieval augmentation, and reinforcement learning from human feedback have been explored, but these often require model retraining, external knowledge bases, or extensive human annotation-limiting their practical deployment in clinical settings. Recent advancements include benchmarks specifically for medical hallucinations and decoding-time methods like dynamic logits calibration. However, these approaches either require training or do not fully address modality-specific aspects in medical imaging.

The fundamental limitation of existing approaches lies in their failure to account for the unique characteristics of medical imaging. Each imaging modality-whether X-ray, CT, MRI, ultrasound, or pathology-has specific physical principles, artifact patterns, and diagnostic features that require tailored treatment. A one-size-fits-all perturbation strategy cannot adequately probe the model's reliance on genuine diagnostic features versus spurious correlations. This gap motivates our development of a modality-aware approach that respects the inherent properties of different medical imaging types.

We propose MedDual, a theoretically grounded framework that synergistically combines our novel MACD with DeCo [12]. Our key contributions are:

- **Modality-Aware Innovation:** We introduce MACD, which applies tailored contrastive strategies for different medical imaging modalities, recognizing that X-rays, CT scans, pathology slides, and ultrasounds have fundamentally different noise characteristics and diagnostic features.
- **Medical Adaptation:** We develop specific perturbation strategies that preserve diagnostic features while disrupting spurious correlations, validated through radiologist evaluation.
- **Rigorous Evaluation:** We conduct comprehensive experiments on three medical VQA benchmarks with systematic hyperparameter analysis, demonstrating consistent improvements across diverse medical tasks while maintaining computational efficiency.

2   Methodology

2.1   Problem Formulation

Given medical image $I \in \mathbb{R}^{H \times W \times C}$ and question Q, a VLM generates response $Y = \{y_1, ..., y_T\}$ by modeling:

$$p(Y|I,Q) = \prod_{t=1}^{T} p(y_t|y_{<t}, I, Q; \theta) \tag{1}$$

We aim to reduce hallucinations while maintaining accuracy, measured through task-specific metrics.

2.2 Modality-Aware Contrastive Decoding (MACD)

Unlike VCD which applies uniform perturbations, our MACD introduces tailored perturbations specific to each medical imaging modality designed to disrupt spurious correlations and identify hallucinations effectively. Given a visual input v, the MACD approach generates a perturbed counterpart v' using modality-specific perturbation strategies informed by extensive research on medical image augmentation techniques:

- **Pathology:** We implement stain color augmentation using HSV(Hue, Saturation, Value) and HED(Hematoxylin, Eosin, DAB) color space transformations [19], which have been proven most effective for histopathology images. Specifically, we apply HSV transformations (hue shift $\pm10\%$, saturation $\pm15\%$, brightness $\pm10\%$) and HED color deconvolution-based augmentation to simulate realistic stain variations across different laboratories while preserving cellular morphology [22].
- **MRI:** We simulate motion artifacts characteristic of MRI acquisition, including ghosting artifacts in the phase-encoding direction and random motion blur using the view2Dmotion framework. Motion parameters include rigid transformations with translation (±2mm) and rotation ($\pm2°$), plus simulated k-space corruption to generate realistic ghosting patterns [20].
- **CT:** We apply beam hardening simulation and Poisson noise modeling that reflects the polychromatic nature of X-ray CT. This includes cupping artifacts (center-to-edge intensity variation $\pm15\%$) and streaking artifacts around high-density objects, implemented using physically-motivated attenuation models [23].
- **X-ray:** We combine traditional geometric augmentations with X-ray specific artifacts including grid artifacts, motion blur (simulating patient movement), and contrast variations ($\gamma \in [0.7, 1.3]$) that preserve bone-tissue contrast ratios critical for diagnosis [24].

Our augmentation strategies are implemented using established open-source frameworks including TorchIO for MRI-specific transforms [21], torchstain for pathology color normalization, and custom implementations based on validated medical image augmentation libraries [19]. The critical insight in MACD is that if the model produces the same diagnosis or prediction despite these targeted perturbations, it indicates a reliance on hallucinated or spurious visual cues rather than genuine diagnostic evidence. Thus, our contrastive decoding formulation explicitly penalizes such robustness to perturbation:

$$p_{\text{MACD}}(y|v, v', x) = \text{softmax}\left[(1 + \alpha)\text{logit}_\theta(y|v, x) - \alpha\text{logit}_\theta(y|v', x)\right] \quad (2)$$

where x represents the textual query, y is the model-generated output, and α modulates the strength of the contrastive term. By encouraging the model's predictions to change appropriately under realistic clinical perturbations, MACD effectively identifies and mitigates hallucinations, enhancing model reliability in medical diagnostics.

2.3 Dynamic Correction Decoding

Following [8], we introduce Dynamic Correction Decoding (DeCo) to mitigate hallucinations by dynamically incorporating intermediate-layer representations into the final decoding step. This conceptually mirrors the residual connections found in ResNet architectures, leveraging earlier network layers to correct the predictions of deeper layers.

Specifically, we identify optimal contrastive layers by computing the Jensen-Shannon Divergence (JSD) between intermediate layer distributions $p^{(l)}$ and the final layer distribution $p^{(L)}$:

$$l^* = \arg \max_{l \in \{1,\dots,L-1\}} \mathrm{JSD}(p^{(l)}, p^{(L)}) \tag{3}$$

Once the optimal intermediate layer l^* is identified, DeCo refines the final output distribution through an additive correction term, thus explicitly correcting and calibrating the model predictions:

$$p_{\mathrm{DeCo}}(y_t) = \mathrm{softmax}\left[\log p^{(L)}(y_t) + \beta(\log p^{(L)}(y_t) - \log p^{(l^*)}(y_t))\right] \tag{4}$$

Here, β controls the extent of correction from the intermediate layer. By effectively integrating earlier-layer knowledge into the final predictions, DeCo improves model robustness against hallucinations, resulting in enhanced accuracy and reliability in medical imaging applications.

2.4 MedDual: Integrated Modality-Aware Dynamic Correction Decoding

To leverage the strengths of both MACD and DeCo, we propose MedDual, an integrated approach that combines modality-specific contrastive perturbations with dynamic layerwise corrections (Fig. 2). This hybrid method first applies DeCo to refine the logits from both the original and perturbed visual inputs, ensuring that intermediate-layer knowledge corrects potential hallucinations in each branch. Subsequently, the corrected logits are contrasted to penalize invariance under perturbations, thereby disrupting spurious correlations while enhancing overall robustness.

Formally, using the DeCo-corrected logits for the original input v and the perturbed input v' (with the same optimal intermediate layer l^* identified from the original input's distributions), the integrated MedDual distribution is obtained by applying the contrastive formulation:

$$p_{\mathrm{MedDual}}(y|v, v', x) = \mathrm{softmax}\left[(1 + \alpha) \cdot \mathrm{logit}_{\mathrm{DeCo}}(y|v, x) - \alpha \cdot \mathrm{logit}_{\mathrm{DeCo}}(y|v', x)\right] \tag{5}$$

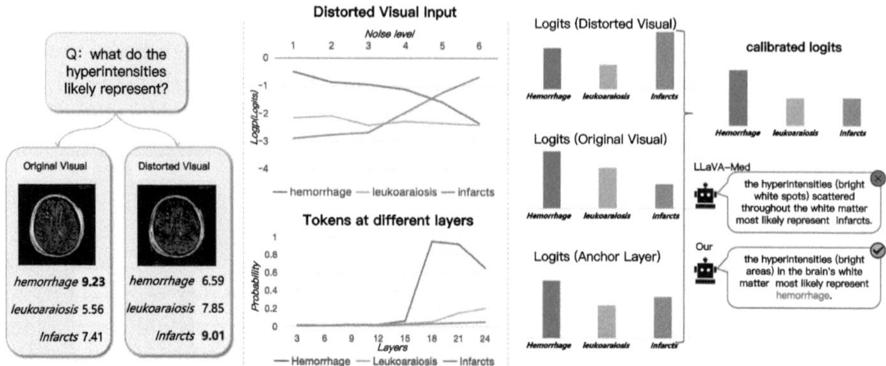

Fig. 2. Illustration of MedDual method for hallucination mitigation in medical VQA. Our approach processes the original image and its modality-specific perturbed counterpart (left). The contrastive decoding analyzes logits changes across noise level and token probabilities across transformer layers(center), where the 'Anker layer' refers to the optimal intermediate layer. The final probability distributions show how Med-Dual recalibrates overconfident predictions, shifting from incorrect "infarcts" to correct "hemorrhage" diagnosis (right), demonstrating effective hallucination detection and mitigation.

This combination organically synergizes the two techniques: DeCo's layer-wise corrections calibrate the model's internal representations against overconfidence and hallucinations, while MACD's targeted perturbations ensure that the refined predictions remain sensitive to genuine diagnostic features rather than artifacts. By integrating these mechanisms, MedDual achieves superior hallucination mitigation and diagnostic reliability in multimodal medical applications, as demonstrated in our experimental evaluations.

3   Experiments

3.1   Experimental Setup

All experiments are conducted on NVIDIA A6000 (48GB) hardware. Following the LLaVA-Med evaluation protocol, we use three datasets: VQA-RAD [10], which consists of 3,515 radiology QA pairs evaluated using accuracy for closed-ended questions by directly comparing predictions with ground truth and recall for open-ended questions by calculating the proportion of ground-truth tokens present in the generated sequences; SLAKE [13], which includes 14,028 medical QA pairs assessed with the same metrics of accuracy for close-ended/multi-choice questions and recall for open-ended questions; and PathVQA [9], comprising 32,799 pathology QA pairs similarly evaluated using accuracy for close-ended questions and recall for open-ended questions.

Table 1. Performance comparison across medical VQA benchmarks. CE: Close-ended accuracy (%), OE: Open-ended recall (%). Best in **bold**.

Method	VQA-RAD		SLAKE		PathVQA	
	OE	CE	OE	CE	OE	CE
LLaVA-Med 7b [3]	61.52	84.19	83.08	85.34	37.95	91.21
+ ITI [18]	61.32	83.78	83.42	85.27	37.78	91.26
+ VCD [11]	63.79	84.11	84.75	85.31	38.24	91.50
+ DoLa [8]	65.02	84.94	89.12	86.09	**38.85**	91.36
+ DeCo [12]	62.88	84.53	85.47	85.68	38.13	91.45
+ MACD	64.26	84.78	87.93	85.89	38.52	91.68
MedDual (Ours)	**65.59**	**85.10**	**89.36**	**86.21**	38.77	**91.92**

We compare against several baselines: LLaVA-Med 7b [3], the base 7B parameter model; LLaVA-Med 7b + ITI [18] with Inference-Time Intervention; LLaVA-Med 7b + VCD [11] with Visual Contrastive Decoding; and LLaVA-Med 7b + DoLa [8] with layer-wise decoding.

3.2 Results

Figure 2 shows MedDual's improved handling of complex queries on VQA-RAD, enabling accurate identification of infarct-like features in renal CT scans, e.g., "The hypodense areas in the kidney are most likely to represent infarcts."

Quantitative comparisons are summarized in Table 1. MedDual outperforms baselines, with generic methods like VCD showing modest gains while our modality-aware approach excels in open-ended tasks.

Key Findings:

- MedDual achieves consistent superiority, with up to 6.3% OE gain on SLAKE, enhancing comprehensive medical descriptions.
- MACD's tailored perturbations outperform VCD's uniform ones, boosting recall by 0.5–3.2% on radiology datasets.
- Balanced improvements: 0.7–0.9% in CE, larger in OE for mitigating descriptive hallucinations.
- On PathVQA, color-aware perturbations aid accuracy despite smaller gains due to microscopic focus.

These validate MedDual's value for reliable clinical deployment.

3.3 Component Analysis

We perform an ablation study based on Table 1 to evaluate component contributions.

- **MACD Alone:** Delivers OE improvements of +2.74% on VQA-RAD, +4.85% on SLAKE, and +0.57% on PathVQA over the base, by using modality-specific perturbations to disrupt spurious visual correlations.
- **DeCo Alone:** Offers OE gains of +1.36% on VQA-RAD, +2.39% on SLAKE, and +0.18% on PathVQA, with average CE increase of +0.31%, through dynamic layer corrections for better textual factuality.
- **MedDual Synergy:** Achieves super-additive effects, e.g., +6.28% OE on SLAKE exceeding the sum of individual gains, particularly benefiting open-ended questions via integrated visual and textual optimization.

This ablation highlights how MACD focuses on visual grounding, DeCo on textual accuracy, and their combination enhances overall hallucination mitigation.

4 Discussion

Our results demonstrate the effectiveness of MedDual in mitigating hallucinations in medical VLMs, particularly through its modality-aware design. The consistent improvements across VQA-RAD, SLAKE, and PathVQA benchmarks highlight the advantages of tailored perturbations over uniform approaches like VCD, enabling better preservation of diagnostic features while disrupting spurious correlations. The synergy between MACD and DeCo not only enhances accuracy in close-ended questions but also significantly boosts recall in open-ended responses, which are critical for comprehensive clinical reporting.

Clinically, these advancements translate to more reliable AI-assisted diagnostics, potentially reducing errors such as misidentifying lesions or omitting key findings. By improving factual grounding without requiring retraining, MedDual offers a practical path toward safer deployment in healthcare settings.

However, limitations persist: performance on rare pathologies is constrained by the base model's training data; designing modality-specific perturbations requires domain expertise; and evaluation is limited to English medical texts.

5 Conclusion

We present MedDual, a dual-decoding framework that integrates Modality-Aware Contrastive Decoding (MACD) with Dynamic Correction Decoding (DeCo) to effectively mitigate hallucinations in medical Vision-Language Models. Our approach achieves substantial gains in accuracy and recall across three benchmarks, balancing performance and efficiency for clinical use.

Future work includes exploring automated perturbation learning via generative models and extending to multilingual applications. Additionally, we plan to construct synthetic datasets by masking lesion regions in medical images and using generative models to inpaint lesion-free versions, inducing controlled hallucinations for fine-tuning-based suppression, thereby creating specialized data to further reduce VLM hallucinations in medical contexts.

Acknowledgments. This work was supported by the National Key R&D Program of China (2022YFA1004203, 2021YFF0501503), the National Natural Science Foundation of China (62125111, 62331028).

References

1. OpenAI: GPT-4 Technical Report. arXiv:2303.08774 (2023)
2. Liu, H., Li, C., Wu, Q., Lee, Y.J.: Visual instruction tuning. In: Advances in Neural Information Processing Systems, vol. 36, pp. 6003–6021 (2023)
3. Li, C., et al.: LLaVA-Med: training a large language-and-vision assistant for biomedicine in one day. In: Advances in Neural Information Processing Systems, vol. 36, pp. 28541–28564 (2023)
4. Wu, C., Zhang, X., Zhang, Y., Wang, Y., Xie, W.: Towards generalist foundation model for radiology by leveraging web-scale 2D&3D medical data. In: The Twelfth International Conference on Learning Representations (ICLR) (2024)
5. Li, Y., Du, Y., Zhou, K., Wang, J., Zhao, X., Wen, J.R.: Evaluating object hallucination in large vision-language models. In: Proceedings of the 2023 Conference on Empirical Methods in Natural Language Processing, pp. 292–305 (2023)
6. Wu, J., Kim, Y., Wu, H.: Hallucination benchmark in medical visual question answering. arXiv:2401.05827 (2024)
7. Pal, A., Sankarasubbu, M.: Gemini goes to med school: exploring the capabilities of multimodal large language models on medical challenge problems & hallucinations. In: Proceedings of the 6th Clinical Natural Language Processing Workshop, pp. 21–46 (2024)
8. Chuang, Y.S., Xie, Y., Luo, H., Kim, Y., Glass, J., He, P.: DoLa: decoding by contrasting layers improves factuality in large language models. In: The Twelfth International Conference on Learning Representations (ICLR) (2024)
9. He, X., Zhang, Y., Mou, L., Xing, E., Xie, P.: PathVQA: 30000+ questions for medical visual question answering. In: MICCAI 2020. LNCS, vol. 12261, pp. 222–231 (2020)
10. Lau, J.J., Gayen, S., Ben Abacha, A., Demner-Fushman, D.: A dataset of clinically generated visual questions and answers about radiology images. Sci. Data **5**, 180251 (2018)
11. Leng, S., Zhang, H., Chen, G., et al.: Mitigating object hallucinations in large vision-language models through visual contrastive decoding. In: Proceedings of the IEEE/CVF Conference on Computer Vision and Pattern Recognition (CVPR), pp. 25952–25961 (2024)
12. Wang, C., et al.: MLLM can see? Dynamic correction decoding for hallucination mitigation. arXiv preprint arXiv:2410.11779 (2024)
13. Liu, B., Zhan, L.M., Wu, X.M.: SLAKE: a semantically-labeled knowledge-enhanced dataset for medical visual question answering. In: ISBI 2021 (2021)
14. Moor, M., et al.: Med-flamingo: a multimodal medical few-shot learner. In: Proceedings of the 3rd Machine Learning for Health Symposium, PMLR, vol. 225, pp. 353–367 (2023)
15. Ouyang, L., Wu, J., Jiang, X., et al.: Training language models to follow instructions with human feedback. In: NeurIPS 2022 (2022)
16. Yao, S., Yu, D., Zhao, J., et al.: Tree of thoughts: deliberate problem solving with large language models. In: Advances in Neural Information Processing Systems, vol. 36, pp. 23512–23535 (2023)

17. Zhang, S., et al.: Large-scale domain-specific pretraining for biomedical vision-language processing. arXiv:2303.00915 (2023)
18. Zhang, S., et al.: Inference-time intervention: eliciting truthful answers from a language model. In: Advances in Neural Information Processing Systems, vol. 36, pp. 20743–20757 (2023)
19. Tellez, D., et al.: Quantifying the effects of data augmentation and stain color normalization in convolutional neural networks for computational pathology. Med. Image Anal. **58**, 101544 (2019)
20. Al-masni, M.A., et al.: Stacked U-Nets with self-assisted priors towards robust correction of rigid motion artifact in brain MRI. Neuroimage **259**, 119411 (2022)
21. Pérez-García, F., Sparks, R., Ourselin, S.: TorchIO: a Python library for efficient loading, preprocessing, augmentation and patch-based sampling of medical images in deep learning. Comput. Methods Programs Biomed. **208**, 106236 (2021)
22. Wagner, S.J., et al.: Structure-preserving multi-domain stain color augmentation using style-transfer with disentangled representations. In: de Bruijne, M., et al. (eds.) MICCAI 2021. LNCS, vol. 12908, pp. 257–266. Springer, Cham (2021). https://doi.org/10.1007/978-3-030-87237-3_25
23. Elbakri, I.A., Fessler, J.A.: Statistical image reconstruction for polyenergetic x-ray computed tomography. IEEE Trans. Med. Imaging **21**(2), 89–99 (2002)
24. Shorten, C., Khoshgoftaar, T.M.: A survey on image data augmentation for deep learning. J. Big Data **6**(1), 1–48 (2019). https://doi.org/10.1186/s40537-019-0197-0

Beyond One Size Fits All: Customization of Radiology Report Generation Methods

Tom van Sonsbeek[(⊠)], Arnaud A. A. Setio, Jung Oh Lee, Junwoo Cho, Junha Kim, Hyeonsoo Lee, Gunhee Nam, Laurent Dillard, and Taesoo Kim

Lunit Inc., Seoul, South Korea
{tom.vansonsbeek,arnaud.setio,jwcho,jhkim,hslee,ghnam,
laurent.dillard,taesoo.kim}@lunit.io

Abstract. Recent advances in generative models have accelerated progress in automated report generation from chest X-rays, which can potentially reduce the workload for clinicians. However, existing methods are often biased to the most prevalent reporting style in their training data, overlooking variations in clinical workflows across regions, institutions, and languages that occur in real-world clinical data. To address these limitations, we introduce a report generation framework that customizes reports through in-context learning from style examples. Our approach includes (i) a style classification metric to quantify customization effectiveness, (ii) a report customization method to adapt generated reports to diverse styles, and (ii) a standalone report generation model that enables multi-style, multilingual report generation. Through our results we show current biases in report generation methods and how our report customization strategies mitigate this. By improving adaptability, our method enhances the possibilities for integration of automated report generation into diverse clinical environments.

Keywords: Radiology report generation · Customization · Chest X-rays

1 Introduction

Chest X-ray imaging is one of the most frequently performed diagnostic procedures, due to its clinical utility, minimal patient burden, and low cost. Radiology reports generated from these images are critical for patient assessment and decision-making, yet their manual creation is time-consuming. Therefore, automated approaches to report generation (RG) could reduce the workload on radiologists and potentially improve patient outcomes [25].

Recent developments in generative language models (LMs) have accelerated the progress of automated generation of reports from X-rays [2,3,6,16, 17,21,26,32]. These methods can effectively learn from large databases of existing radiology reports, accurately inferring key findings from X-rays. However,

© The Author(s), under exclusive license to Springer Nature Switzerland AG 2026
J. Qiu et al. (Eds.): Agentic AI 2025/CMLLMs 2025/CREATE 2025, LNCS 16147, pp. 259–268, 2026.
https://doi.org/10.1007/978-3-032-06004-4_26

we observe that existing methods are biased towards the dominant reporting style of the most prevalent data source. This bias overlooks the variability in clinical workflows and reporting standards, which can vary based on regional, demographic, and institution-specific practices-ranging from punctuation, abbreviations and reporting language [19] to placement of key information [27]. Because these workflows are already well-established and often practically not feasible for modification, a flexible solution is necessary to accommodate diverse reporting styles.

A reason for the bias in generated reports is the scarcity of publicly available datasets with paired chest X-rays and reports, e.g., MIMIC-CXR [14], IU-X-Ray [8], or the recently introduced (reports of) CheXpert-Plus [5]. As model development is typically performed on these same datasets, the risk of overfitting towards specific reporting styles increases. This is a known, and persisting problem in computer vision [18]. Moreover, an often overlooked aspect in RG is language bias. Approximately ~85% of research papers on RG use English for the reports, despite English accounting for only about ~50% of radiology reports worldwide [19]. We argue that creating RG methods that can generate reports in different writing styles and languages can help elevate the field of RG. Using advances in generative LMs, especially in-context learning [7], we aim to construct a more flexible and adaptive RG method that functions well under data-constrained settings.

In this paper we show (i) that RG methods are biased toward the training corpus in terms of report style, by introducing a style classification metric. Next, we introduce (ii) a report adaptation method to mitigate style bias by incorporating in-context style examples, which can be integrated with any existing RG method. Lastly, we introduce (iii) a stand-alone RG method for generation of customizable reports, that encodes images into a general content representation, which can then be used to generate reports in any style and language.

2   Related Works

In-Context-Learning and **Retrieval Augmented Generation** are learning strategies that are known to work well on generative tasks in the general [15], but also the medical domain. Several studies found that RG can be boosted by using additional (retrieved) information, such as disease documentation [1], image or text embeddings [20,22], or similar report texts [23].

Generation of radiology reports in specific styles is mostly unexplored. With increased capabilities of generative models, the following studies made a contribution towards report customization: Hartsock *et al.* [12] explored a method to adapt reports so their length and conciseness matches reference reports. Yan *et al.* [27] use radiologist-specific graphs of disease-terminology that is used for in-context adaptation with LMs of reports to match the writing style of that specific radiologist. We propose a simple and more adaptable customization method that can be applied to a wide range of customization scenarios (Fig. 1).

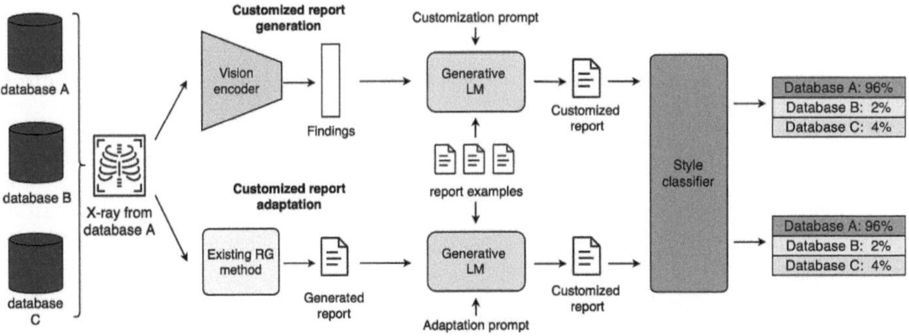

Fig. 1. Method overview, showing the proposed customized report adaptation, report generation and style prediction towards possible source database styles, with example output.

3   Methods

3.1   Problem Statement

We formulate two distinct customization scenarios: customized Report Adaptation (RA) and customized RG. Both customization methods can be formalized as: $R_c = C(R, p|i)$, with R the original report or input data representation, C a generative LM and p being the customization prompt which contains examples that correspond to the style identifier i. R_c is the customized report. We create a style classifier $S(R_c)$, which predicts style identifier i from a predefined set of learned styles. We use a cross-entropy based measure to assess successful customization through the F1 score over S, called **Style-F1**.

3.2   Style-F1: Measuring Style Disentanglement

The style classifier S is designed to categorize reports into a predefined set of report styles. In this paper, we define each dataset as a distinct report style. The training splits of these datasets are used to train S, which is implemented as a Multi-layer Perceptron (MLP). The classifier uses text embeddings from R_c as input and predicts the source dataset i for R_c. The Style-F1 score predicts the source dataset of generated reports from the test split. This metric quantifies how well the output of the customization aligns with dataset-specific stylistic characteristics.

3.3   Customized Report Adaption

We customize the style of a generated report g as follows in a token-by-token manner: $(g_1, g_2, \ldots, g_j) \xrightarrow{C(g, p|i)} (r_1^c, r_2^c, \ldots, r_k^c)$, with randomly selected examples from the training corpus of i embedded in prompt p. To verify whether the customization does not alter the meaning of the report, we compare R_c with g through existing RG metrics that measure content.

3.4 Customized Report Generation

We propose a stand-alone RG method that produces an interpretable intermediate state that summarizes the findings of a given X-ray, generated by a pre-trained vision encoder.

We used **vision-language pre-training (VLP)** for X-ray interpretation, leveraging a Vision Transformer (ViT-L) backbone [9]. The model is trained using a combination of contrastive and captioning objectives on a large-scale dataset comprising pairs of X-rays and radiology reports. This pre-training strategy enables the model to learn representations that capture both visual and textual semantics.

Following pre-training, we fine-tuned the model on a manually annotated dataset to predict radiological findings. We apply 1×1 convolutional heads on top of the frozen VLP backbone, each dedicated to a specific finding. These heads generate feature maps, which are subsequently aggregated using max-pooling to produce a list of positive findings.

In a similar manner to customized RA, we generate a customized report with a LM, guided by a prompt embedded with randomly selected samples: $(t_1, t_2, \ldots, t_j) \xrightarrow{C(t,p|i)} (r_1^c, r_2^c, \ldots, r_k^c)$ where t is structured as an itemized list of positive findings. Alternatively, instead of using random samples, the relevant examples can be selected through kNN retrieval over the VLP embeddings from the training corpus of i.

4 Experimental Setup

Datasets. The three most widely used X-ray-report datasets are used for our method: MIMIC-CXR [14] (228k pairs), CheXpert-Plus [5] (187k pairs) and IU-X-Ray [8] (4k pairs). Next to that, to test multi-linguality of our customization method, we used PadChest [4] (160k pairs) which contains reports written in Spanish. The official train/validation/test split is used for each dataset.

Evaluation. Alongside our style classification F1, we measure performance of English report content with RadGraph [13], RaTEScore [31] and 1/RadCliQ-v1 (so higher scores are better for all metrics) [29]. Semantic similarity is measured with BLEU-2. This is in line with RG metrics used in the ReXrank leaderboard [30]. We apply customization to three existing RG methods: RGRG [24], MAIRA-2 [2] and Med-Gemini [28], selected based on performance and availability of a public code implementation.

Implementation Details. The style classifier S is implemented using Sentence Transformer embeddings from PubMedBERT [11], followed by a MLP classifier. The sizes of the MLP classifier layers are 768, 128, 64, l_i, with $l_i = 4$ the four public datasets we evaluate on are used to train S for five epochs with a batch size of 128.

Table 1. Abstractions of the prompts used for customized report adaption (a) and generation (b). The full prompts can be found in the public code base[1].

(a) **Customized Report Adaptation**
Adjust the style of the following radiology report (**original report**) based on the style of the given example reports: (**examples**). Follow the phrasing and language of the examples. Maintain the content and length of the report

---

(b) **Customized Report Generation**
Generate a radiology report for the given chest X-ray, based on the following structured data (**X-ray findings**). Use the following examples as style reference: (**examples**). Follow the phrasing and language of the examples

---

For customized RG, the ViT is pre-trained with an internal dataset of 5.5M paired X-rays and reports in three epochs with a batch size of 12. Subsequently, fine-tuning on 5.7k images with manual annotations was done, where three radiologists annotated each image to identify 57 findings. A majority vote was applied to determine a single label for each finding. The fine-tuning was performed for 60 epochs with a batch size of 32.

For C, the customization LM, The 'Instruct' versions of Llama3.1-8B and Llama3.3-70B [10] are used. Their multi-linguality and additional instruct-tuning steps make them suitable for our customization tasks. A maximum of 4000 output tokens is set. In case fine-tuning is applied, LoRA ($r = 8, \alpha = 16$) is used for 5000 steps and a batch size of 12.

Training of the style classifier and inference of all models <70B is done on 4 NVIDIA RTX 4090 GPUs. Inference of models >70B, VLP training and LM fine-tuning is done on 8 NVIDIA H100 GPUs.

Table 1 shows abstractions of the prompts used for customized RA and RG. By default they are used with ten style examples. Full prompts are available in the projects code base[1].

5   Results

Customization Improves Style Disentanglement. When applied to ground-truth reports, Style-F1 accurately predicts the dataset source of those reports, demonstrating its ability to distinguish and identify dataset-specific stylistic features (Fig. 2). However, reports generated by existing RG methods are predominantly predicted as having the style of MIMIC-CXR, as shown by the confusion matrices of S. This confirms that there is an existing bias in RG. After applying customized RA or RG methods, the generated reports exhibit improved style alignment. This also extends to the PadChest dataset, which contains reports written in Spanish. Existing RG methods generate reports in English, leading to mis-classification by the style classifier. After customization, reports are generated in Spanish, and are easily aligned with the PadChest report characteristics.

---

[1] github.com/tjvsonsbeek/customized-radiology-report-generation/.

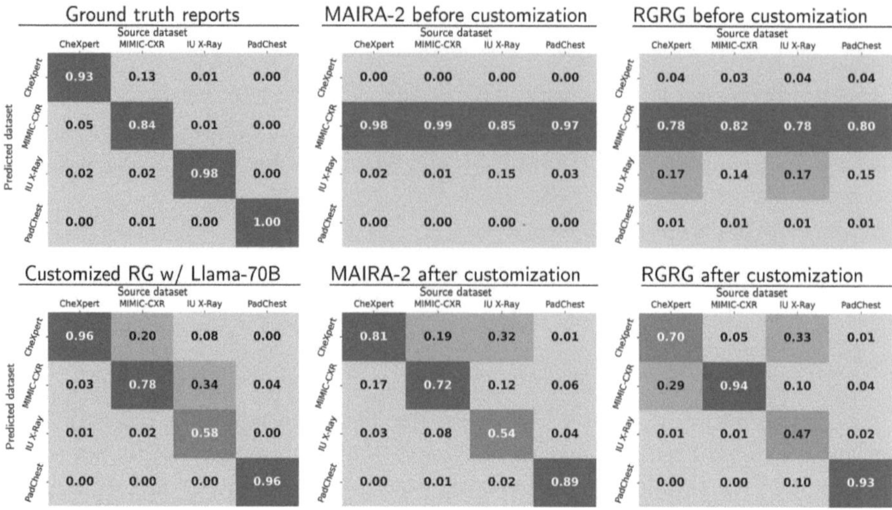

Fig. 2. Confusion matrices showing accuracy of source dataset prediction (with the test split), indicating a positive effect of customization on style disentanglement. Style-F1 is calculated based on these scores. The style of MIMIC-CXR is largely assumed by existing RG methods. After customization the style of a report becomes more similar to their source dataset.

Report Adaptation Improves Style Without Content Performance Loss. The effectiveness of customized RA is shown in Table 2. It shows a consistent improvement in Style-F1 (also illustrated in Fig. 2), with only minimal performance fluctuations on report content metrics. This is most pronounced in CheXpert-Plus and IU-X-Ray. These findings support the hypothesis that RG methods have a bias towards the style of MIMIC-CXR, a large dataset extensively used for training and development over the years. It is likely that this bias extends to unseen X-rays from other data sources. This further motivates the need to consider style in RG. We observed no benefit of using a customization LM larger than 8B parameters.

Customized Report Generation Is Competitive with Existing Methods. The lower half of Table 2 shows the performance of our proposed method to generate customized reports. Using even off-the-shelf LMs, competitive performance can be reached, specifically on CheXpert-Plus, a dataset not part of the dominant training corpus (e.g., MIMIC-CXR). Directly generating customized reports offers a more efficient approach compared to RA.

Fine-Tuning or Retrieving Style Examples Have Minimal Impact on Customization Quality. Results on customized RG in Table 2 show that using style examples retrieved based on VLP embeddings, or using fine-tuning has little effect on report quality, both in style and content metrics. This aligns with the

Table 2. Comparison of customized report adaptation (top) showing improvements style disentanglement after customization with ten random examples from the source dataset. For customized report generation (bottom) results are shown for fine-tuning (FT) and inclusion of random or retrieved style examples.

		MIMIC-CXR					CheXpert-plus					IU-X-ray					PadChest	
		1/RadCliQ-v1	Radgraph	BLEU	RATE score	Style-F1	1/RadCliQ-v1	Radgraph	BLEU	RATE score	Style-F1	1/RadCliQ-v1	Radgraph	BLEU	RATE score	Style-F1	BLEU	Style-F1
Customized report adaptation																		
Med-Gemini [21]	Original reports	**0.68**	**0.24**	0.13	**0.51**	0.44	0.46	0.07	0.12	**0.47**	0.24	**1.16**	**0.34**	**0.28**	**0.64**	0.09	-	-
	w/ Llama3.1-8B	0.63	**0.24**	0.14	0.51	0.57	0.52	**0.14**	0.14	0.45	0.67	0.81	0.33	0.18	0.64	0.19	-	-
RGRG [24]	Original reports	0.63	0.22	**0.15**	0.48	0.66	0.44	0.06	0.11	0.43	0.00	0.29	0.12	0.07	0.46	0.06	0.01	0.00
	w/ Llama3.1-8B	0.61	0.22	**0.15**	0.48	**0.81**	0.48	0.12	0.13	0.42	0.67	0.52	0.15	0.09	0.54	0.68	0.06	0.75
MAIRA-2 [2]	Original reports	0.58	0.16	0.07	0.46	0.49	0.43	0.04	0.06	0.42	0.07	0.70	0.23	0.11	0.47	0.23	0.00	0.00
	w/ Llama3.1-8B	0.58	0.18	0.09	0.47	0.72	0.50	0.13	0.12	0.41	0.70	0.74	0.29	0.16	0.55	0.66	0.08	0.84
Customized report generation																		
Llama3.1-8B	Random examples	0.52	0.10	0.06	0.40	0.54	0.47	0.07	0.07	0.41	0.75	0.65	0.19	0.14	0.39	0.54	0.06	0.88
	Retrieved examples	0.54	0.13	0.08	0.45	0.56	0.52	0.12	0.10	0.43	0.74	0.69	0.21	0.15	0.45	0.59	0.08	0.93
	Retrieved examples + FT	0.54	0.14	0.08	0.44	0.56	0.53	0.12	**0.16**	0.44	0.75	0.69	0.21	0.16	0.46	0.58	0.08	0.93
Llama3.3-70B	Random examples	0.55	0.13	0.10	0.43	0.66	0.50	0.09	0.08	0.40	0.81	0.81	0.29	0.21	0.55	0.68	0.07	0.94
	Retrieved examples	0.56	0.15	0.11	0.47	0.66	0.56	0.13	0.10	0.43	**0.82**	0.87	0.30	0.23	0.56	**0.73**	**0.10**	**0.96**
	Retrieved examples + FT	0.55	0.17	0.12	0.47	0.66	**0.57**	0.13	0.10	0.44	0.81	0.88	0.32	0.22	0.60	**0.73**	0.09	**0.96**

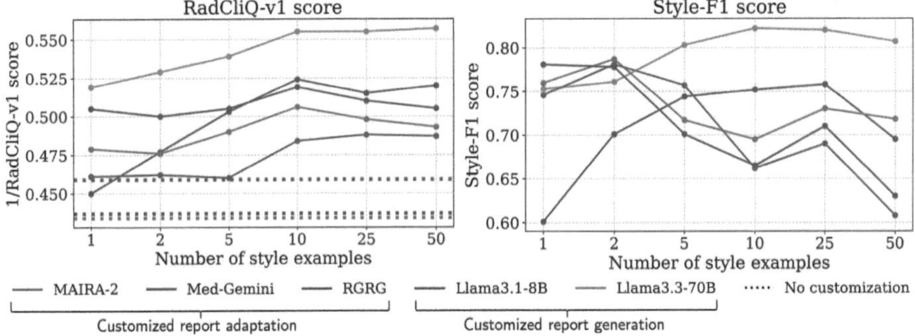

Fig. 3. The effect of the number of in-context examples on style and content metrics for CheXpert-Plus, together with baseline performance for customization methods. More in-context style examples improve content metric performance, while it has a slight negative effect on Style-F1.

design of our method, which requires no domain-specific or formatting knowledge, as this information is provided directly through the in-context examples.

More Style Examples Are Not Always Better. The presence of more style examples has a positive effect on the content quality of reports, while it has a negative effect on the Style-F1 scores as shown in Fig. 3 for CheXpert-Plus. It is likely that individual report examples can skew content in the wrong direction. While using a larger amount of style examples can mitigate this issue, the

Table 3. Examples of customization on IU-X-Ray (top) and PadChest (bottom). Inclusion of key words or phrases that show the stylistic customization towards the original report are highlighted.

Ground truth report	MAIRA-2	MAIRA-2 with customization	Llama3.3-70B customized RG
The cardiac contour is widened. The costophrenic angles are blunted, suggesting small pleural effusions. The lungs are clear without evidence of focal consolidation or pneumothorax. The bony structures are intact.	There is mild bibasilar atelectasis and mild pulmonary edema. An enlarged cardiac silhouette is observed.	There is mild bibasilar atelectasis and mild pulmonary edema. There is no pneumothorax. An enlarged cardiac contour is observed.	Chest radiograph demonstrates severe cardiomegaly, as well as small bilateral pleural effusions. A more confluent opacity is seen in the right middle lobe. There is no pneumothorax.
Presencia de enfisema y EPOC.	Signs of COPD with chronic inflammatory changes mainly in the right upper lobe.	Signos radiológicos de EPOC. Tractos pleuroparenquimatosos parahiliares derechos.	Hallazgos radiológicos característicos en la EPOC.

added complexity of using many examples (e.g., >25) reduces style performance. Therefore ten style examples is seen as the optimal option.

What Is Being Customized? Table 3 presents examples of customization in English and in Spanish. The adapted MAIRA-2 report incorporates stylistic elements such as the naming conventions for the cardiac silhouette and the explicit mention of absent findings. The second example demonstrates the correct translation of COPD into its Spanish equivalent. Since PadChest reports are generally more concise than those in other datasets, condensing information to retain only key details is preferable. This is achievable through customized RG, but not through RA, as the latter is bound to original report content.

Limitations on Customization Scenarios. The customized methods proposed in this paper have broader clinical use-cases than simply adapting towards style of specific datasets. Think of the inclusion of specific phrases, emphasis on findings that are especially relevant (by institution or regional factors) or adaptations to more languages than just Spanish. This is achievable because our customization method only requires a few stylistic examples. However, a limiting factor in exploring these scenarios is the ability to quantify and validate clinical performance, for which reviews from radiologists are needed.

6   Conclusion

In this paper, we address the overlooked issue of bias in automatically generated radiology reports. While existing approaches generate coherent and clinically accurate content, they inherently reflect biases from training data. Our proposed methods make use of in-context learning to guide the style of radiology reports. First, we show how the style or language of an existing report can be modified while preserving the content. Secondly, we show that it is possible to generate customized reports from scratch. By addressing these biases in RG, this paper highlights a critical yet under-recognized challenge that can impact advancement and real-world integration of automated radiology reporting systems.

References

1. Alam, H.M.T., Srivastav, D., Kadir, M.A., Sonntag, D.: Towards interpretable radiology report generation via concept bottlenecks using a multi-agentic RAG. arXiv preprint arXiv:2412.16086 (2024)
2. Bannur, S., et al.: MAIRA-2: grounded radiology report generation. arXiv preprint arXiv:2406.04449 (2024)
3. Bu, S., Li, T., Yang, Y., Dai, Z.: Instance-level expert knowledge and aggregate discriminative attention for radiology report generation. In: CVPR, pp. 14194–14204 (2024)
4. Bustos, A., Pertusa, A., Salinas, J.M., Iglesia-Vaya, M.: Padchest: a large chest X-ray image dataset with multi-label annotated reports. Med. Image Anal. **66**, 101797 (2020)
5. Chambon, P., et al.: Chexpert plus: hundreds of thousands of aligned radiology texts, images and patients. arXiv e-prints pp. arXiv–2405 (2024)
6. Chen, Z., Song, Y., Chang, T.H., Wan, X.: Generating radiology reports via memory-driven transformer. EMNLP (2020)
7. Davidson, E.M., et al.: The reporting quality of natural language processing studies: systematic review of studies of radiology reports. BMC Med. Imaging **21**, 1–13 (2021)
8. Demner-Fushman, D., et al.: Preparing a collection of radiology examinations for distribution and retrieval. JAMIA **23**(2), 304–310 (2016)
9. Dosovitskiy, A., et al.: An image is worth 16x16 words: transformers for image recognition at scale. In: ICLR (2021)
10. Dubey, A., et al.: The Llama 3 herd of models. arXiv preprint arXiv:2407.21783 (2024)
11. Gu, Y., et al.: Domain-specific language model pretraining for biomedical natural language processing. ACM Health **3**(1), 1–23 (2021)
12. Hartsock, I., Araujo, C., Folio, L., Rasool, G.: Improving radiology report conciseness and structure via local large language models. arXiv preprint arXiv:2411.05042 (2024)
13. Jain, S., et al.: Radgraph: extracting clinical entities and relations from radiology reports. NeurIPS Benchmarks Track (Round 1) (2021)
14. Johnson, A.E., et al.: MIMIC-CXR, a de-identified publicly available database of chest radiographs with free-text reports. Sci. Data **6**(1), 317 (2019)
15. Lewis, P., et al.: Retrieval-augmented generation for knowledge-intensive NLP tasks. NeurIPS **33**, 9459–9474 (2020)
16. Li, Y., Wang, Z., Liu, Y., Wang, L., Liu, L., Zhou, L.: Kargen: knowledge-enhanced automated radiology report generation using large language models. In: MICCAI, pp. 382–392. Springer (2024)
17. Liu, Z., Zhu, Z., Zheng, S., Zhao, Y., He, K., Zhao, Y.: From observation to concept: a flexible multi-view paradigm for medical report generation. IEEE Trans. Multimedia (2023)
18. Liu, Z., He, K.: A decade's battle on dataset bias: are we there yet? ICLR (2025)
19. Meddeb, A., et al.: Large language model ability to translate CT and MRI free-text radiology reports into multiple languages. Radiology **313**(3), e241736 (2024)
20. Ranjit, M., Ganapathy, G., Manuel, R., Ganu, T.: Retrieval augmented chest X-ray report generation using OpenAI GPT models. In: Machine Learning for Healthcare Conference, pp. 650–666. PMLR (2023)

21. Saab, K., et al.: Capabilities of gemini models in medicine. arXiv preprint arXiv:2404.18416 (2024)
22. van Sonsbeek, T., Worring, M.: X-TRA: Improving chest X-ray tasks with cross-modal retrieval augmentation. In: IPMI, pp. 471–482. Springer (2023)
23. Sun, L., Zhao, J., Han, M., Xiong, C.: Fact-aware multimodal retrieval augmentation for accurate medical radiology report generation. NAACL (2025)
24. Tanida, T., Müller, P., Kaissis, G., Rueckert, D.: Interactive and explainable region-guided radiology report generation. In: CVPR, pp. 7433–7442 (2023)
25. Tanno, R., et al.: Collaboration between clinicians and vision–language models in radiology report generation. Nat. Med. 1–10 (2024)
26. Wang, X., et al.: Activating associative disease-aware vision token memory for LLM-based X-ray report generation. arXiv preprint arXiv:2501.03458 (2025)
27. Yan, B., et al.: Style-aware radiology report generation with radgraph and few-shot prompting. EMNLP Findings (2023)
28. Yang, L., et al.: Advancing multimodal medical capabilities of gemini. arXiv preprint arXiv:2405.03162 (2024)
29. Yu, F., et al.: Evaluating progress in automatic chest X-ray radiology report generation. Patterns **4**(9) (2023)
30. Zhang, X., et al.: ReXrank: a public leaderboard for AI-powered radiology report generation. arXiv preprint arXiv:2411.15122 (2024)
31. Zhao, W., Wu, C., Zhang, X., Zhang, Y., Wang, Y., Xie, W.: Ratescore: a metric for radiology report generation. EMNLP (2024)
32. Zhou, H.Y., Adithan, S., Acosta, J.N., Topol, E.J., Rajpurkar, P.: A generalist learner for multifaceted medical image interpretation. arXiv preprint arXiv:2405.07988 (2024)

Predictive Multimodal Modeling of Diagnoses and Treatments in EHR

Cindy Shih-Ting Huang[✉], Clarence Boon Liang Ng, and Marek Rei

Imperial College London, London, UK
{cindy.huang23,clarence.ng21}@alumni.imperial.ac.uk,
marek.rei@imperial.ac.uk

Abstract. While the ICD code assignment problem has been widely studied, most works have focused on post-discharge document classification. Models for early forecasting of this information could be used for identifying health risks, suggesting effective treatments, or optimizing resource allocation. To address the challenge of predictive modeling using the limited information at the beginning of a patient stay, we propose a multimodal system to fuse clinical notes and tabular events captured in electronic health records. The model integrates pre-trained encoders, feature pooling, and cross-modal attention to learn optimal representations across modalities and balance their presence at every temporal point. Moreover, we present a weighted temporal loss that adjusts its contribution at each point in time. Experiments show that these strategies enhance the early prediction model, outperforming the current state-of-the-art systems.

Keywords: Multimodality · Cross-modal application · Cross-modal information extraction

1 Introduction

Electronic health records (EHR) are comprehensive repositories of patient information, encompassing clinical notes, laboratory tests, diagnostic imaging, and other data sources that collectively document the medical trajectory of a given patient. Much of the research on EHR documents has focused on assigning accurate International Classification of Diseases (ICD) codes based on discharge summaries, which are written at the end of a hospital stay and contain a textual description of the relevant diagnoses and treatments [13,18]. Although recent work has shown that multimodal features [19] and earlier clinical notes [14] can provide additional useful context for this task, these studies have primarily aimed to automate the retrospective analysis of individual documents.

While ICD code classification during discharge has useful applications, the rich temporal structure of EHR has further potential. Systems for jointly modeling and predicting the overall health trajectory of a patient during hospitalization could potentially be used for identifying health risks, suggesting timely treatments, or optimizing healthcare workflow efficiency. The early assignment of

J. Qiu et al. (Eds.): Agentic AI 2025/CMLLMs 2025/CREATE 2025, LNCS 16147, pp. 269–279, 2026.
https://doi.org/10.1007/978-3-032-06004-4_27

diagnoses and treatments is a key factor in improving the effectiveness of patient care, yet very few works on EHR so far have explored prospective models that provide earlier prognostic estimates to allow for integration into clinical pipelines [2]. Furthermore, no prior research has examined the impacts of incorporating multimodal information on the performance for this early-stage prediction task.

In this work, we investigate the use of multimodal learning to predict the diagnoses and treatments that patients will encounter. The system performs ICD code forecasts at various stages of the hospital stay, with the predictions continuously updated as more data becomes available. We design the **M**ultimodal **I**ntegrated **H**ierarchical **S**equence **T**ransformer (MIHST) architecture for augmenting the information in clinical notes with additional data sources, as these may reveal early indicators and complementary features which are not yet captured by textual reports. The model integrates pre-trained encoders, feature pooling, and cross-modal attention to learn optimal representations across modalities and balance their presence at every temporal point.

Experiments show that this additional information is necessary for early prediction, as MIHST with textual and tabular data outperforms all existing models at any time cutoff prior to the final discharge summary. Cross-modal causal attention together with feature pooling is shown to be the best combination, as it allows the architecture to dynamically adjust to the shifting significance of each data source over time, eliminating the need for constant data availability or paired multimodal records. A novel loss function in the model also enhances early predictions by balancing the performance across multiple temporal points. Code for the model and experiments is available at our repository.[1]

2   Related Work

The discharge summary has been a primary focus of research for automating ICD code assignments at the end of a stay. Initial models were based on convolutional neural networks (CNNs) [12,13] and long short-term memory (LSTM) [18,21]. Later, transformer approaches like the Pre-trained Language Model-ICD (PLM-ICD) [8] and the Hierarchical Transformer for Document Sequences (HTDS) model [14] surpassed their performance by dividing long documents into smaller sequences ("chunks") and retaining all token embeddings encoded to represent a document. Notably, HTDS also established the significance of including earlier clinical documents for improved ICD code classification, as these provide additional context for diagnoses and treatments.

Researchers have also investigated multimodal fusion to improve clinical task performance. Early fusion methods textualize other data types with associated source tags [15] or inject token embeddings into the prompt via modal-specific encoders [1]. Recent work has also explored framing ICD code classification as a text-to-text task [3], yet performance still lags behind state-of-the-art. These studies reveal limitations of early fusion, where textualization can obscure data properties and the relative priority of modalities. On the other hand, late fusion

---

[1] https://github.com/cindyellow/ehr-predictive-multimodal-modeling.

frameworks lack information flow between modalities, as seen by Xu et al. [19], who predict ICD-10 codes by averaging the outputs from separate models for notes and tabular events, relying on text availability when other modalities are missing.

These approaches were designed to output their prediction based on the discharge summary at the end of the hospital stay. In contrast, recent work has argued that for practical downstream applications, such code classification should instead be performed on earlier medical notes [5]. The Label-Attentive Hierarchical Sequence Transformer (LAHST) [4] introduced temporal ICD code prediction using causal and label-wise attention for generating predictions at any time point, focusing only on the textual notes as input. Our proposed approach combines both textual and tabular information into a multimodal framework that allows for making real-time predictions throughout the hospital stay, improving performance during the crucial early stages with limited available evidence.

3   Proposed Framework

3.1   Multimodal Representations

Tabular Feature Selection: To assess the benefits of additional modalities, tabular events – specifically laboratory measurements – are examined in this work as they embody diverse information that can reveal valuable insights into disease progression, complementary to those mentioned in textual notes. These entries are represented as name-value pairs, where feature names denote event types and values are the corresponding measurements. Typically, measurement units and event entry time are also provided. We apply feature selection using the training and development sets to identify laboratory events most closely associated with ICD codes.

First, lab feature values undergo Yeo-Johnson transformation with standardization to ensure a uniform scale and Gaussian-like distribution. Missing values are imputed with the mean from the training set. We employ an iterative process of training a logistic regression model with an L1 penalty term for each ICD code. Models are trained on two variables per lab feature: the average measurement and the average difference between consecutive measurements in a stay. We start with features measured in more than 5,000 admissions. For each model, the 10 variables with the highest absolute coefficients are identified. We count how often each lab feature appears across all models and select those important for 20 or more ICD codes. ICD codes with micro-F1 scores (rescaled to 0–100) below 30 are retrained with an expanded variable set, achieved by lowering the admission threshold for filtering lab features to 2,000. The list of significant variables is updated for that code if its score improves. A second retraining phase targets codes with scores under 20, further reducing the threshold to 500 admissions. We do not further retrain to prevent overfitting to rare event types. The final list of 22 laboratory features used in the main model includes laboratory tests that are important for 10 or more ICD codes, as well as those significant for at least 5 labels among codes with scores less than 30.

Tabular Representation: To encode laboratory data, we employ the Tabular Prediction adapted BERT approach (TP-BERTa) [20] pre-trained on classification tasks for a large tabular database. Measurements for events in the list of selected laboratory test features are normalized and discretized with a quantile bin value between 1 and 256 to align with the foundation model. Each bin value has been registered as `mask_token_id + bin_value` in the model vocabulary. Both the feature name and bin value are encoded, yielding an embedding matrix:

$$\mathbf{U}_i = [\mathbf{E}_{\mathrm{CLS}}, \mathbf{E}^i_{\mathrm{name}}, \mathbf{E}^i_{\mathrm{value}}] \in \mathbb{R}^{(F+2) \times D_{\mathrm{tabular}}} \tag{1}$$

where F is the maximum number of tokens used to represent the feature name. $\mathbf{E}^i_{\mathrm{value}}$ is scaled by the normalized lab measurement value.

After intra-feature attention, the embedding at the [CLS] position is used as the final representation of this feature, denoted $\hat{\mathbf{u}}_i \in \mathbb{R}^{1 \times D_{\mathrm{tabular}}}$. The tensor of feature embeddings for all laboratory events is $\mathbf{E}_{\mathrm{tabular}} = [\hat{\mathbf{u}}_1, \ldots, \hat{\mathbf{u}}_M] \in \mathbb{R}^{M \times D_{\mathrm{tabular}}}$, where M is the number of lab events for the admission and D_{tabular} is the hidden dimension of TP-BERTa.

Document Representation: Documents are divided into chunks of T tokens that can be encoded by a pre-trained language model (PLM). During training, a maximum of N chunks are selected to limit resource usage. The resulting matrix $\mathbf{S} \in \mathbb{R}^{N \times T}$ contains all the chunk tokens that serve as input to the model. $\mathbf{S}$ is passed through the PLM to obtain token embeddings, where the tensor at the [CLS] position is selected to represent the document chunk, yielding $\mathbf{E}_{\mathrm{note}} \in \mathbb{R}^{N \times D_{\mathrm{textual}}}$. D_{textual} is the hidden dimension of the PLM. We use `RoBERTa-base-PM-M3-Voc` as the PLM since it was pre-trained on abstracts and full-text content of biomedical works on PubMed and physician notes from MIMIC-III [11].

3.2   Model Design

We develop a model that integrates information from multiple modalities without requiring paired multimodal data for predictions during a patient stay. The model encodes chunks of medical documents and tabular event records, pooling tabular embeddings by timestamp. Textual and tabular representations are merged and sorted chronologically, then given as input to a hierarchical transformer. Next, a causally masked label-wise attention network extracts relevant information for each label up to that time point. Finally, label-specific embeddings are processed by a projection layer to generate temporal predictions for each ICD code. We refer to the model as the **Multi-modal Integrated Hierarchical Sequence Transformer** (MIHST), illustrated in Fig. 1.

Step 1: Clinical event encoding. As described in Sect. 3.1, clinical events are encoded by either the PLM or the tabular foundation model, depending on the event type. Two embeddings are obtained: $\mathbf{E}_{\mathrm{note}}$ for document chunks and $\mathbf{E}_{\mathrm{tabular}}$ for laboratory measurements.

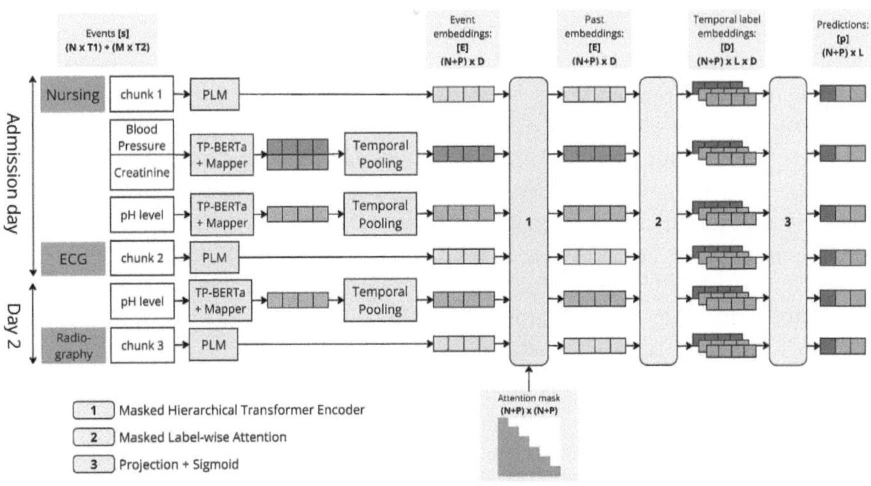

Fig. 1. An overview of the model architecture. Document chunks are encoded by a pre-trained language model (PLM) while tabular measurements are encoded with TP-BERTa, followed by a modality mapper. The resulting tabular representations are pooled based on their timestamps. Updated encodings are passed through the masked hierarchical transformer and label-wise attention. Finally, a projection and sigmoid module outputs the label predictions.

Step 2: Modality mapper. To align the tabular and textual vector spaces, $\mathbf{E}_{\text{tabular}}$ is updated with a trainable mapping network [16] consisting of a linear layer that transforms the tabular dimension D_{tabular} to the textual dimension D_{textual}, followed by a LeakyReLU.

Step 3: Feature pooling. Feature pooling enables the model to manage large volumes of tabular events without increasing computational resources or complexity. Embeddings with the same timestamp are pooled, condensing the tabular dimension into representations comparable to document chunk embeddings, which encapsulate textual information at a given time. For $p \in [1, \ldots, P]$, P being the total number of unique temporal points for tabular events, feature pooling for the pth temporal position is performed on $\mathbf{w}(p)$, the set of tabular embeddings with time p. The entry time of the mth event is denoted $Time(m)$.

$$\mathbf{E}_{\text{pooled}}^{p} = \max_{m \in \mathbf{w}(p)} \left(\mathbf{E}_{\text{tabular}}^{m} \right)$$
$$\mathbf{w}(p) = \{m \in [1, \ldots, M] \mid Time(m) = p\}$$

$$(2)$$

Step 4: Causal Attention. $\mathbf{E} = [\mathbf{E}_{\text{note}}, \mathbf{E}_{\text{pooled}}] \in \mathbb{R}^{(N+P) \times D_{\text{textual}}}$ is obtained by merging and sorting $\mathbf{E}_{\text{note}}$ and $\mathbf{E}_{\text{pooled}}$ by event timestamp. A hierarchical transformer with causal attention [6] refines event embeddings with information from prior events. A masked attention block ensures each position accesses only past information. This generates an embedding matrix $\mathbf{H} \in \mathbb{R}^{(N+P) \times D_{\text{textual}}}$, where $H_i = \text{CausalAttn}(e_1, \ldots, e_i), i \in [1, \ldots, N + P]$.

Step 5: Masked label-wise attention. Label-wise attention network [13] is utilized to prevent any predictions based on future events. The mask at temporal point t, denoted a_t, is constant in the label dimension and nullifies events beyond t. Multi-head attention [17] with learnable label embeddings $\mathbf{Q} \in \mathbb{R}^{L \times D_{\text{textual}}}$ is applied. Linear projections of the key, query, and value embeddings $e_{k,i} = \mathbf{H}\mathbf{W}_i^K$, $e_{q,i} = \mathbf{Q}\mathbf{W}_i^Q$, $e_{v,i} = \mathbf{H}\mathbf{W}_i^V$ are used for each head. The output is $\mathbf{D}_t = \text{MultiHeadAttn}(\mathbf{Q}, \mathbf{H}, \mathbf{H}, a_t) \in \mathbb{R}^{1 \times L \times D_{\text{textual}}}$, which are label-specific embeddings for L labels at each time point $t \in [1, \dots, N + P]$.

Lastly, $\mathbf{D}_{t,\ell} \in \mathbb{R}^{D_{\text{textual}} \times 1}$ is passed through a projection layer followed by the sigmoid function: $p_{t,\ell} = \text{Sigmoid}(\mathbf{W}_\ell \cdot \mathbf{D}_{t,\ell})$. This represents the probability for the lth label at time t. Masking in the preceding modules guarantees that each output is computed using only embeddings of past events.

Training: The same training scheduler and hyperparameters as LAHST [4] are used, tuning only the temporal loss weights. We similarly apply the Extended Context Algorithm (ECA) for textual documents to accommodate indefinite document length by randomly sampling a maximum of N_{max} text chunks during training. Tabular events do not face the same restraint at the encoding step, so we retain all of them to minimize information loss.

During inference, ECA is adapted for multimodality. N_{total} textual note chunks are processed in batches of size N_{max}. Tabular events between the earliest and latest notes in the batch are also included, with M_i denoting the number of tabular entries in batch i. The model encodes both inputs into batch embeddings, which are concatenated to form the embedding for the entire sequence of clinical events $h \in \mathbb{R}^{(N_{\text{total}}+M) \times D_{\text{textual}}}$, $M = \sum M_i$. This is passed to the masked multi-head label attention module for predictions based on the entire event sequence. If no lab records are present for a sample, the model proceeds under the unimodal setting using textual information.

Model training employs binary cross-entropy loss, computed per label l among L total labels by comparing predictions p_l with ground truth y_l. To enhance early performance, we consider the loss across a set of temporal points C, where labels are compared against predictions based on events up to each $t \in C$. We further propose a weighted temporal loss to adjust the contribution of each time point to gradient propagation: $\mathcal{L}_{\text{w}} = -\sum_{t \in C} w_t [\frac{1}{L} \sum_{\ell=1}^{L} (y_\ell \cdot logp_{t,\ell}) + ((1 - y_\ell) \cdot log(1 - p_{t,\ell}))]$, where the weights w_t sum to 1. C in our setup includes 5 temporal points: 2, 5, 13 days after admission, the time point right before the discharge summary, and the entry time of the summary – these temporal positions are also used in evaluation. Experiments showed the best results when the last temporal point is given the highest weight (0.6) and others assigned 0.1.

Training experiments are conducted on Nvidia RTX 6000 GPUs (72 GB RAM). 4 samples are processed per second, yielding a total training time of 11.5 h.

4   Experiment Set-Up and Results

Dataset: For this study, we use the MIMIC-III [10] dataset, which contains de-identified multimodal health records from patients admitted to crit-

Table 1. Evaluation on the test set for early ICD code prediction using all data until each of the specified temporal cutoffs. TrLDC performance is from the original paper [7]. PubMedBERT-Hier (PMB-H; [9]), HTDS [14], and LAHST results are from [4]. HTDS* is a variation of HTDS with similar computation requirements as LAHST and MIHST. Results for all models except TrLDC are averaged across 3 runs with random seeds. Standard deviations for MIHST results are < 0.3.

Model	Last day			0–13 days			0–5 days			0–2 days		
	F1	AUC	P@5	F1	AUC	P@5	F1	AUC	P@5	F1	AUC	P@5
TrLDC	70.1	93.7	65.9	–	–	–	–	–	–	–	–	–
PMB-H	67.2	91.5	63.0	30.7	68.0	30.2	31.3	68.4	31.0	31.7	68.7	31.5
HTDS	**73.3**	**95.2**	**68.1**	49.7	82.1	47.6	47.5	80.6	45.9	44.5	78.7	43.6
HTDS*	70.7	93.8	66.2	48.6	82.0	47.0	46.7	80.7	45.5	43.6	78.7	43.3
LAHST	70.4	94.7	67.6	52.9	87.0	52.8	50.3	85.3	50.5	46.1	82.9	46.9
MIHST	66.0	93.5	65.0	**53.1**	**88.1**	**54.2**	**51.6**	**86.9**	**52.5**	**48.2**	**85.0**	**49.9**

ical care units (ICU) between 2001 and 2012 at the Beth Israel Deaconess Medical Center in Boston, Massachusetts. We align data preprocessing, train/development/test splits, and label space with previous studies [4,13] for comparability, using the top 50 most frequent codes for modeling and evaluation. Note that patients are not excluded if they lack measurements for laboratory tests.

Evaluation Framework: In this task, we define temporal cutoffs at 2 days, 5 days and 13 days for standardized comparison with LAHST [4]. For instance, in the 5-day setting, the model predicts ICD codes based on textual and laboratory events occurring within the first 5 days of admission. We also report the model performance using all events up to (but excluding) the discharge summary to test the model without it, and with all events including the summary. Metrics follow standard conventions in the ICD coding task [13].

We compare MIHST to existing baselines for real-time prediction. TrLDC [7], PMB-H [9], and HTDS [14] are the best-performing models for the post-discharge task, while LAHST [4] serves as the state-of-the-art for early predictions during hospitalization. As shown in Table 1, MIHST consistently outperforms LAHST in early prediction settings, achieving higher Micro-F1, Micro-AUC, and Precision@5 scores. By integrating multimodal representations, MIHST leverages both textual and non-textual information to make more accurate early predictions – this is especially important in the early stages of the hospital stay, as each individual modality contains very limited information.

For post-discharge predictions, other approaches outperform MIHST, likely due to the trade-off from optimizing across multiple time points and modalities. This indicates that optimal model choice depends on the required application: MIHST excels in all early-stage prediction settings, while unimodal models learning from the discharge summary may be more effective for post-discharge assignments.

Table 2. Micro-F1 scores computed on the development set, comparing with feature pooling ablated and different weight schemes. Values are averaged across 3 runs, with the standard deviation shown in the subscript.

	Model	Abl. Pooling	Equal	First	None
2 day	**49.6**$_{\pm 0.1}$	48.8$_{\pm 0.5}$	49.4$_{\pm 0.3}$	48.0$_{\pm 0.3}$	47.8$_{\pm 0.5}$
5 day	**54.0**$_{\pm 0.3}$	52.8$_{\pm 0.6}$	52.8$_{\pm 0.2}$	51.0$_{\pm 0.4}$	52.2$_{\pm 0.5}$
13 day	**56.1**$_{\pm 0.3}$	54.7$_{\pm 0.5}$	54.7$_{\pm 0.2}$	52.1$_{\pm 0.3}$	54.7 $_{\pm 0.6}$
Excl. DS	**56.4**$_{\pm 0.4}$	55.0$_{\pm 0.5}$	55.0$_{\pm 0.3}$	52.2$_{\pm 0.4}$	55.0$_{\pm 0.5}$
Last day	68.6$_{\pm 0.3}$	67.4$_{\pm 0.5}$	64.9$_{\pm 0.2}$	61.1$_{\pm 0.3}$	**69.5**$_{\pm 0.3}$

Ablation Experiments: Table 2 presents key ablation results. Removing pooling lowers performance, indicating its role in preserving the strongest signals and preventing overfitting. We also examine the impact of weighted temporal loss across three settings: "Equal" assigns uniform weights to all time points; "First" assigns the largest weight (0.6) to the first cutoff and 0.1 to the rest; and "None" removes temporal loss, optimizing only for last-day predictions. The "Equal" setting maintains early performance but decreases the last-day score, likely due to reduced emphasis on the discharge summary, which only appears in the last temporal point and is the most relevant evidence for ICD codes. The low scores in the "First" setup further highlight the value of the summary in complementing other clinical records. Nevertheless, the "None" setting shows that distributing weights across all time points is crucial to enhance early prediction.

In Table 3, we evaluate the importance of cross-modal interaction by comparing it to other fusion strategies. These experiments use a model variation that excludes temporal loss and feature pooling to eliminate interfering effects. The "Cross-Modal" setting mirrors the MIHST architecture, where tabular and note embeddings are combined before passing through the masked hierarchical transformer. In other settings, modalities are merged *after* the transformer. "Intra-Modal" allows interaction within each modality by using separate masked transformers, while "None" removes causal attention for tabular features.

Table 3. Micro-F1 scores computed on the development set for models with cross-modal, intra-modal, and no interaction between tabular and text embeddings. Values are averaged across 3 runs, with the standard deviation shown in the subscript.

	2 day	5 day	13 day	Excl. DS	Last day
None	45.1 $_{\pm 0.3}$	48.9 $_{\pm 0.1}$	51.2 $_{\pm 0.1}$	51.8 $_{\pm 0.1}$	69.0 $_{\pm 0.3}$
Intra-Modal	46.1 $_{\pm 0.0}$	49.8 $_{\pm 0.2}$	52.1 $_{\pm 0.1}$	52.8 $_{\pm 0.2}$	69.6 $_{\pm 0.2}$
Cross-Modal	**47.1** $_{\pm 0.2}$	**51.0** $_{\pm 0.1}$	**52.9** $_{\pm 0.1}$	**53.5** $_{\pm 0.2}$	**69.7** $_{\pm 0.1}$

In particular, we observe notable gains in 2-day prediction performance driven by enhanced interactions between tabular and textual embeddings. Introducing causal attention for tabular data yields a 1.0 improvement in Micro-F1, illustrating the value of modeling temporal relationships. Further integration of cross-modal interactions contributes an additional 1.0 gain, which highlights the influence of identifying multimodal patterns on overall model capabilities.

5   Conclusion

This study leverages multimodal data to predict ICD codes at various points during hospitalization, with an emphasis on early prediction of diagnoses and treatments for a given patient. Strengthening prediction quality at the beginning of a hospital stay has the potential to aid clinicians in improving patient outcomes and planning resources. MIHST utilizes pre-trained foundation models for meaningful textual and tabular encodings, which then interact in a causal attention module that updates each representation based on previous information. A weighted temporal loss helps achieve an optimal balance between predictions at temporal points.

Experiments demonstrated that multimodality benefits predictions when unimodal data offers weaker evidence, notably soon after admission. The novel weighted temporal loss aligns optimizations across temporal positions, while feature pooling moderates modality presence to emphasize the most informative features. This yields a system that surpasses the state-of-the-art for early predictions.

MIHST is agnostic to the pre-trained encoder choice and easily extends to new modalities via modality-specific encoders. Its cross-modal interaction design adapts to varying data availability and alignment. As more powerful PLMs emerge, MIHST can incorporate a wider range of data sources to improve diagnoses and treatment decisions. The results highlight the potential for mining rich multimodal EHR data to advance prospective applications in clinical practices.

References

1. Belyaeva, A., et al.: Multimodal LLMs for health grounded in individual-specific data. LNCS (including subseries Lecture Notes in Artificial Intelligence and Lecture Notes in Bioinformatics), vol. 14315, pp. 86–102 (2024). https://doi.org/10.1007/978-3-031-47679-2_7/TABLES/3
2. Ben-Israel, D., et al.: The impact of machine learning on patient care: a systematic review. Artif. Intell. Med. **103**, 101785 (2020). https://doi.org/10.1016/J.ARTMED.2019.101785
3. Boyle, J.S., Kascenas, A., Lok, P., Liakata, M., O'Neil, A.Q.: Automated clinical coding using off-the-shelf large language models. In: Deep Generative Models for Health Workshop NeurIPS 2023 (2023). https://openreview.net/forum?id=mqnR8rGWkn
4. Caralt, M.H., Boon, C., Ng, L., Rei, M.: Continuous predictive modeling of clinical notes and ICD codes in patient health records. In: Demner-Fushman, D., Ananiadou, S., Miwa, M., Roberts, K., Tsujii, J. (eds.) Proceedings of the 23rd Workshop on Biomedical Natural Language Processing, pp. 243–255. Association for Computational Linguistics (2024). https://doi.org/10.18653/v1/2024.bionlp-1.19

5. Cheng, H., et al.: MDACE: MIMIC documents annotated with code evidence. In: Proceedings of the 61st Annual Meeting of the Association for Computational Linguistics (Volume 1: Long Papers), vol. 1, pp. 7534–7550. Association for Computational Linguistics (2023). https://doi.org/10.18653/V1/2023.ACL-LONG.416

6. Choromanski, K., et al.: From block-Toeplitz matrices to differential equations on graphs: towards a general theory for scalable masked transformers. In: International Conference on Machine Learning, vol. 162, pp. 3962–3983. PMLR (2022). https://doi.org/10.48550/arXiv.2107.07999

7. Dai, X., Chalkidis, I., Darkner, S., Elliott, D.: Revisiting transformer-based models for long document classification. In: Findings of the Association for Computational Linguistics: EMNLP 2022, pp. 7212–7230. Association for Computational Linguistics (2022). https://doi.org/10.18653/v1/2022.findings-emnlp.534

8. Huang, C.W., Tsai, S.C., Chen, Y.N.: PLM-ICD: automatic ICD coding with pre-trained language models. In: Proceedings of the 4th Clinical Natural Language Processing Workshop, pp. 10–20. Association for Computational Linguistics (ACL) (2022). https://doi.org/10.18653/V1/2022.CLINICALNLP-1.2

9. Ji, S., Höltä, M., Marttinen, P.: Does the magic of BERT apply to medical code assignment? a quantitative study. Comput. Biol. Med. **139**, 104998 (2021). https://doi.org/10.1016/J.COMPBIOMED.2021.104998

10. Johnson, A.E., et al.: MIMIC-III, a freely accessible critical care database. Sci. Data **3**(1), 1–9 (2016). https://doi.org/10.1038/sdata.2016.35

11. Lewis, P., Ott, M., Du, J., Stoyanov, V.: Pretrained language models for biomedical and clinical tasks: Understanding and extending the state-of-the-art. In: Proceedings of the 3rd Clinical Natural Language Processing Workshop, pp. 146–157. Association for Computational Linguistics, Online (2020). https://doi.org/10.18653/v1/2020.clinicalnlp-1.17

12. Liu, Y., Cheng, H., Klopfer, R., Gormley, M.R., Schaaf, T.: Effective convolutional attention network for multi-label clinical document classification. In: Proceedings of the 2021 Conference on Empirical Methods in Natural Language Processing, Online and Punta Cana, Dominican Republic, pp. 5941–5953. Association for Computational Linguistics (2021). https://doi.org/10.18653/v1/2021.emnlp-main.481

13. Mullenbach, J., Wiegreffe, S., Duke, J., Sun, J., Eisenstein, J.: Explainable prediction of medical codes from clinical text. In: Proceedings of the 2018 Conference of the North American Chapter of the Association for Computational Linguistics: Human Language Technologies, Volume 1 (Long Papers), New Orleans, Louisiana, pp. 1101–1111. Association for Computational Linguistics (2018). https://doi.org/10.18653/v1/N18-1100, https://aclanthology.org/N18-1100

14. Ng, C.B.L., Santos, D., Rei, M.: Modelling temporal document sequences for clinical ICD coding. In: Vlachos, A., Augenstein, I. (eds.) Proceedings of the 17th Conference of the European Chapter of the Association for Computational Linguistics, pp. 1640–1649. Association for Computational Linguistics (2023). https://doi.org/10.18653/V1/2023.EACL-MAIN.120

15. Niu, S., Ma, J., Bai, L., Wang, Z., Guo, L., Yang, X.: EHR-KnowGen: knowledge-enhanced multimodal learning for disease diagnosis generation. Inf. Fusion **102**, 102069 (2024). https://doi.org/10.1016/J.INFFUS.2023.102069

16. Ramos, R., Bugliarello, E., Martins, B., Elliott, D.: PAELLA: parameter-efficient lightweight language-agnostic captioning model. In: Duh, K., Gomez, H., Bethard, S. (eds.) Findings of the Association for Computational Linguistics: NAACL 2024, pp. 3549–3564. Association for Computational Linguistics (2024). https://doi.org/10.18653/V1/2024.FINDINGS-NAACL.225

17. Vaswani, A., et al.: Attention is all you need. In: Advances in Neural Information Processing Systems, vol. 30. Curran Associates, Inc. (2017)
18. Vu, T., Nguyen, D.Q., Nguyen, A.: A label attention model for ICD coding from clinical text. In: Bessiere, C. (ed.) Proceedings of the Twenty-Ninth International Joint Conference on Artificial Intelligence, IJCAI-2020, vol. 2021-January, pp. 3335–3341. International Joint Conferences on Artificial Intelligence Organization (2020). https://doi.org/10.24963/ijcai.2020/461
19. Xu, K., et al.: Multimodal machine learning for automated ICD coding. In: Proceedings of the 4th Machine Learning for Healthcare Conference, vol. 106, pp. 197–215. PMLR (2019). https://proceedings.mlr.press/v106/xu19a.html
20. Yan, J., et al.: Making pre-trained language models great on tabular prediction. In: The Twelfth International Conference on Learning Representations (2024). https://openreview.net/forum?id=anzIzGZuLi
21. Yuan, Z., Tan, C., Huang, S.: Code synonyms do matter: multiple synonyms matching network for automatic ICD coding. In: Proceedings of the 60th Annual Meeting of the Association for Computational Linguistics (Volume 2: Short Papers), Dublin, Ireland, pp. 808–814. Association for Computational Linguistics (2022). https://doi.org/10.18653/v1/2022.acl-short.91

Adapting and Evaluating Multimodal Large Language Models for Adolescent Idiopathic Scoliosis Self-management: A Divide and Conquer Framework

Zhaolong Wu, Pu Luo, Jason Pui Yin Cheung, and Teng Zhang[✉]

University of Hong Kong, Hong Kong, China
{wuzl101,luopudent}@connect.hku.hk, {cheungjp,tgzhang}@hku.hk

Abstract. This study presents the first comprehensive evaluation of Multimodal Large Language Models (MLLMs) for Adolescent Idiopathic Scoliosis (AIS) self-management. We constructed a database of approximately 3,000 anteroposterior X-rays with diagnostic texts and evaluated five MLLMs through a 'Divide and Conquer' framework consisting of a visual question-answering task, a domain knowledge assessment task, and a patient education counseling assessment task. Our investigation revealed limitations of MLLMs' ability in interpreting complex spinal radiographs and comprehending AIS care knowledge. To address these, we pioneered enhancing MLLMs with spinal keypoint prompting and compiled an AIS knowledge base for retrieval augmented generation (RAG), respectively. Results showed varying effectiveness of visual prompting across different architectures, while RAG substantially improved models' performances on the knowledge assessment task. Our findings indicate current MLLMs are far from capable in realizing personalized assistant in AIS care. The greatest challenge lies in their abilities to obtain accurate detections of spinal deformity locations (best accuracy: 0.55) and directions (best accuracy: 0.13).

Keywords: Multimodal Large Language Models · Scoliosis · X-ray

1 Introduction

Adolescent Idiopathic Scoliosis (AIS) is the most common spinal deformity in pediatric populations, affecting up to 4.8% of adolescents [5,9]. This condition predominantly occurs during growth spurts between ages 11 and 14. Without timely intervention, spinal deformities may progressively worsen, leading to serious consequences such as back pain and impaired pulmonary function, reducing patients' quality of life [3].

While treatments, surgical-interventions and clinical management are key in AIS care, patient self-management plays an essential role in the recovery from AIS, which could have paramount effects on outcomes, quality of life and long-term well-being [7,13]. This includes exercise and physical therapy, compliance with treatment, emotion and mental health as well as monitoring and reporting [10,17,19].

MLLMs such as LLaVA-Med have demonstrated powerful capabilities in medical image analysis, detecting lesions, analyzing conditions, and providing medical advice [16,21,26]. Although these models excel in chest radiographs and fundus image analysis, research targeting spinal diseases, particularly AIS, remains scarce due to the limited availability of specialized spinal data. Furthermore, it is unclear whether and how well current state-of-the-art open-source/weight MLLMs are at realizing a personal assistant for AIS self-management, which requires capabilities beyond medical imaging comprehension.

To address this gap, we constructed a database of approximately 3,000 anteroposterior X-rays with corresponding diagnostic texts. We evaluated leading MLLMs through a comprehensive framework that breaks down the AIS self-management requirement into three downstream tasks: a visual spinal assessment task (the ability of x-ray analysis for disease progression), a domain knowledge assessment task (the understanding of the disease and managements), and a patient education counseling assessment task (the ability to do personalized assistant given patient's x-ray and open-ended queries).

The main contributions of this paper can be summarized as follows.

1. We conducted the first comprehensive study on using MLLMs for AIS self-management through a divide-and-conquer framework, which is aligned well with clinical practice, separating complex AIS care into specialized tasks.
2. For adapting MLLMs for AIS care, we integrated a spinal keypoint detection model with MLLMs for improving their ability on analyzing spinal deformities for the image modality. To improve MMLMs in comprehending AIS knowledge, we compiled an AIS knowledge base and implemented a knowledge auugmented generation approach.
3. We constructed a large-scale specialized image-text database for AIS, the largest to date to the best of our knowledge, providing foundational resources for future related research.

2   Related Work

Recent years have witnessed significant advances in applying MLLMs to medical tasks. These models successfully integrate various modalities to perform complex medical tasks, such as medical image-text generation, disease diagnosis, and automated report generation [4,8,14,18]. However, current research exhibits a notable data bias toward certain anatomical regions.

Current MLLMs, including medical-specific models like LLaVA-Med and general models like Qwen-VL 2.5 and InternVL-3 [1,16,27], have developed expertise primarily in chest radiograph analysis due to the predominance of chest X-ray datasets [11,12,22]. This has resulted in limited capabilities for spinal disorders like AIS, which requires precise assessment of curve patterns and specialized knowledge of classification systems and progression risk factors.

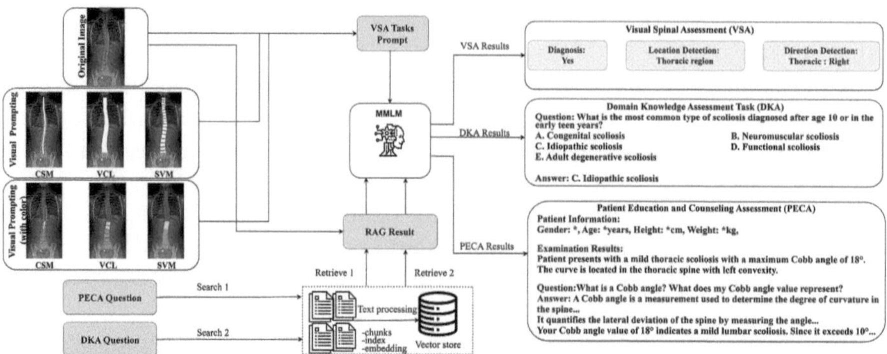

Fig. 1. Divide and Conquer Framework for Adapting and Evaluating MLLMs for AIS Care.

To address this gap, we introduce a novel framework that decomposes the complex AIS care process into distinct, manageable components. This "Divide and Conquer" approach allows us to systematically evaluate and adapt MLLMs for specific aspects of AIS management, from visual diagnosis to patient education. We construct a specialized spinal database and develop targeted assessment methods for each component, while also employing a keypoint detection model to supply critical anatomical landmarks and investigate what missing information leads to diagnostic failures.

3   The 'Divide and Conquer' Evaluation Framework

Figure 1 shows our multi-granularity clinical assessment framework for AIS analysis. We designed three comprehensive evaluation tasks to assess MLLMs' effectiveness in analyzing spinal deformities from AP spine X-rays across multiple clinical perspectives. Additionally, we integrated a spinal keypoint detection model to investigate whether diagnostic errors stem from MLLMs' inability to identify critical spinal landmarks.

Visual Spinal Assessment (VSA). When diagnosing AIS in clinical practice, clinicians primarily focus on examining images to analyze spinal deformities, their location, and curve patterns. We decomposed this complex visual reasoning process into three sequential tasks of increasing granularity:

AIS Diagnosis (AD): This binary classification task requires MLLMs to generate a potential diagnosis based on spinal X-ray images and textual prompts, determining whether AIS is present or absent.

Spinal Deformity Location Detection (SDLD): This multi-class classification task focuses on localizing spinal deformities. Building on the diagnosis from AD, MLLMs must identify whether the spinal curvature occurs in the thoracic, thoracolumbar, or lumbar segments based on X-ray images and textual prompts. This localization helps determine the type of spinal deformity curve.

Spinal Deformity Direction Detection (SDDD): This multi-class classification task determines the direction of spinal deformity. Following AD and SDLD tasks, MLLMs must assess whether the spinal curvature is directed leftward or rightward, distinguishing between left-convex and right-convex curvatures.

Domain Knowledge Assessment Task (DKA). The second major component of our framework, DKA, addresses the professional knowledge dimension of AIS care. This multiple-choice task evaluates whether MLLMs possess sufficient domain knowledge of AIS. We designed questions across six categories: basic knowledge, etiology and pathophysiology, clinical presentation and diagnosis, assessment and monitoring, treatment options, and complications and prognosis. This comprehensive assessment examines the models' understanding of AIS from fundamental concepts to diagnosis, patient management, and treatment strategies.

Patient Education and Counseling Assessment (PECA). This patient-oriented question-answering task evaluates MLLMs' ability to provide accurate, accessible information to patients with varying degrees of spinal deformity. We developed 161 questions across five categories: disease explanation, treatment options, daily life management, long-term prognosis, and follow-up and monitoring. Each question was stratified by three severity levels (mild, moderate, severe) to assess how well models adapt their responses to different clinical scenarios. This task specifically measures the models' capabilities in translating complex medical knowledge into patient-appropriate explanations while maintaining clinical accuracy.

4 Method

4.1 Datasets

In this study, we collected 3,683 AP spinal X-rays with corresponding radiological reports from 3,022 patients at X Hospital between December 2019 and July 2023, with all data collection protocols approved by the Institutional Review Board (IRB). The dataset was randomly partitioned into training, validation, and testing sets following an 8:1:1 ratio, resulting in 2,946 samples for training, 368 samples for validation, and 369 samples for testing. To address the inherent class imbalance common in medical datasets, we implemented a stratified sampling approach that maintained consistent distribution of scoliosis severity across all partitions, ensuring that the proportion of each scoliosis category (normal, mild, moderate, and severe) remained constant across all sets. To minimise input image variance, all X-rays were standardized by cropping to a uniform size of 896 × 448 pixels, ensuring that the entire spinal column was captured within each image.

4.2 Visual Prompting Strategies

We established MLLMs zero-shot predictions as baselines and investigated how performance is affected by visual prompts provided by spine keypoint detection models. As shown in Fig. 1, we designed three visual prompting strategies:

Curved Spine Midline (CSM), Vertebral Connection Line (VCL), and Segmented Vertebrae Marks (SVM). These visual prompts were designed to provide models with critical information about spinal curvature and structure. For SDLD and SDDD tasks, we differentiated thoracic, thoracolumbar, and lumbar regions using distinct colors to provide anatomical localization information.

4.3 Retrieval-Augmented Generation (RAG) Approach

To enhance model performance on knowledge-intensive tasks, we implemented a specialized RAG framework utilizing an AIS-specific knowledge database. This database was constructed by integrating authoritative sources including clinical practice guidelines, research publications from PubMed, and patient education resources from organizations such as the Scoliosis Research Society (SRS) [2,6,15,23,24]. To optimize information retrieval, we employed Gemini to generate structured knowledge graphs that capture key relationships between AIS concepts, treatments, and outcomes [25].

4.4 DKA and PECA Data Population

Our evaluation datasets were designed to assess different aspects of AIS understanding. The DKA dataset consists of multiple-choice questions targeting professional medical knowledge, while the PECA dataset simulates patient-centered scenarios requiring both clinical accuracy and appropriate communication. To ensure comprehensive coverage, we developed questions spanning diagnosis, treatment options, management strategies, and daily living accommodations for patients with varying degrees of spinal deformity. A rigorous quality control process involving two junior doctors and verification by a senior physician was implemented to ensure clinical relevance, accuracy, and appropriate difficulty levels across all questions.

4.5 Evaluation

For the VSA task, we addressed dataset class imbalance by employing a comprehensive evaluation approach combining F1 score, AUC, and accuracy metrics. This multi-metric strategy ensures balanced assessment of model performance across both majority and minority classes. For the DKA task, we used accuracy as our primary performance metric, as it directly measures the models' proficiency in selecting correct answers within the multiple-choice format. In evaluating the PECA task, we implemented a structured human evaluation protocol wherein three junior physicians independently assessed model responses using a five-dimensional framework on a 5-point Likert scale. The detailed assessment criteria, outlined in Table 1, were specifically designed to evaluate both clinical accuracy and communication effectiveness, providing comprehensive insight into models' capabilities in addressing patient concerns across varying degrees of spinal deformity.

5   Experiments

5.1   Experimental Setup

All experiments were conducted using 4 NVIDIA GeForce RTX 3090 GPUs. Table 1 presents the models used for comparison. We primarily utilized open-source general-domain models, selecting five multimodal large language models from four different companies, with parameter sizes ranging from 4.2 B to 14 B. To ensure experimental reproducibility and consistency, we standardized all MLLM configurations with a temperature parameter of 0, bfloat16 quantization, and disabled flash attention. For the PECA Task, we set the maximum response length to 300 tokens. These settings minimized model generation randomness and ensured result reliability. For keypoint detection, we employed SpineHR-Net+, a model specifically trained on spine data based on HRNet and UNet architectures [20].

Table 1. Multimodal Large Language Models Description.

Model	Vision Encoder	Backbone LLM	Connector	Release dates
Qwen2.5-VL-7B	Redesigned ViT	Qwen2.5-7B	MLP	2025.01
InternVL3-8B	InternViT-300M-448px-V2_5	Qwen2.5-7B	MLP	2024.04
InternVL3-14B	InternViT-300M-448px-V2_5	Qwen2.5-14B	MLP	2025.04
Llama 3.2-Vision	CLIP ViT-H/14	Llama 3.1-8B	MLP	2024.07
Phi-3-Vision	CLIP ViT-L/14	Phi-3 Mini	Projection	2024.05

5.2   Results and Discussion

Table 2. Visual Spinal Assessment Results. Thor. = Thoracic, TL = Thoracolumbar, Lum. = Lumbar, OA = Overall accuracy. Baseline shows results without visual prompts. "Color" indicates region-specific color encoding in visual prompts.

Model	Method	Task 1: AIS F1	Task 1: AIS AUC	Task 2 No Color OA	Task 2 No Color Thor./TL/Lum.	Task 2 With Color OA	Task 2 With Color Thor./TL/Lum.	Task 3 No Color OA	Task 3 No Color Thor./TL/Lum.	Task 3 With Color OA	Task 3 With Color Thor./TL/Lum.
Qwen2.5-VL-7B	Baseline	0.83	0.74	0.15	0.19,0.52/0.22,0.52/0.51,0.58	–	–	0.09	0.37/0.45/0.43	–	–
	CSM	0.93	0.79	0.33	0.68,0.64/0.25,0.57/0.61,0.62	0.31	0.79,0.69/0.17,0.47/0.71,0.57	0.09	0.49/0.51/0.36	0.13	0.58/0.53/0.30
	VCL	0.96	0.71	0.43	0.81,0.67/0.15,0.49/0.71,0.58	0.44	0.81,0.77/0.23,0.55/0.80,0.64	0.08	0.58/0.66/0.24	0.08	0.64/0.64/0.22
	SVM	0.92	0.52	0.35	0.81,0.55/0.19,0.51/0.75,0.48	0.40	0.78,0.57/0.11,0.47/0.77,0.54	0.03	0.52/0.67/0.17	0.12	0.51/0.69/0.32
InternVL3-8B	Baseline	0.94	0.50	0.10	0.38,0.59/0.24,0.54/0.20,0.50	–	–	0.03	0.42/0.22/0.32	–	–
	CSM	0.94	0.50	0.07	0.82,0.56/0.23,0.51/0.02,0.50	0.06	0.80,0.50/0.23,0.50/0.00,0.50	0.01	0.53/0.07/0.36	0.00	0.49/0.05/0.36
	VCL	0.94	0.50	0.07	0.82,0.56/0.23,0.51/0.02,0.50	0.05	0.81,0.53/0.23,0.50/0.00,0.50	0.01	0.53/0.07/0.36	0.00	0.51/0.05/0.36
	SVM	0.94	0.50	0.07	0.83,0.62/0.23,0.50/0.02,0.50	0.06	0.80,0.50/0.23,0.50/0.00,0.50	0.01	0.56/0.05/0.36	0.00	0.49/0.05/0.36
InternVL3-14B	Baseline	0.60	0.57	0.26	0.57,0.56/0.00,0.50/0.48,0.53	–	–	0.10	0.46/0.87/0.30	–	–
	CSM	0.95	0.65	0.17	0.82,0.55/0.00,0.50/0.02,0.50	0.28	0.79,0.77/0.00,0.50/0.43,0.54	0.11	0.52/0.87/0.36	0.16	0.61/0.87/0.34
	VCL	0.96	0.83	0.21	0.84,0.63/0.00,0.50/0.02,0.50	0.32	0.83,0.80/0.04,0.51/0.49,0.53	0.14	0.57/0.87/0.36	0.21	0.65/0.87/0.35
	SVM	0.92	0.84	0.22	0.83,0.67/0.00,0.50/0.25,0.52	0.50	0.82,0.74/0.00,0.50/0.79,0.66	0.16	0.60/0.87/0.37	0.38	0.62/0.87/0.57
Llama 3.2-Visio	Baseline	0.94	0.50	0.05	0.81,0.53/0.24,0.55/0.80,0.58	–	–	0.02	0.50/0.17/0.17	–	–
	CSM	0.94	0.50	0.03	0.80,0.50/0.23,0.52/0.74,0.54	0.01	0.81,0.52/0.23,0.50/0.79,0.51	0.01	0.49/0.09/0.17	0.00	0.50/0.05/0.10
	VCL	0.80	0.77	0.10	0.75,0.67/0.23,0.54/0.70,0.63	0.10	0.74,0.66/0.23,0.54/0.73,0.64	0.09	0.57/0.38/0.27	0.09	0.56/0.38/0.27
	SVM	0.94	0.50	0.06	0.80,0.50/0.23,0.50/0.34,0.44	0.01	0.80,0.50/0.23,0.50/0.78,0.50	0.00	0.49/0.05/0.26	0.00	0.49/0.05/0.09
Phi-3-Vision	Baseline	0.84	0.57	0.11	0.00,0.50/0.00,0.50/0.00,0.50	–	–	0.11	0.33/0.87/0.36	–	–
	CSM	0.94	0.50	0.11	0.00,0.50/0.00,0.50/0.00,0.50	0.11	0.00,0.50/0.00,0.50/0.00,0.50	0.11	0.33/0.87/0.36	0.11	0.33/0.87/0.36
	VCL	0.67	0.74	0.11	0.00,0.50/0.00,0.50/0.00,0.50	0.11	0.00,0.50/0.00,0.50/0.00,0.50	0.11	0.33/0.87/0.36	0.11	0.33/0.87/0.36
	SVM	0.94	0.60	0.11	0.00,0.50/0.00,0.50/0.00,0.50	0.11	0.00,0.50/0.00,0.50/0.00,0.50	0.11	0.33/0.87/0.36	0.11	0.33/0.87/0.36

Impact of Different Visual Prompts on AIS Diagnosis. Table 2 shows significant variations in the performance of AIS diagnosis between models with different visual prompts. Structured visual cues generally improved diagnostic accuracy, with VCL achieving high F1 scores for Qwen2.5 (0.96) and InternVL3-14B (0.96). CSM boosted InternVL3-14B's F1 score from 0.60 to 0.95. Notably, InternVL3-8B and most Llama-3.2 configurations showed AUC=0.50, indicating they function as constant positive predictors rather than discriminative classifiers. Only VCL enabled meaningful discrimination in Llama-3.2 (AUC=0.77), while Phi-3.5 showed improved discrimination with VCL (AUC=0.74) and SVM (0.60). These findings demonstrate that visual prompts' effectiveness varies by model architecture, with certain prompting strategies uniquely enabling discriminative capabilities absent in baseline conditions.

Impact of Different Visual Prompts on SDLD. Qwen2.5's accuracy improved significantly from a baseline of 0.15 to 0.43 with VCL prompting, while InternVL3-14B initially declined from 0.26 to 0.17 (CSM) and 0.22 (SVM). Color encoding produced mixed overall effects but dramatically enhanced InternVL3-14B with SVM (0.22→ 0.50). Regional performance was significantly impacted by color enhancement: for thoracic detection, Qwen2.5 maintained strong F1 scores (CSM: 0.68→0.79, VCL: 0.81→0.81, SVM: 0.81 → 0.78), while InternVL3-14B showed impressive gains (CSM: 0.52→0.79, SVM: 0.63→0.82). Thoracolumbar detection remained challenging with modest improvements for Qwen2.5 (F1 scores 0.55–0.67), while lumbar region detection showed no significant improvement across models with color enhancement. Results indicate that despite interventions successfully improving some models, others like InternVL3-8B, Phi-3.5, and certain Llama-3.2 configurations remained unimproved (OA 0.10–0.11, F1 = 0, AUC = 0.50), likely due to insufficient domain knowledge in these models. This highlights the importance of foundational model capabilities.

Impact of Different Visual Prompts on SDDD. The best overall performance without color comes from InternVL3-14B+SVM (0.16), while with color enhancement, this same configuration dramatically improves to 0.38. Qwen2.5 shows moderate performance, with its best configuration being VCL without color for regional detection (0.58/0.66/0.24) but lower overall accuracy (0.08), suggesting it can identify individual region directions but struggles to integrate these into correct overall bending states. Color enhancement generally improves Qwen2.5's performance, particularly with CSM prompting (0.10→0.13). In stark contrast, several models demonstrate consistently poor performance regardless of prompting or color enhancement: InternVL3-8B achieves near-zero overall accuracy with color (0.0000 across all prompting strategies) and very poor thoracolumbar detection (0.04–0.06); Llama3.2 with SVM prompting similarly achieves 0.00 overall accuracy with both color versions; and multiple configurations show minimal response to different prompting strategies. Regional analysis reveals an important pattern: the seemingly high thoracolumbar accuracy (0.87) for InternVL3-14B and Phi-3.5 is misleading, as Task 2 results indicate these models are consistently outputting negative results rather than actually detecting deformities.

Enhancement Models Through RAG. DKA and PECA results consistently demonstrate significant improvements through RAG implementation across all models. In DKA, InternVL 2.5-14B achieved the highest accuracy (0.97 with RAG), while Phi 3.5-Vision showed the largest improvement (+0.20). Similarly, for PECA, RAG substantially enhanced Medical Accuracy (+1.06 to +1.09) and Safety (+0.94 to +1.02) across all models, though Communication Clarity saw more modest gains (+0.49 to +0.68). This suggests that while RAG effectively addresses knowledge limitations in specialized domains like AIS, extensive retrieved content may occasionally impact narrative coherence. Notably, performance gaps between models narrowed with RAG implementation, with smaller models showing proportionally greater improvements. These findings confirm that retrieval augmentation effectively compensates for limited parameters in specialized medical applications, enabling smaller models to approach the performance of larger counterparts (Table 3).

Table 3. MLLMs Performance Comparison on DKA and PECA Tasks With and Without RAG. "w/o RAG" = Without RAG, "w/ RAG" = With RAG, "Imp" = Improvement.

Model	DKA (Acc)			PECA (Acc)														
	w/o RAG	w/ RAG	Imp	Medical Accuracy			Response Completeness			Communication Clarity			Response Personalization			Safety		
				w/o RAG	w/ RAG	Imp	w/o RAG	w/ RAG	Imp	w/o RAG	w/ RAG	Imp	w/o RAG	w/ RAG	Imp	w/o RAG	w/ RAG	Imp
InternVL 2.5-14B	0.82	0.97	+0.16	2.96	3.95	+0.98	3.20	3.93	+0.73	3.38	3.88	+0.50	3.13	3.84	+0.71	2.87	3.88	+1.02
InternVL 2.5-8B	0.77	0.91	+0.14	2.88	3.84	+0.96	3.11	3.95	+0.84	3.30	3.86	+0.56	3.08	3.75	+0.67	2.87	3.84	+0.97
Llama 3.2-Vision	0.77	0.94	+0.17	2.84	3.90	+1.06	3.15	3.89	+0.74	3.30	3.80	+0.50	3.20	3.79	+0.59	2.84	3.78	+0.94
Phi 3.5-Vision	0.70	0.90	+0.20	2.80	3.88	+1.08	2.89	3.88	+0.99	3.20	3.88	+0.68	2.88	3.67	+0.79	2.84	3.82	+0.98
Qwen 2.5VL-7B	0.78	0.94	+0.16	2.86	3.95	+1.09	3.09	3.91	+0.82	3.27	3.76	+0.49	2.89	3.72	+0.83	2.88	3.84	+0.96

6 Conclusion

This research introduces a novel Divide and Conquer framework that breaks down complex AIS analysis into distinct, evaluable stages. Findings demonstrate that current MLLMs remain insufficient for implementing automated AIS patient self-management systems, despite showing promise in specialized tasks. While anatomical guidance improved quantification performance for models with adequate baseline capabilities, RAG significantly enhanced models' capabilities in specialized AIS knowledge domains, particularly overcoming knowledge limitations in patient education and domain knowledge tasks. This systematic evaluation approach provides a roadmap for targeted improvements, suggesting that as MLLMs advance in both foundational capabilities and specialized medical understanding, they will increasingly support clinical practice without yet replacing human expertise in AIS care.

Disclosure of Interests. The authors have no competing interests to declare that are relevant to the content of this article.

References

1. Bai, S., et al.: Qwen2. 5-vl technical report. arXiv preprint arXiv:2502.13923 (2025)
2. Berdishevsky, H., et al.: Physiotherapy scoliosis-specific exercises–a comprehensive review of seven major schools. Scoliosis Spinal Disorders **11**, 1–52 (2016)
3. Cheung, J.P.Y., Cheung, P.W.H., Samartzis, D., Luk, K.D.K.: Curve progression in adolescent idiopathic scoliosis does not match skeletal growth. Clin. Orthopaedics Related Res.® **476**(2), 429–436 (2018)
4. Davis, A., Souza, R., Lim, J.H.: Knowledge-augmented language models interpreting structured chest x-ray findings. arXiv preprint arXiv:2505.01711 (2025)
5. De Sèze, M., Cugy, E.: Pathogenesis of idiopathic scoliosis: a review. Ann. Phys. Rehabil. Med. **55**(2), 128–138 (2012)
6. Dimitrijević, V., Rašković, B., Popović, M., Viduka, D., Nikolić, S., Drid, P., Obradović, B.: Treatment of idiopathic scoliosis with conservative methods based on exercises: a systematic review and meta-analysis. Front. Sports Active Living **6**, 1492241 (2024)
7. Dufvenberg, M., et al.: Six-month results on treatment adherence, physical activity, spinal appearance, spinal deformity, and quality of life in an ongoing randomised trial on conservative treatment for adolescent idiopathic scoliosis (contrais). J. Clin. Med. **10**(21), 4967 (2021)
8. Fan, Z., Liang, C., Wu, C., Zhang, Y., Wang, Y., Xie, W.: Chestx-reasoner: advancing radiology foundation models with reasoning through step-by-step verification. arXiv preprint arXiv:2504.20930 (2025)
9. Fong, D.Y., et al.: A population-based cohort study of 394,401 children followed for 10 years exhibits sustained effectiveness of scoliosis screening. Spine J. **15**(5), 825–833 (2015)
10. Holt, C.J., McKay, C.D., Truong, L.K., Le, C.Y., Gross, D.P., Whittaker, J.L.: Sticking to it: a scoping review of adherence to exercise therapy interventions in children and adolescents with musculoskeletal conditions. J. Orthop. Sports Phys. Ther. **50**(9), 503–515 (2020)
11. Irvin, J., et al.: Chexpert: a large chest radiograph dataset with uncertainty labels and expert comparison. In: Proceedings of the AAAI Conference on Artificial Intelligence, vol. 33, pp. 590–597 (2019)
12. Johnson, A.E., et al.: Mimic-cxr, a de-identified publicly available database of chest radiographs with free-text reports. Sci. Data **6**(1), 317 (2019)
13. Karol, L.A., Virostek, D., Felton, K., Wheeler, L.: Effect of compliance counseling on brace use and success in patients with adolescent idiopathic scoliosis. JBJS **98**(1), 9–14 (2016)
14. Kim, Y., Wu, J., Abdulle, Y., Gao, Y., Wu, H.: Enhancing human-computer interaction in chest x-ray analysis using vision and language model with eye gaze patterns. In: International Conference on Medical Image Computing and Computer-Assisted Intervention, pp. 184–194. Springer (2024)
15. Kuznia, A.L., Hernandez, A.K., Lee, L.U.: Adolescent idiopathic scoliosis: common questions and answers. Am. Fam. Phys. **101**(1), 19–23 (2020)
16. Li, C., et al.: Llava-med: training a large language-and-vision assistant for biomedicine in one day. Adv. Neural. Inf. Process. Syst. **36**, 28541–28564 (2023)
17. Li, J., et al.: "Am i different?" coping and mental health among teenagers with adolescent idiopathic scoliosis: a qualitative study. J. Pediatr. Nurs. **75**, e135–e141 (2024)

18. Li, Q., et al.: Aor: anatomical ontology-guided reasoning for medical large multi-modal model in chest x-ray interpretation. arXiv preprint arXiv:2505.02830 (2025)
19. Marchese, R.: World-wide variation in Schroth therapists' clinical reasoning and exercise prescription for adolescents with idiopathic scoliosis. Ph.D. thesis, Macquarie University (2023)
20. Meng, N., et al.: An artificial intelligence powered platform for auto-analyses of spine alignment irrespective of image quality with prospective validation. EClinicalMedicine **43** (2022)
21. Moor, M., et al.: Med-flamingo: a multimodal medical few-shot learner. In: Machine Learning for Health (ML4H), pp. 353–367. PMLR (2023)
22. Nguyen, H.Q., et al.: Vindr-cxr: an open dataset of chest x-rays with radiologist's annotations. Sci. Data **9**(1), 429 (2022)
23. Roye, B.D., et al.: Establishing consensus on the best practice guidelines for the use of bracing in adolescent idiopathic scoliosis. Spine Deformity **8**, 597–604 (2020)
24. Seifert, J., Thielemann, F., Bernstein, P.: Adolescent idiopathic scoliosis: guideline for practical application. Orthopade **45**, 509–517 (2016)
25. Team, G., et al.: Gemini 1.5: unlocking multimodal understanding across millions of tokens of context. arXiv preprint arXiv:2403.05530 (2024)
26. Tu, T., et al.: Towards generalist biomedical ai. Nejm Ai **1**(3), AIoa2300138 (2024)
27. Zhu, J., et al.: Internvl3: exploring advanced training and test-time recipes for open-source multimodal models. arXiv preprint arXiv:2504.10479 (2025)

MedM-VL: What Makes a Good Medical LVLM?

Yiming Shi[1], Shaoshuai Yang[1], Xun Zhu[1], Haoyu Wang[1], Xiangling Fu[4],
Miao Li[1(✉)], and Ji Wu[1,2,3(✉)]

[1] Department of Electronic Engineering, Tsinghua University, Beijing 100084, China
{miao-li,wuji_ee}@tsinghua.edu.cn
[2] College of AI, Tsinghua University, Beijing 100084, China
[3] Beijing National Research Center for Information Science and Technology,
Tsinghua University, Beijing 100084, China
[4] School of Computer Science, Beijing University of Posts and Telecommunications,
Beijing 100876, China

Abstract. Medical image analysis is essential in modern healthcare. Deep learning has redirected research focus toward complex medical multimodal tasks, including report generation and visual question answering. Traditional task-specific models often fall short in handling these challenges. Large vision-language models (LVLMs) offer new solutions for solving such tasks. In this study, we build on the popular LLaVA framework to systematically explore **model architectures** and **training strategies** for both 2D and 3D medical LVLMs. We present extensive empirical findings and practical guidance. To support reproducibility and future research, we release a modular codebase, **MedM-VL**, and two pre-trained models: **MedM-VL-2D** for 2D medical image analysis and **MedM-VL-CT-Chest** for 3D CT-based applications. The code is available at: https://github.com/MSIIP/MedM-VL.

Keywords: 2D Medical LVLMs · 3D Medical LVLMs

1 Introduction

Medical image analysis plays a critical role in modern healthcare. As clinical needs become increasingly complex, research has shifted toward multimodal tasks [12], including medical visual question answering (VQA) [6,10], report generation [14,18], and diagnostic reasoning [21]. These tasks require joint understanding of 2D images (e.g., X-rays), 3D volumes (e.g., CT and MRI), and associated textual information. Early methods rely on task-specific designs or simple fusion, but struggle with complex multimodal integration, cross-modal reasoning, and scalability [7,23]. These limitations hinder practical deployment in real-world workflows.

To address these limitations, large vision-language models (LVLMs) have emerged as a promising direction. By leveraging advances in both computer vision and natural language processing, LVLMs aim to unify multimodal understanding within a single framework. These models are particularly well-suited to handle the diversity and complexity of medical data, enabling integrated analysis

J. Qiu et al. (Eds.): Agentic AI 2025/CMLLMs 2025/CREATE 2025, LNCS 16147, pp. 290–299, 2026.
https://doi.org/10.1007/978-3-032-06004-4_29

across 2D and 3D modalities alongside rich textual context [2]. Their potential to support open-ended, instruction-driven medical tasks makes them a promising candidate for next-generation medical AI systems.

With the success of ChatGPT-4 [1], large language models (LLMs) [4,24] have fundamentally reshaped the landscape of deep learning, enabling powerful capabilities in natural language understanding and generation. Building on this progress, LVLMs [5,8] have emerged as a unified framework for handling complex multimodal tasks that were previously beyond the reach of traditional models. Among them, LLaVA [19] stands out for its architectural simplicity and effectiveness. It adopts a standard encoder-connector-LLM architecture and demonstrates that even a linear layer can achieve robust alignment between visual and textual modalities.

Meanwhilie, LVLMs in the medical domain have experienced rapid development [20,22,28], aiming to bridge the gap between complex medical image data and clinical language understanding. LLaVA-Med [16], extending the standard LLaVA framework, leverages ChatGPT-4 to generate large-scale instruction-following data, significantly improving the model's ability to handle diverse medical prompts and multimodal tasks. RadFM [25] further advances the filed by training on web-scale multimodal data, encompassing both 2D and 3D medical images, to build a powerful and generalizable foundation model. In parallel, models such as M3D-LaMed [3] and CT-CHAT [9] focus specifically on 3D CT-based multimodal understanding, addressing the challenges of volumetric data representation and task diversity in diagnostic scenarios.

Although both 2D and 3D medical LVLMs generally adopt the LLaVA-style framework, their model architectures and training strategies differ substantially. A systematic and fair comparison among them remains unexplored. In this study, we introduce MedM-VL, a modular and extensible framework built upon TinyLLaVA Factory [13], tailored for both 2D and 3D medical LVLMs. Based on MedM-VL, we conduct a comprehensive study on how different model architectures and training strategies affect medical LVLM performance. We provide extensive engineering findings and release two pre-trained models: MedM-VL-2D and MedM-VL-CT-Chest. Our main contributions are summarized as follows:

- We systematically explore **model architectures** and **training strategies** for 2D and 3D medical LVLMs, offering extensive empirical findings and practical guidance.
- We release an open-source, modular framework, **MedM-VL**, to facilitate the training, evaluation, and development of medical LVLMs.
- We provide two pre-trained model weights: **MedM-VL-2D** for 2D medical image analysis and **MedM-VL-CT-Chest** for 3D CT-based applications.

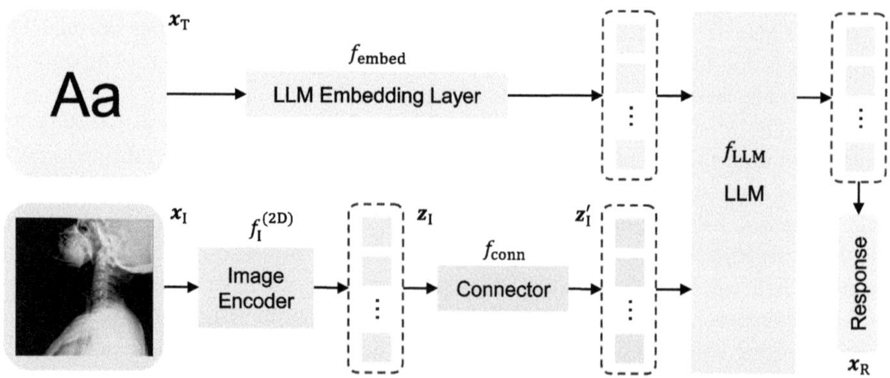

Fig. 1. Model architecture of MedM-VL-2D. Following the popular LLaVA framework [19], MedM-VL-2D consists of three main components: an image encoder, a connector, and an LLM. It takes a 2D medical image and a textual prompt as input, and generates a textual response, supporting a wide range of medical multimodal tasks.

2   Model Architecture

2.1   2D Medical LVLMs

The architecture of 2D medical LVLMs generally follow LLaVA [19], which is one of the most widely adopted architectures for LVLMs. As illustrated in Fig. 1, it consists of three main components: an image encoder, a connector, and an LLM.

Given a 2D medical image $\mathbf{x}_I$ and a textual prompt $\mathbf{x}_T$, the LVLM employs the 2D image encoder $f_I^{(2D)}$ to extract a sequence of image features $\mathbf{z}_I$:

$$\mathbf{z}_I = f_I^{(2D)}(\mathbf{x}_I) \in \mathbb{R}^{L_I^{(2D)} \times D_I}, \tag{1}$$

where $L_I^{(2D)}$ is the sequence length and D_I is the embedding dimension.

Next, the connector f_{conn} projects the image features into the input space of the LLM:

$$\mathbf{z}_I' = f_{\text{conn}}(\mathbf{z}_I) \in \mathbb{R}^{L_I^{(2D)} \times D_T}, \tag{2}$$

where D_T denotes the embedding dimension of the LLM.

The LLM f_{LLM} then integrates the image features $\mathbf{z}_I'$ with the textual input $\mathbf{x}_T$ and generates the textual response $\mathbf{x}_R$:

$$\mathbf{x}_R = f_{\text{LLM}}(\text{concat}(\mathbf{z}_I', f_{\text{embed}}(\mathbf{x}_T))), \tag{3}$$

where the features are concatenated along the sequence (token) dimension, while the embedding dimension remains unchanged.

This LLaVA-style architecture supports multimodal input (image and text) and textual output, enabling a wide range of medical multimodal tasks such as image classification, report generation, VQA, referring expression comprehension (REC), and referring expression generation (REG).

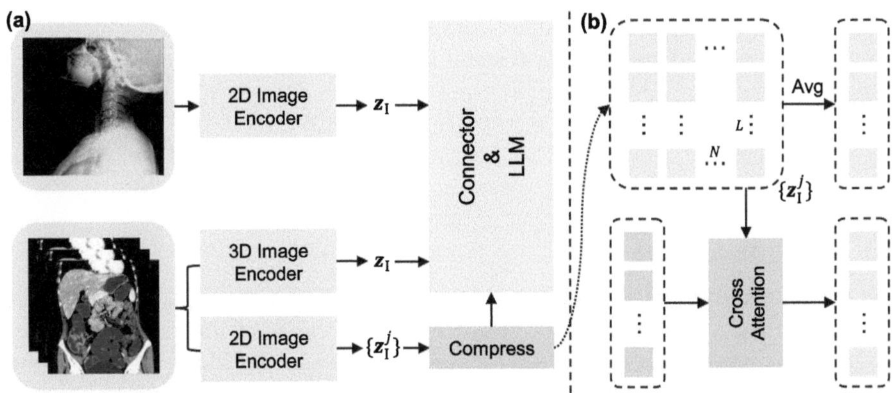

Fig. 2. (a) Overview of the 3D medical LVLM architecture. Compared to the 2D case, 3D LVLMs adopt either a 3D encoder or apply a 2D image encoder to each slice. Using 2D encoders leads to long feature sequences that require compression. (b) Two compression strategies are explored: a cross-attention module that reduces the sequence to a fixed length, and average pooling across all slice features.

2.2 3D Medical LVLMs

The architecture of 3D medical LVLMs largely follows LLaVA [19]. However, due to the dimensional differences between 3D and 2D medical images, specialized strategies for feature extraction are required, as illustrated in Fig. 2 (a).

One approach is to directly process the volumetric input $\mathbf{x}_I^{(3D)}$ using a 3D image encoder $f_I^{(3D)}$:

$$\mathbf{z}_I = f_I^{(3D)}(\mathbf{x}_I) \in \mathbb{R}^{L_I^{(3D)} \times D_I}. \tag{4}$$

Alternatively, a 2D image encoder $f_I^{(2D)}$ can be applied independently to each of the N slices:

$$\mathbf{z}_I^j = f_I^{(2D)}(\mathbf{x}_I^j) \in \mathbb{R}^{L_I^{(2D)} \times D_I}, \quad j = 1, 2, \ldots, N, \tag{5}$$

resulting in a long feature sequence of length $N \times L_I^{(2D)}$. To reduce the sequence length for efficient computation, we explore two compression strategies, as illustrated in Fig. 2 (b).

In the first strategy, all slice-level features are concatenated into a single sequence and passed through a cross-attention module to compress them into a fixed-length sequence of $L_I^{(\mathrm{Attn})}$ tokens:

$$\mathbf{z}_I' = f_{\mathrm{conn}}\left(\mathrm{concat}(\mathbf{z}_I^1, \mathbf{z}_I^2, \ldots, \mathbf{z}_I^N)\right) \in \mathbb{R}^{L_I^{(\mathrm{Attn})} \times D_T}. \tag{6}$$

In the second strategy, an average pooling operation is applied across all slices to produce a compact representation:

$$\mathbf{z}_I' = f_{\mathrm{conn}}\left(\frac{1}{N}\sum_{j=1}^{N} \mathbf{z}_I^j\right) \in \mathbb{R}^{L_I^{(2D)} \times D_T}. \tag{7}$$

Following feature extraction and compression, the LLM integrates image and textual features and generates a textual response:

$$\mathbf{x}_R = f_{\mathrm{LLM}}(\mathrm{concat}(\mathbf{z}_I', f_{\mathrm{embed}}(\mathbf{x}_T))). \tag{8}$$

3   Training

3.1   Data Preparation

2D Datasets for Multi-task Learning. The 2D dataset is constructed by integrating multiple public sources to support a wide range of medical multimodal tasks. It includes image classification samples from MedMNIST v2 [27], report generation data from MIMIC-CXR [15] and MPx-Single [25], and VQA samples from Path-VQA [11] and Slake-VQA [17]. For REC and REG, we use data from SA-Med2D-20M [29]. To further enhance instruction diversity, we incorporate 60K instruction-tuning samples from the LLaVA-Med [16] training set.

3D Dataset Based on Chest CT. Due to the limited availability of 3D medical data, we adopt CT-RATE [9], a large-scale 3D medical multimodal dataset consisting of 50K non-contrast chest CT scans from 21K patients. CT-RATE supports various VQA, including long answer, short answer, multiple choice, and report generation. We follow the official training and validation split provided by CT-RATE to ensure consistency.

3.2   Training Strategy

We adopt a standard two-stage training strategy [19] for both 2D and 3D medical LVLMs. The training process consists of a pre-training stage followed by the instruction-tuning stage.

Pre-training Stage. In the pre-training stage, the goal is to align the visual and textual modalities. To achieve this, the image encoder and LLM are kept frozen, and only the connector is trained. The training data consists solely of image-caption pairs.

For the 2D medical LVLM, we use the LLaVA [19] pre-training dataset, which contains 558K image-text pairs from the general domain. For the 3D medical LVLM, we use the report generation subset of CT-RATE [9].

Instruction-Tuning Stage. The instruction-tuning stage focuses on learning to follow diverse prompts and perform various tasks, aiming to enhance the task generalization ability of LVLMs. In this stage, all model components, including the image encoder, connector, LLM, are trained jointly. The training data includes a wide range of tasks and instruction formats.

As detailed in Sect. 3.1, the 2D medical LVLM is trained on a multi-task dataset covering image classification, report generation, VQA, REC and REG. The 3D medical LVLM is trained on the full CT-RATE dataset [9], which supports long answer, short answer, multiple choice, and report generation.

4 Experiment

4.1 Overall Performance

Table 1. Training configurations for 2D and 3D medical LVLMs. Values before and after the slash (/) indicate settings for 2D and 3D models.

	Pre-training stage	Instruction-tuning stage
Sequence length	2048	2048
Epoch	1	3/1
Batch size	64/16	16/8
Learning rate	1e–3	2e–5

For language compatibility and computational efficiency, we adopt Qwen2.5-3B-Instruct [26] as the LLM. All images or slices are resized to a uniform resolution of 256×256. Training is performed using bf16 precision and ZeRO-3 optimization on two NVIDIA A800 GPUs (80 GB).

Table 2. Performance of 2D medical LVLMs across multiple benchmarks.

Method	MedMNIST	MedPix	MIMIC-CXR	PathVQA	SAMed	SLAKE
Med-Flamingo [20]	0.089	0.081	**0.233**	0.334	–	0.215
LLaVA-Med [16]	0.668	**0.151**	0.204	0.378	0.458	0.337
RadFM [25]	0.189	–	0.068	0.248	–	0.817
MedM-VL-2D	**0.808**	0.126	0.199	**0.634**	**0.693**	**0.841**

For MedM-VL-2D, we use SigLIP [30] as the image encoder and a two-layer MLP as the connector. As shown in Table 2, MedM-VL-2D achieves either the best or highly competitive performance across multiple benchmarks, demonstrating its effectiveness in various 2D medical multimodal tasks.

Table 3. Performance of 3D medical LVLMs on CT-RATE [9].

Method	Long	Short	Choice	RG
CT-CHAT [9]	0.482	0.274	0.838	0.395
MedM-VL-CT-Chest (3D)	0.619	0.658	**0.924**	0.419
MedM-VL-CT-Chest (2D+Avg)	0.622	0.664	0.920	**0.441**
MedM-VL-CT-Chest (2D+Attn)	**0.623**	**0.667**	**0.924**	0.439

We compare different strategies for 3D image feature extraction in MedM-VL-CT-Chest. The 3D image encoder is initialized with the pre-trained M3D-CLIP [3], while the 2D encoder uses the pre-trained SigLIP [30]. All CT volumes are uniformly resized to $32 \times 256 \times 256$.

As shown in Table 3, despite being pre-trained on large-scale CT datasets covering multiple anatomical regions, the LVLM using the M3D-CLIP encoder underperforms compared to the one using the general-purpose 2D encoder SigLIP. This highlights the strong generalization ability of high-capacity 2D encoders, even when applied to 3D medical volumetric data.

For LVLMs based on the 2D encoder, we evaluate two token compression methods, attention pooling and average pooling. Both approaches achieve comparable results, with attention pooling showing a slight advantage.

4.2   Ablation Study

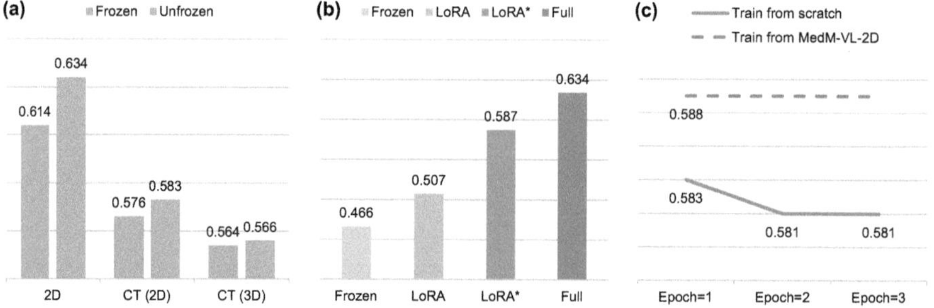

Fig. 3. (a) Impact of encoder training strategies: frozen vs. unfrozen. (b) Comparison of LLM training strategies: frozen, LoRA, and full-parameter tuning. LoRA* indicates training with a higher learning rate (2e-4). (c) Comparison between training 3D LVLMs from scratch and initializing from pre-trained 2D LVLMs.

Unfreezing General-Purpose Encoders Improves LVLM Performance. As shown in Fig. 3 (a), unfrozen general-purpose image encoders like SigLIP yield notable gains during instruction tuning, whereas domain-specific encoders such as M3D-CLIP benefit less.

We attribute this difference to the domain gap between the pre-training and fine-tuning datasets. SigLIP, pre-trained on natural images, requires adaptation to medical data, while M3D-CLIP, pre-trained on CT images, is already domain-aligned. This highlights the importance of considering domain alignment when deciding whether to freeze or fine-tune pre-trained encoders in LVLM training.

Full-Parameter Tuning of the LLM Achieves the Best Performance.
We compare different LLM training strategies during instruction-tuning, including freezing, LoRA-based tuning, and full-parameter tuning. As shown in Fig. 3 (b), full-parameter tuning yields the best performance, while keeping the LLM frozen results in the worst.

Interestingly, the performance of LoRA is highly sensitive to the learning rate. Increasing the learning rate from 2e-5 to 2e-4 significantly improves results, indicating that appropriate hyperparameter tuning is essential for achieving competitive performance with parameter-efficient methods.

These findings suggest that, while full-parameter tuning remains the most effective, carefully optimized parameter-efficient fine-tuning strategies like LoRA can serve as viable alternatives under resource-constrained settings.

Table 4. Comparison of one-stage and two-stage training strategies. Two-stage training leads to better overall performance.

Method	MedMNIST	MedPix	MIMIC-CXR	PathVQA	SAMed	SLAKE
One-stage Training	0.712	**0.140**	0.131	0.604	0.657	0.817
Two-stage Training	**0.808**	0.126	**0.199**	**0.634**	**0.693**	**0.841**

Two-Stage Training Improves LVLM Performance. We compare a one-stage instruction-tuning strategy with a two-stage approach that adds a pre-training stage. As shown in Table 4, the two-stage strategy achieves superior overall performance. This result highlights the importance of pre-training in aligning visual and textual modalities before exposing the LVLM to diverse task instructions. Without this alignment, instruction-tuning alone may be insufficient for the LVLM to fully exploit the image features, especially in complex medical scenarios.

3D LVLMs Benefit from 2D LVLM Initialization. We compare two strategies for training 3D medical LVLMs: training from scratch and fine-tuning based on a pre-trained 2D medical LVLM. As shown in Fig. 3 (c), initializing the 3D LVLM with weights from a 2D LVLM leads to better performance. Notably, simply extending the training duration of the 3D LVLM does not result in consistent improvement and can even cause performance degradation. This further supports the effectiveness of transferring knowledge from 2D LVLMs, which likely provide a better starting point for multimodal alignment and instruction following.

5   Conclusion

In this work, we conduct a comprehensive study of medical LVLMs based on the LLaVA framework. Through systematic comparisons of model architectures and training strategies, we provide practical insights into designing effective medical

LVLMs. To support extensibility and reproducibility, we introduce MedM-VL, a modular and flexible codebase that facilitates easy integration and replacement of image encoders, connectors, and LLMs. We hope that our findings and open-source resources will support further research and practical deployment of LVLMs in real-world medical applications.

The field of medical multimodal AI is advancing at a remarkable pace. A diverse array of medical multimodal datasets, encompassing both 2D and 3D images, is continuously emerging. Concurrently, foundational models from the general domain are undergoing rapid progress. Therefore, we believe the pivotal direction for future development lies in how to efficiently leverage existing open-source models and data within resource-constrained environments. Effectively transferring and adapting cutting-edge technologies from the general domain to specific clinical applications will be the key to unlocking the next wave of innovation in the field of medical multimodal intelligence.

Acknowledgments. This study was funded by Noncommunicable Chronic Diseases-National Science and Technology Major Project (Grant No. 2023ZD0506501), Beijing Natural Science Foundation NO. L251072 and Beijing Natural Science Foundation NO. 4252046.

Disclosure of Interests. The authors have no competing interests to declare that are relevant to the content of this article.

References

1. Achiam, J., et al.: Gpt-4 technical report. arXiv preprint arXiv:2303.08774 (2023)
2. Acosta, J.N., Falcone, G.J., Rajpurkar, P., Topol, E.J.: Multimodal biomedical ai. Nat. Med. **28**(9), 1773–1784 (2022)
3. Bai, F., Du, Y., Huang, T., Meng, M.Q.H., Zhao, B.: M3d: advancing 3d medical image analysis with multi-modal large language models. arXiv preprint arXiv:2404.00578 (2024)
4. Bai, J., et al.: Qwen technical report. arXiv preprint arXiv:2309.16609 (2023)
5. Bai, J., et al.: Qwen-vl: a versatile vision-language model for understanding, localization, text reading, and beyond. arXiv preprint arXiv:2308.12966 (2023)
6. Ben Abacha, A., Hasan, S.A., Datla, V.V., Demner-Fushman, D., Müller, H.: Vqa-med: overview of the medical visual question answering task at imageclef 2019. In: Proceedings of CLEF (Conference and Labs of the Evaluation Forum) 2019 Working Notes. 9–12 September 2019 (2019)
7. Bian, Y., Li, J., Ye, C., Jia, X., Yang, Q.: Artificial intelligence in medical imaging: from task-specific models to large-scale foundation models. Chin. Med. J. **138**(06), 651–663 (2025)
8. Chen, Z., et al.: Internvl: scaling up vision foundation models and aligning for generic visual-linguistic tasks. In: Proceedings of the IEEE/CVF Conference on Computer Vision and Pattern Recognition, pp. 24185–24198 (2024)
9. Hamamci, I.E., et al.: Developing generalist foundation models from a multimodal dataset for 3d computed tomography (2024), https://arxiv.org/abs/2403.17834
10. Hartsock, I., Rasool, G.: Vision-language models for medical report generation and visual question answering: a review. Front. Artif. Intell. **7**, 1430984 (2024)

11. He, X., Zhang, Y., Mou, L., Xing, E., Xie, P.: Pathvqa: 30000+ questions for medical visual question answering. arXiv preprint arXiv:2003.10286 (2020)
12. Huang, S.C., Pareek, A., Seyyedi, S., Banerjee, I., Lungren, M.P.: Fusion of medical imaging and electronic health records using deep learning: a systematic review and implementation guidelines. NPJ Digit. Med. 3(1), 136 (2020)
13. Jia, J., et al.: Tinyllava factory: a modularized codebase for small-scale large multimodal models. arXiv preprint arXiv:2405.11788 (2024)
14. Jin, H., Che, H., Lin, Y., Chen, H.: Promptmrg: diagnosis-driven prompts for medical report generation. In: Proceedings of the AAAI Conference on Artificial Intelligence, vol. 38, pp. 2607–2615 (2024)
15. Johnson, A.E., et al.: Mimic-cxr, a de-identified publicly available database of chest radiographs with free-text reports. Sci. Data 6(1), 317 (2019)
16. Li, C., et al.: Llava-med: training a large language-and-vision assistant for biomedicine in one day. Adv. Neural. Inf. Process. Syst. 36, 28541–28564 (2023)
17. Liu, B., Zhan, L.M., Xu, L., Ma, L., Yang, Y., Wu, X.M.: Slake: a semantically-labeled knowledge-enhanced dataset for medical visual question answering. In: 2021 IEEE 18th international symposium on biomedical imaging (ISBI), pp. 1650–1654. IEEE (2021)
18. Liu, G., et al.: Clinically accurate chest x-ray report generation. In: Machine Learning for Healthcare Conference, pp. 249–269. PMLR (2019)
19. Liu, H., Li, C., Wu, Q., Lee, Y.J.: Visual instruction tuning. Adv. Neural. Inf. Process. Syst. 36, 34892–34916 (2023)
20. Moor, M., et al.: Med-flamingo: a multimodal medical few-shot learner. In: Machine Learning for Health (ML4H), pp. 353–367. PMLR (2023)
21. Savage, T., Nayak, A., Gallo, R., Rangan, E., Chen, J.H.: Diagnostic reasoning prompts reveal the potential for large language model interpretability in medicine. NPJ Digit. Med. 7(1), 20 (2024)
22. Shi, Y., Zhu, X., Hu, Y., Guo, C., Li, M., Wu, J.: Med-2e3: A 2d-enhanced 3d medical multimodal large language model. arXiv preprint arXiv:2411.12783 (2024)
23. Sopruchi, A., Anguzu, R., University II, K.I.: The integration of ai-driven decision support systems in healthcare: enhancements, challenges, and future directions. Idosr J. Comput. Appl. Sci. 9, 17–25 (2024). https://doi.org/10.59298/JCAS/2024/92.1725
24. Touvron, H., et al.: Llama: open and efficient foundation language models. arXiv preprint arXiv:2302.13971 (2023)
25. Wu, C., Zhang, X., Zhang, Y., Wang, Y., Xie, W.: Towards generalist foundation model for radiology. arXiv preprint arXiv:2308.02463 (2023)
26. Yang, A., et al.: Qwen2. 5 technical report. arXiv preprint arXiv:2412.15115 (2024)
27. Yang, J., et al.: Medmnist v2-a large-scale lightweight benchmark for 2d and 3d biomedical image classification. Sci. Data 10(1), 41 (2023)
28. Yang, L., et al.: Advancing multimodal medical capabilities of gemini. arXiv preprint arXiv:2405.03162 (2024)
29. Ye, J., et al.: Sa-med2d-20m dataset: segment anything in 2d medical imaging with 20 million masks. arXiv preprint arXiv:2311.11969 (2023)
30. Zhai, X., Mustafa, B., Kolesnikov, A., Beyer, L.: Sigmoid loss for language image pre-training. In: Proceedings of the IEEE/CVF International Conference on Computer Vision, pp. 11975–11986 (2023)

KokushiMD-10: Benchmark for Evaluating Large Language Models on Ten Japanese National Healthcare Licensing Examinations

Junyu Liu[1], Lawrence K.Q. Yan[2], Tianyang Wang[3], Qian Niu[4], Momoko Nagai-Tanima[1], and Tomoki Aoyama[1](✉)

[1] Kyoto University, Kyoto, Japan
aoyama.tomoki.4e@kyoto-u.ac.jp
[2] The Hong Kong University of Science and Technology, Hong Kong, China
[3] Xi'an Jiaotong-Liverpool University, Xi'an, China
[4] The University of Tokyo, Tokyo, Japan

Abstract. Recent advances in large language models (LLMs) have demonstrated notable performance in medical licensing exams. However, comprehensive evaluation of LLMs across various healthcare roles, particularly in high-stakes clinical scenarios, remains a challenge. Existing benchmarks are typically text-based, English-centric, and focus primarily on medicines, which limits their ability to assess broader healthcare knowledge and multimodal reasoning. To address these gaps, we introduce KokushiMD-10, the first multimodal benchmark constructed from ten Japanese national healthcare licensing exams. This benchmark spans multiple fields, including Medicine, Dentistry, Nursing, Pharmacy, and allied health professions. It contains over 11588 real exam questions, incorporating clinical images and expert-annotated rationales to evaluate both textual and visual reasoning. We benchmark over 30 state-of-the-art LLMs, including GPT-4o, Claude 3.5, and Gemini, across both text and image-based settings. Despite promising results, no model consistently meets passing thresholds across domains, highlighting the ongoing challenges in medical AI. KokushiMD-10 provides a comprehensive and linguistically grounded resource for evaluating and advancing reasoning-centric medical AI across multilingual and multimodal clinical tasks.

Keywords: Reasoning Medical AI · Multimodal Benchmark · Japanese Licensing Examinations

1 Introduction

Healthcare licensing examinations offer a rigorous and standardized framework for evaluating the clinical reasoning capabilities of large language models (LLMs) in realistic healthcare contexts [14]. These exams test not only factual recall but also complex diagnostic reasoning, making them valuable foundations for

---

J. Liu, L.K.Q. Yan and T. Wang—Equal contribution.

© The Author(s), under exclusive license to Springer Nature Switzerland AG 2026
J. Qiu et al. (Eds.): Agentic AI 2025/CMLLMs 2025/CREATE 2025, LNCS 16147, pp. 300–309, 2026.
https://doi.org/10.1007/978-3-032-06004-4_30

benchmark construction. For example, questions may involve interpreting laboratory results, imaging findings, or patient histories to arrive at differential diagnoses and appropriate treatment plans [3,19]. Accordingly, many benchmark datasets-such as MedQA [8], MedMCQA [16], IgakuQA [9], and YakugakuQA [20] have been developed from national licensing exams to support research in medical question answering and clinical decision-making. These datasets usually comprise real-exam multiple-choice questions spanning internal medicine, pharmacology, surgery, and public health. However, existing benchmarks often remain limited in scope, constrained by profession, modality, language, or depth of reasoning, which restricts their value for evaluating general-purpose medical AI systems across the full spectrum of healthcare domains.

While benchmarks such as MedMCQA [16] and YakugakuQA [20] have significantly expanded the scale and subject diversity of medical QA datasets, critical gaps remain. For example, MedMCQA spans 21 subjects from real-world Indian postgraduate medical entrance exams (AIIMS and NEET PG), yet focuses exclusively on Medicine-related content and excludes image-based questions, limiting its utility for multimodal assessment. Similarly, YakugakuQA is derived from Japan's National Pharmacist Licensing Examinations and introduces profession-specific coverage beyond medicine. However, it only includes text-based questions and centers on pharmacy-related knowledge [21]. Moreover, most existing benchmarks are either monolingual-typically in English [10,23]-or narrowly target a single healthcare profession (e.g., dentists [14] or pharmacists [20]), leaving important domains such as nursing, dentistry, and rehabilitation underrepresented. In addition, few benchmarks provide detailed reasoning annotations, such as chain-of-thought (CoT) explanations [22], which are essential for investigating the underlying decision-making behavior of LLMs.

To address these limitations, we introduce KokushiMD-10[1], a large-scale multimodal benchmark constructed from ten Japanese national licensing examinations across diverse healthcare professions, including Medicine, Dentistry, Nursing, Physical Therapy, Occupational Therapy, Midwifery, Public Health Nursing, Radiologic Technology, Optometry, and Pharmacy. This broad coverage enables the evaluation of both domain-specific knowledge and cross-disciplinary generalization for LLMs. In addition to text-based questions, KokushiMD-10 includes a substantial number of image-based items (e.g., radiographs, clinical photographs), supporting robust assessment of visual reasoning in clinical contexts. All content is written in Japanese, addressing the linguistic gap in existing medical QA benchmarks. Furthermore, six of them are paired with detailed CoT explanations, enabling fine-grained analysis of model reasoning beyond mere accuracy. Our contributions are three-fold:

- To the best of our knowledge, we present the first benchmark spanning ten distinct healthcare professions, allowing a fine-grained evaluation of LLMs in both the core medical and allied health domains.

---

[1] https://huggingface.co/datasets/humanalysis-square/KokushiMD-10.

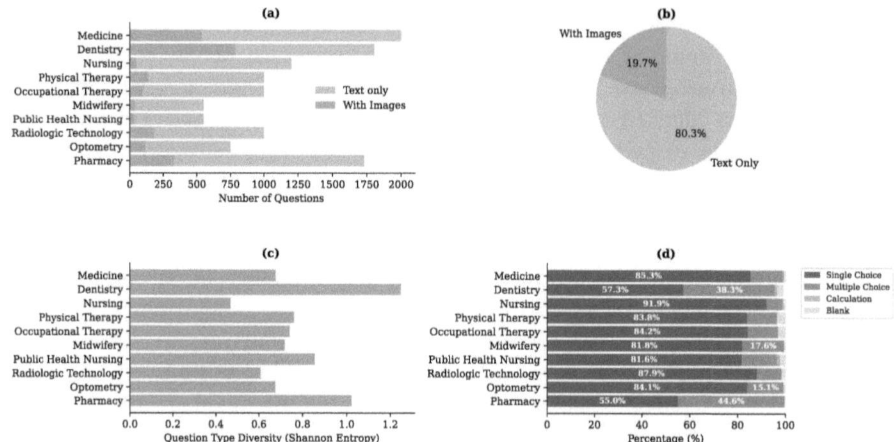

Fig. 1. Dataset statistics across ten healthcare professions: (a) Total and image-based question counts per exam; (b) Overall proportion of image vs. text-only questions; (c) Question type diversity measured by Shannon entropy; (d) Distribution of question types (single-choice, multiple-choice, calculation, blank).

- We release a large-scale QA dataset in Japanese that incorporates both textual and visual modalities, facilitating a comprehensive evaluation of vision-language understanding.
- We benchmark a wide range of open and closed source LLMs on KokushiMD-10. The results show that even state-of-the-art models like GPT-4o [15] passed 70% exams while many models passed no exams, underscoring the challenges of safe and accurate medical reasoning.

2 Methodology

2.1 Benchmark Overview

KokushiMD-10 is a multimodal benchmark constructed from PDFs of official Japanese healthcare licensing examination issued by the Ministry of Health, Labor and Welfare between 2020 and 2024 [12]. The benchmark covers ten healthcare professions, including Medicine, Dentistry, Nursing, Pharmacy, Midwifery, Public Health Nursing, Physical Therapy, Occupational Therapy, Optometry, and Radiologic Technology, and 6 of them comprise both text-only and image-based multiple-choice questions.

As shown in Fig. 1, the dataset presents profession-level variation in question quantity, modality, and type. Image-based questions account for 19.7% overall, with Dentistry showing the highest ratio. Medicine and Dentistry lead in total volume, while Nursing primarily features text-only items. Single-choice questions dominate in Medicine and Nursing, whereas Pharmacy and Dentistry contain

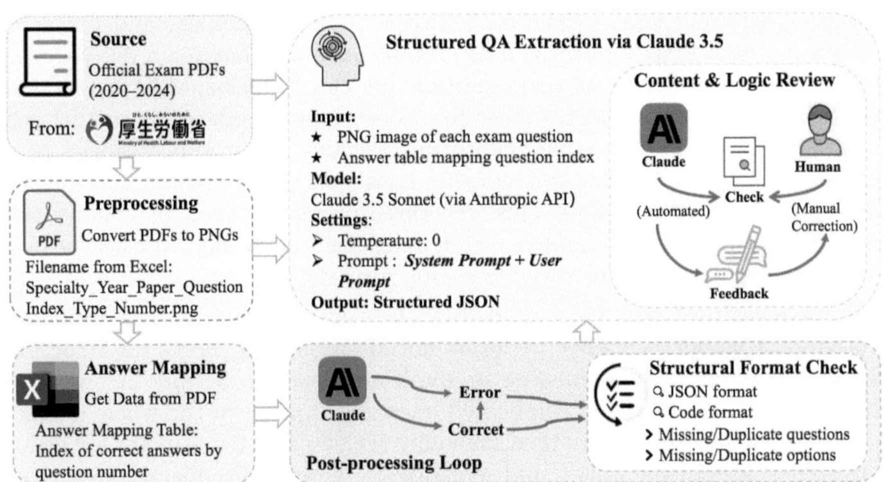

Fig. 2. The construction pipeline of KokushiMD-10, including PDF-to-image preprocessing, structured QA extraction, answer alignment, human-in-the-loop correction, and structural format verification.

more multiple-choice formats. Diversity in question types, quantified via Shannon entropy [18], is highest in Dentistry and lowest in Radiologic Technology.

2.2 Benchmark Construction

To ensure structural consistency and scalability, we employ a multi-stage pipeline (Fig. 2) for benchmark construction.

Data Processing and Answer Mapping. Official PDF exam papers are first converted into high-resolution PNG images using Adobe Acrobat [2]. Each filename encodes key metadata such as the specialty, year, paper section, and question index in the format

`Specialty_Year_Paper_Index_Type_Number.png`. Answer mapping tables are extracted from the PDFs using Excel's built-in table recognition and aligned with the question index to serve as the ground truth for downstream evaluation.

Structured QA Extraction. Each PNG question image is fed into Claude-3.5-Sonnet [4] via the Anthropic API using carefully engineered system-level and user-level prompts, with the temperature fixed at 0 to ensure deterministic outputs. The system prompt defines Claude as a data scientist responsible for organizing Japanese healthcare QA datasets, specifying a required JSON schema with fields for metadata (`year`, `section`, `index`), question components (`question`, `content`, options a–e), and format indicators (`answer`, `answer_sub`, `text_only`). The prompt includes comprehensive processing rules for sequential question handling, multimodal content detection (automatically setting `text_only` to `false` when figures or tables are present), and quality control measures requiring cross-validation between question numbers and extracted indices. For each PNG input, the user prompt integrates metadata derived from the filename (specialty, exam year, question index) with task-specific instructions and references to the correct answer from our mapping table. The model generates structured JSON outputs capturing domain-specific subcategories (e.g., `corrected_question_index` for indexing consistency and `answer_sub` for specialized fields like Pharmacy). This structured prompting strategy ensures that extracted QA entries are complete, valid, and semantically aligned with the original exam logic, even under varied question formats.

Post-processing Loop. Generated JSON files undergo a multi-stage validation procedure. Claude-3.5-Sonnet performs initial automated checks for structural format violations such as missing or duplicated entries. Scripts are made to check if there are missing or duplicated questions, images in the whole dataset. A team of human annotators then manually reviews selected samples for logical consistency, formatting issues, and semantic correctness. Feedback is iteratively incorporated through a looped refinement process. Special formatting, such as underlined text (<u>... </u>), chemical ions (e.g., HCO_3^-), and isotope symbols (e.g., ^{99m}Tc), is normalized to enhance parsing reliability for LLMs. This hybrid approach ensures the production of high-quality QA records aligned with medical-domain expectations. CoT explanations for six licensing examinations are constructed entirely from publicly available governmental and instructional sources [1,5,13]. Additional exam preparation platforms were also referenced for explanatory consistency and structure [7,11,17]. English version of all the information in the benchmark is also provided.

3 Experiments and Results

3.1 Baselines

We adopt publicly available passing criteria from ten Japanese national healthcare licensing exams as reference points for human-level performance [12]. For example, the National Medical Licensing Examination in Japan requires candidates to achieve at least 80% accuracy on designated essential questions and to meet the overall score thresholds. In addition, it includes "forbidden options" that, if selected, result in automatic failure-allowing us to benchmark high-risk decision errors. While some professions, such as dentistry and midwifery, apply composite scoring across general and scenario-based sections, others use varying sectional thresholds or qualitative grading. These diverse standards provide profession-specific baselines to assess model reliability in real-world clinical contexts.

3.2 Implementation Details

A total of 33 models were tested in text-only mode and 11 in multimodal mode on the KokushiMD-10 benchmark. Closed-source models from OpenAI, Anthropic, and Google Gemini [6] were accessed through their public APIs, whereas each

Table 1. Number of models that passed at least one healthcare licensing exam, failed all exams, or scored zero in all exams.

Model Type	Passed ≥1 Exam (out of total)	Failed All Exams	Scored Zero in All Exams
Text-only	12/33	21	6
Multimodal	6/11	5	0

open-source model was executed from its Hugging Face release on a single compute node equipped with eight NVIDIA A100 GPUs.

Every query was issued with a two-part prompt. The system prompt, originally in Japanese, begins with the instruction, "As an expert [role], please answer the following [exam_type] question." It subsequently defines the required output format and directs the model to provide only that formatted answer. The user prompt contains the question text and, for multimodal runs, the associated images. All generated responses were recorded for subsequent evaluation.

3.3   Evaluation Protocol and Metrics

Questions are categorized into four types: *Single Answer*, *Multiple Choice*, *Numerical Calculation*, and *Blank*. Each type is scored with strict correctness rules-for example, multiple-choice questions are marked incorrect unless the selected options exactly match the correct set. Scenario-based questions in nursing and midwifery are evaluated with higher weight to reflect real-world decision-making. We report per-type accuracy, as well as domain-specific thresholds aligned with official exam standards. A small subset of blank or invalid questions is excluded following real-world practice. In total, we evaluate 33 models in text-only mode and 11 models in multimodal mode, across 50 licensing exams spanning ten professions and five years. A model is considered to have passed an exam if its score meets the official passing threshold for that exam. We record both the number of passed exams per model and the raw scores for each model in each exam[2].

3.4   Benchmark Performance

Pass/Fail Evaluation under Official Thresholds. We evaluate each model on 50 full licensing exams-five years of test data across ten healthcare professions-using official criteria for each. As shown in Table 1, 33 models were tested in the Text-Only mode and 11 in the Multimodal mode. In the text-only setting, 12 models passed at least one exam, while 6 failed all exams with zero correct answers. In the multimodal setting, 6 models passed at least one exam, and all models scored non-zero. Figure 3 shows the cumulative number of passed exams per model, with colors indicating per-domain contributions. Only models with non-zero total scores are included.

---

[2] Scripts for evaluation is available at: https://github.com/jamed-llm/KokushiMD-10.

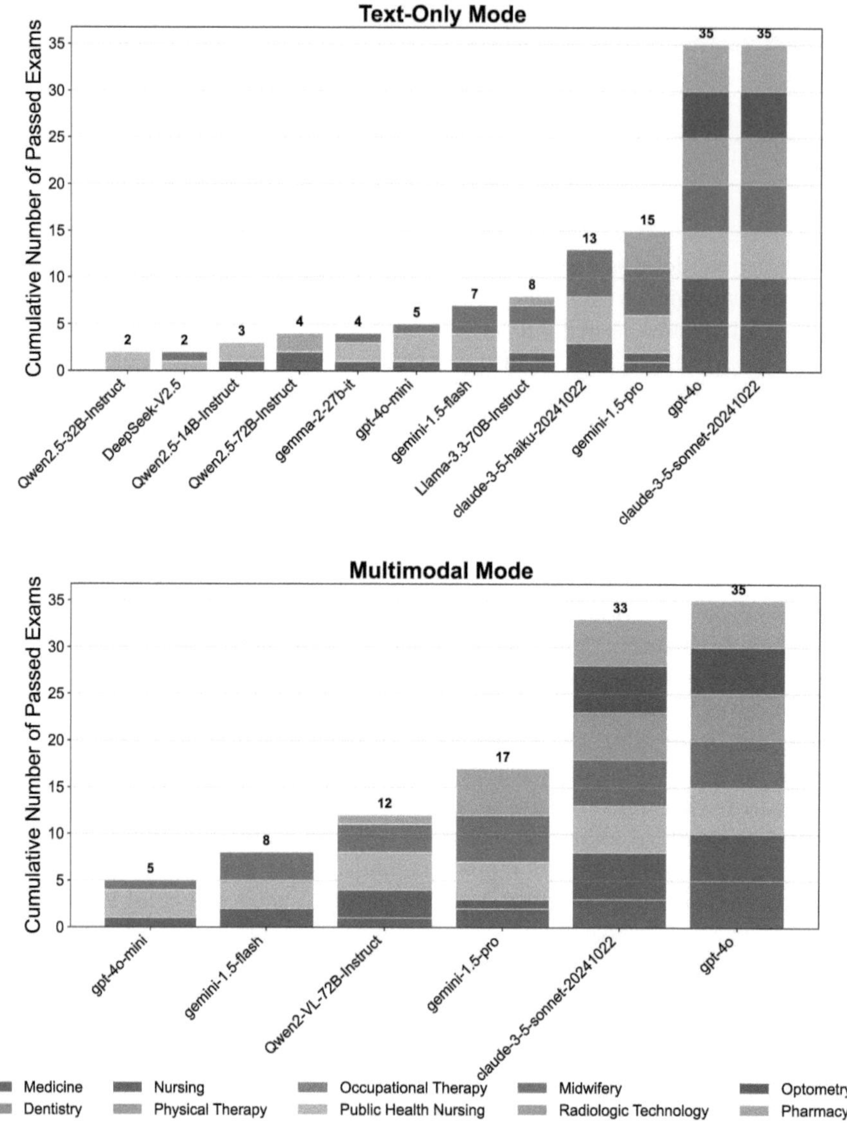

Fig. 3. Cumulative number of passed exams per model across 50 licensing exams (five years per healthcare profession).

Quantitative Score Analysis Per Exam. Table 2 reports scores across ten healthcare specialties in text-only mode. Among 27 models with non-zero scores, GPT-4o leads in four specialties (Medicine, Physical Therapy, Occupational Therapy, Optometry), while Claude-3.5-Sonnet dominates five areas (Nursing, Public Health Nursing, Midwife, Radiological Technology, Pharmacy). Notably, Pharmacy examinations yield the highest absolute scores (Claude: 536.8 ± 18.70

Table 2. Mean ± standard deviation of scores per healthcare specialty over five years (Text-Only Mode). The highest score in each subject is highlighted in bold. (Six models getting zero score in all tests were excluded, leaving 27 text-only models).

Model	Medicine	Dentistry	Nursing	Physical Therapy	Occupational Therapy	Public Health Nursing	Midwife	Radiological Technology	Optometry	Pharmacy
Gemma-2-2B-IT	0.00 ± 0.00	0.00 ± 0.00	0.20 ± 0.40	0.00 ± 0.00	0.40 ± 0.49	0.00 ± 0.00	0.00 ± 0.00	0.00 ± 0.00	0.20 ± 0.40	0.00 ± 0.00
Phi-3.5-Vision-Instruct	1.20 ± 0.98	0.00 ± 0.00	1.40 ± 1.36	1.00 ± 0.89	0.20 ± 0.40	0.80 ± 1.17	1.00 ± 0.63	1.60 ± 0.80	0.20 ± 0.40	3.20 ± 3.25
QwQ-32B-Preview	2.40 ± 2.06	1.60 ± 0.80	8.40 ± 2.87	5.60 ± 2.65	4.80 ± 2.79	3.20 ± 2.04	2.00 ± 1.67	4.40 ± 2.33	2.60 ± 1.62	13.60 ± 5.71
GPT-3.5-Turbo	11.40 ± 3.44	5.00 ± 2.61	0.80 ± 0.75	2.20 ± 2.04	2.60 ± 1.74	1.40 ± 1.50	0.40 ± 0.49	0.20 ± 0.40	0.00 ± 0.00	14.00 ± 5.06
Qwen2.5-0.5B-Instruct	22.00 ± 3.69	3.80 ± 2.99	8.40 ± 7.00	0.80 ± 1.17	3.00 ± 1.79	3.60 ± 2.24	2.20 ± 2.99	0.00 ± 0.00	0.00 ± 0.00	4.00 ± 2.83
Llama-3.2-1B-Instruct	27.40 ± 6.53	20.80 ± 7.28	17.60 ± 3.44	8.40 ± 5.08	10.60 ± 3.77	8.80 ± 2.93	6.00 ± 2.68	3.40 ± 1.62	6.20 ± 1.83	17.20 ± 5.31
Qwen2.5-1.5B-Instruct	69.00 ± 8.81	3.60 ± 5.82	2.20 ± 3.49	0.40 ± 0.80	0.40 ± 0.80	0.00 ± 0.00	0.00 ± 0.00	0.00 ± 0.00	2.20 ± 2.48	12.80 ± 11.57
Llama-3.2-3B-Instruct	142.00 ± 10.66	119.20 ± 5.98	88.40 ± 12.77	46.80 ± 7.14	48.00 ± 3.85	34.20 ± 6.31	36.40 ± 6.47	29.60 ± 4.32	25.40 ± 6.09	123.60 ± 5.71
Phi-4	168.40 ± 12.66	114.40 ± 11.59	116.20 ± 9.52	68.80 ± 3.76	76.60 ± 4.59	47.40 ± 6.89	29.80 ± 3.54	35.40 ± 3.07	30.40 ± 4.08	150.80 ± 9.52
Llama-3.1-8B-Instruct	169.20 ± 14.30	127.40 ± 10.35	84.20 ± 7.08	32.40 ± 3.77	34.20 ± 5.34	35.40 ± 3.93	29.00 ± 4.60	22.80 ± 2.86	18.60 ± 3.83	123.60 ± 5.28
Qwen2.5-3B-Instruct	197.80 ± 9.74	92.00 ± 9.38	114.00 ± 4.98	43.00 ± 14.63	52.80 ± 10.11	52.40 ± 6.53	45.60 ± 6.28	24.60 ± 4.22	20.40 ± 4.22	123.60 ± 4.45
Gemma-2-9B-IT	255.60 ± 11.11	188.20 ± 18.68	169.20 ± 13.56	91.00 ± 9.08	99.40 ± 14.01	68.40 ± 7.71	65.20 ± 2.79	51.20 ± 7.52	44.60 ± 4.50	261.20 ± 12.43
Claude-3.5-Haiku	264.20 ± 14.61	183.20 ± 12.87	231.40 ± 9.67	185.80 ± 7.19	191.60 ± 9.81	102.20 ± 7.00	95.00 ± 2.10	111.20 ± 0.98	78.60 ± 4.13	379.60 ± 27.93
Qwen2.5-7B-Instruct	264.60 ± 7.81	187.60 ± 18.94	168.60 ± 6.18	100.20 ± 14.48	112.40 ± 8.73	67.20 ± 5.23	60.60 ± 8.87	57.40 ± 6.15	55.40 ± 8.21	266.40 ± 15.36
Gemma-2-27B-IT	311.40 ± 12.04	227.00 ± 26.33	206.00 ± 9.17	130.40 ± 5.46	141.20 ± 9.68	85.60 ± 9.91	77.60 ± 5.57	73.40 ± 8.66	57.00 ± 6.96	349.20 ± 14.23
Qwen2.5-14B-Instruct	314.20 ± 8.52	232.80 ± 25.81	193.60 ± 11.57	121.20 ± 8.75	128.60 ± 14.64	81.00 ± 11.78	73.00 ± 6.90	68.40 ± 7.63	65.60 ± 3.44	340.80 ± 11.57
GPT-4o-Mini	327.40 ± 8.82	209.80 ± 14.50	205.80 ± 9.33	147.40 ± 8.19	161.40 ± 10.80	80.20 ± 12.24	81.20 ± 4.75	83.20 ± 2.79	65.80 ± 5.42	376.00 ± 21.35
DeepSeek-V2.5	335.40 ± 11.74	230.00 ± 10.00	197.60 ± 13.50	138.60 ± 8.16	151.20 ± 9.30	77.00 ± 8.51	74.40 ± 7.66	94.40 ± 7.26	68.20 ± 6.55	351.60 ± 7.31
DeepSeek-V2.5-1210	344.80 ± 8.28	235.40 ± 19.74	192.00 ± 10.45	134.80 ± 7.08	141.40 ± 9.48	71.80 ± 6.62	69.20 ± 10.24	91.00 ± 5.76	63.20 ± 7.36	360.00 ± 12.84
Gemini-1.5-Flash	345.00 ± 7.16	252.80 ± 11.30	214.40 ± 3.26	157.00 ± 7.90	163.40 ± 6.59	90.80 ± 9.79	87.20 ± 3.54	98.20 ± 8.93	76.40 ± 4.03	397.60 ± 21.78
Llama-3.1-70B-Instruct	347.00 ± 8.72	230.40 ± 14.77	180.60 ± 11.50	117.40 ± 8.89	145.80 ± 9.02	77.60 ± 4.72	66.00 ± 5.25	67.00 ± 7.07	55.80 ± 4.71	374.80 ± 12.17
Qwen2.5-32B-Instruct	353.40 ± 14.65	263.60 ± 22.02	195.60 ± 12.60	138.60 ± 2.42	145.80 ± 9.02	79.20 ± 12.02	73.20 ± 8.57	82.60 ± 5.57	67.00 ± 6.78	390.40 ± 25.97
Llama-3.3-70B-Instruct	366.80 ± 9.41	238.00 ± 12.71	215.00 ± 7.48	159.20 ± 9.33	166.00 ± 2.10	86.20 ± 12.22	83.40 ± 7.26	101.20 ± 3.71	71.40 ± 7.76	406.40 ± 22.18
Gemini-1.5-Pro	384.60 ± 11.74	301.60 ± 16.72	221.20 ± 8.47	177.00 ± 8.92	182.20 ± 8.13	96.20 ± 9.68	95.00 ± 4.60	108.60 ± 7.00	90.40 ± 3.98	465.20 ± 27.03
Qwen2.5-72B-Instruct	385.40 ± 5.89	274.20 ± 24.10	187.20 ± 13.26	140.20 ± 11.79	146.00 ± 12.54	71.60 ± 7.86	74.60 ± 3.83	84.60 ± 7.47	66.80 ± 5.04	422.80 ± 14.46

Table 3. Mean ± standard deviation of scores per healthcare specialty over five years (Multimodal Mode). The highest score in each subject is highlighted in bold.

Model	Medicine	Dentistry	Nursing	Physical Therapy	Occupational Therapy	Public Health Nursing	Midwife	Radiological Technology	Optometry	Pharmacy
QVQ-72B-preview	0.40 ± 0.49	0.20 ± 0.40	0.00 ± 0.00	0.40 ± 0.49	0.00 ± 0.00	0.00 ± 0.00	0.40 ± 0.80	0.00 ± 0.00	0.00 ± 0.00	0.80 ± 0.98
Phi-3.5-vision-instruct	2.00 ± 1.10	1.60 ± 1.96	1.80 ± 2.14	1.00 ± 0.89	0.20 ± 0.40	1.20 ± 1.17	1.00 ± 0.63	2.60 ± 1.36	2.20 ± 2.04	13.20 ± 5.46
Qwen2-VL-2B-Instruct	85.00 ± 6.29	67.80 ± 6.05	68.60 ± 6.59	33.40 ± 5.71	37.60 ± 3.98	28.20 ± 5.91	29.60 ± 3.67	18.40 ± 4.32	21.00 ± 2.53	139.60 ± 10.31
Pixtral-12B-2409	168.00 ± 20.65	122.80 ± 15.51	105.60 ± 9.89	37.20 ± 5.19	42.40 ± 5.75	45.20 ± 4.58	39.60 ± 10.61	31.20 ± 2.64	25.00 ± 3.46	139.60 ± 10.31
Qwen2-VL-7B-Instruct	217.20 ± 16.99	140.40 ± 4.03	162.00 ± 14.71	88.00 ± 6.20	100.00 ± 3.41	66.40 ± 9.46	59.20 ± 7.57	58.60 ± 4.54	42.00 ± 3.03	189.20 ± 8.82
gpt-4o-mini	326.20 ± 5.15	210.00 ± 14.79	207.00 ± 8.65	149.60 ± 7.55	164.00 ± 10.86	81.00 ± 11.88	81.40 ± 7.00	86.60 ± 3.38	65.20 ± 2.32	382.80 ± 20.73
gemini-1.5-flash	340.60 ± 4.27	251.60 ± 27.95	215.20 ± 3.19	157.80 ± 7.88	163.80 ± 7.41	91.40 ± 9.56	87.40 ± 4.54	97.40 ± 6.50	74.80 ± 5.15	401.60 ± 19.41
Qwen2-VL-72B-Instruct	371.40 ± 10.37	242.00 ± 28.37	231.80 ± 9.68	181.20 ± 7.14	182.20 ± 4.96	98.00 ± 9.88	92.80 ± 7.86	112.00 ± 7.04	86.40 ± 9.13	426.80 ± 14.89
gemini-1.5-pro	383.60 ± 13.37	296.40 ± 17.41	223.00 ± 8.27	177.20 ± 8.98	182.00 ± 9.70	97.40 ± 10.31	96.40 ± 5.16	111.80 ± 6.85	89.00 ± 6.23	478.80 ± 24.55
claude-3-5-sonnet	389.80 ± 13.98	297.40 ± 26.51	**268.40 ± 2.42**	223.60 ± 7.61	228.80 ± 5.64	**125.80 ± 5.64**	**116.20 ± 3.43**	**150.20 ± 4.66**	110.40 ± 5.31	**551.20 ± 18.96**
gpt-4o	**437.20 ± 5.84**	**329.40 ± 14.62**	265.00 ± 5.25	**227.40 ± 7.06**	**231.80 ± 11.23**	112.80 ± 6.40	112.80 ± 4.31	145.60 ± 4.18	**114.60 ± 6.15**	526.80 ± 7.96

vs GPT-4o: 508.4 ± 11.41), whereas Dentistry consistently shows the lowest performance across all models. Gemini-1.5-Pro maintains competitive scores but achieves no category leadership.

The multimodal setting (Table 3) preserves the same performance hierarchy with minimal score variations. GPT-4o and Claude-3.5-Sonnet maintain their respective domain leadership, with Claude showing slight improvements in Pharmacy (551.2 ± 18.96) and Public Health Nursing categories. Cross-modal consistency indicates that visual information provides limited additional benefit for healthcare licensing performance, as rank ordering remains stable between text-only and multimodal evaluations across all tested models.

4 Conclusion

In this work, we introduced KokushiMD-10, a comprehensive multimodal benchmark constructed from ten Japanese national healthcare licensing examinations. The dataset spans a wide range of healthcare professions and includes both textual and visual questions, enabling thorough evaluation of LLMs across multiple clinical domains. Our evaluation of over 30 state-of-the-art models shows that, while some models perform reasonably well in isolated tasks, none consistently achieve the performance required for real-world clinical use, particularly in visual reasoning and complex decision-making. These findings highlight the need for more robust, generalizable, and multimodal medical AI systems. KokushiMD-10 lays the groundwork for their development by offering a rigorous and diverse evaluation framework. Future efforts will focus on strengthening clinical alignment and expanding real-world applicability.

References

1. Academy, T.: Qualification exam information center. https://www.tokyo-ac.jp/qualification/info-exam/, Accessed 05 June 2025
2. Adobe Inc.: Adobe acrobat pro. Computer software (2024), https://acrobat.adobe.com, version 2024.004.2054, Accessed 8 June 2025
3. Al-Khater, K.M.K.: Comparative assessment of three ai platforms in answering usmle step 1 anatomy questions or identifying anatomical structures on radiographs. Clin. Anat. **38**(2), 186–199 (2025)
4. Anthropic: Claude 3 (opus release) (2025), https://www.anthropic.com/claude, Accessed 8 June 2025
5. Dental, I.: Overview and trends in dental national examinations. https://www.ishiyaku-dental.jp/, Accessed 05 June 2025
6. Google DeepMind: Gemini 1.5 pro (2025), https://gemini.google.com, Accessed 8 June 2025
7. Healthcare, M.: National exam strategy and career information. https://nurse.mynavi.jp/conts/kokushi_kouryaku/, Accessed 25 June 2025
8. Jin, D., Pan, E., Oufattole, N., Weng, W.H., Fang, H., Szolovits, P.: What disease does this patient have? a large-scale open domain question answering dataset from medical exams. Appl. Sci. **11**(14), 6421 (2021)
9. Kasai, J., Kasai, Y., Sakaguchi, K., Yamada, Y., Radev, D.: Evaluating gpt-4 and chatgpt on japanese medical licensing examinations. arXiv preprint arXiv:2303.18027 (2023)
10. Liu, J., et al.: Benchmarking large language models on cmexam-a comprehensive chinese medical exam dataset. Adv. Neural. Inf. Process. Syst. **36**, 52430–52452 (2023)
11. Media, C.: Overview of radiologic technologist national examination. https://contact.ne.jp/media/?p=170, Accessed 05 June 2025
12. Ministry of Health, Labour and Welfare: Qualification exam results announcement (2025), https://www.mhlw.go.jp/kouseiroudoushou/shikaku_shiken/goukaku.html, Accessed 07 June 2025
13. Ministry of Health, Labour and Welfare of Japan: National licensing examination overview and results. https://www.mhlw.go.jp/kouseiroudoushou/shikaku_shiken/ (2024), Accessed 05 June 2025
14. Morishita, M., et al.: Comparison of the performance on the japanese national dental examination using gpt-3.5 and gpt-4. JJDEA **40**, 3–10 (2024)
15. OpenAI: ChatGPT (gpt-4o model) (2025), https://openai.com/chatgpt, Accessed 8 June 2025

16. Pal, A., Umapathi, L.K., Sankarasubbu, M.: Medmcqa: a large-scale multi-subject multi-choice dataset for medical domain question answering. In: Conference on Health, Inference, and Learning, pp. 248–260. PMLR (2022)
17. Platform, G.J.: National exam overview for public health roles. https://www.guppy.jp/, Accessed 05 June 2025
18. Shannon, C.E.: A mathematical theory of communication. Bell Syst. Tech. J. **27**(3), 379–423 (1948)
19. Shieh, A., Tran, B., He, G., Kumar, M., Freed, J.A., Majety, P.: Assessing chatgpt 4.0's test performance and clinical diagnostic accuracy on usmle step 2 ck and clinical case reports. Sci. Rep. **14**(1), 9330 (2024)
20. Sukeda, I., Fujii, T., Buma, K., Sasaki, S., Ono, S.: A japanese language model and three new evaluation benchmarks for pharmaceutical nlp. arXiv preprint arXiv:2505.16661 (2025)
21. Takagi, S., Watari, T., Erabi, A., Sakaguchi, K., et al.: Performance of gpt-3.5 and gpt-4 on the japanese medical licensing examination: comparison study. JMIR Med. Educ. **9**(1), e48002 (2023)
22. Wei, J., et al.: Chain-of-thought prompting elicits reasoning in large language models. Adv. Neural. Inf. Process. Syst. **35**, 24824–24837 (2022)
23. Xie, Y., et al.: Medtrinity-25m: a large-scale multimodal dataset with multigranular annotations for medicine. arXiv preprint arXiv:2408.02900 (2024), https://arxiv.org/abs/2408.02900, Accessed 8 June 2025

DuEL-Med: A Dual-Path Enhanced Language Model for Clinically-Aware Radiology Report Generation

Jin Kim[1,2], Matthew S. Brown[1,2], and Dan Ruan[3,4(✉)]

[1] Department of Radiology, University of California, Los Angeles, CA, USA
kimjin116@g.ucla.edu, mbrown@mednet.ucla.edu
[2] Center for Computer Vision and Imaging Biomarkers, University of California, Los Angeles, CA, USA
[3] Department of Radiation Oncology, University of California, Los Angeles, CA, USA
druan@mednet.ucla.edu
[4] Department of Bioengineering, University of California, Los Angeles, CA, USA

Abstract. Automatic generation of radiology reports from medical imaging has significant clinical utility. This task requires precise pathology identification and coherent and clinically relevant descriptions. This paper introduces **Du**al-path **E**nhanced **L**anguage for **M**edicine, **DuEL-Med**, a novel architecture that integrates a vision encoder-fed large language model (LLM) with an explicit pathology classification informatics path. A lightweight classification head is fine-tuned via LoRA to produce probabilistic pathological predictions to guide and refine the direct LLM-generated text to produce the final report. Experiments on the large-scale MIMIC-CXR and the smaller IU X-Ray datasets demonstrate DuEL-Med's adaptive benefits. In MIMIC-CXR, it improves the GREEN score from 0.262 to 0.273 by adding correct findings. In IU X-Ray, it demonstrates robustness gain in correcting hallucinated findings from the baseline model, leading to a 15.4% improvement in high-confidence negative cases. This study underscores the value of integrating explicit clinical classification to create more reliable and context-aware automated reporting systems. The training weight and code is available at https://github.com/jink-ucla/DuEL-Med.

1 Introduction

The generation of accurate medical reports from radiological images is crucial for clinical decision-making and efficient healthcare delivery [1,2]. It is challenging due to the complexity of medical knowledge and the need for precise interpretation [3,4]. Large vision-language models (VLMs) and encoder-decoder architectures show promise, but still require improvements in clinical reliability [5–7]. It has been reported that models using vision encoders and language decoders fail to meet accuracy requirements, as in the case of reading chest X-rays (CXRs) [8]. Although integrating large language models (LLM) with domain-specific vision encoders, such as MambaXray-VL [9] and MAIRA-1 [10], can improve fluency and coverage, the lack of explicit mechanisms to account for intrinsic pathology

© The Author(s), under exclusive license to Springer Nature Switzerland AG 2026
J. Qiu et al. (Eds.): Agentic AI 2025/CMLLMs 2025/CREATE 2025, LNCS 16147, pp. 310–319, 2026.
https://doi.org/10.1007/978-3-032-06004-4_31

can result in textually fluent yet clinically ambiguous reports. This limitation is evident in evaluations using medical benchmarks such as CheXpert labeling [11–13].

Recent work has explored classifiers to guide the generation of reports by converting labels into discrete prompts [14], or integrating structured findings for broader dialog capabilities [15]. These methods inspire us to consider a dual-path approach, to integrate explicit pathology classification with language generation to improve clinical relevance and awareness. Specifically, we propose a DuEL-Med architecture, consisting of a LoRA-fine-tuned classification head to predict pathology classes with soft probability quantification and a text decoder to generate unstructured text simultaneously. These paths are integrated with a language processor to produce the final report. This work contributes to automated radiology report generation as follows:

- **DuEL-Med Architecture:** A novel architecture combining a vision-language model for report generation with a LoRA-fine-tuned pathology classification pathway, enabling free-text generation and pathology identification. The fusion mechanism integrates structured, probabilistic findings into the free-text report, ensuring accuracy and confidence.
- **Probabilistic Classification Integration:** We use the classifier's full probabilistic output, preserving uncertainty for nuanced report guidance. This differentiates our approach from methods that use discrete classification tokens or binary labels, providing a more granular and uncertainty-aware integration of clinical findings.
- **Demonstrated Robustness:** We show that DuEL-Med delivers performance gain across datasets of different scales, enhancing reports on the large MIMIC-CXR dataset and correcting baseline model hallucinations on the smaller, more challenging IU X-Ray dataset.

2   Approach

Figure 1 illustrates the overall system design. The DuEL-Med model consists of two paths: (1) a domain transfer path where images are processed by LoRA-adapted vision and language coders to generate a preliminary text output and (2) a classification path using encoded image features to predict probabilities for pathology. Results from the text path and classification path are fused using an LLM to generate the final report.

2.1   Vision Encoding

An input image $\mathbf{x}$ is first processed by a vision encoder pretrained on chest X-rays $\mathcal{F}(\cdot)$ to extract a sequence of visual embeddings:

$$\mathbf{E}_{\text{img}} = \mathcal{F}(\mathbf{x}) \in \mathbb{R}^{N \times D}, \tag{1}$$

where N is the number of visual tokens and D is the embedding dimension [9].

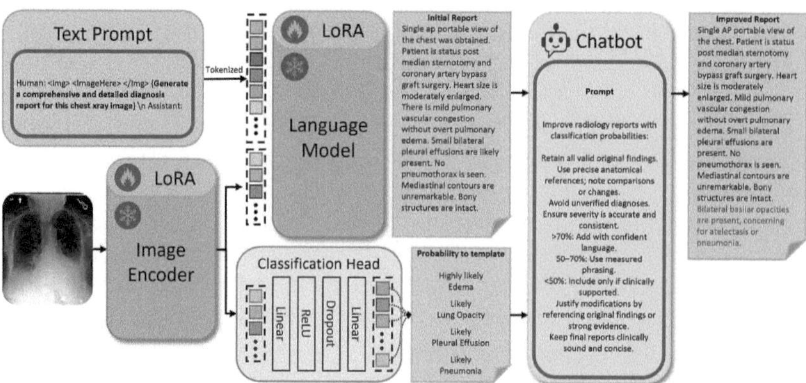

Fig. 1. The proposed DuEL-Med architecture.

We have chosen Mamba because its near-linear complexity is advantageous for high-resolution medical images, enabling efficient patch embedding, and it is compatible with pretraining [16].

2.2　Language Generation Path

In the text path, the visual embeddings are converted into a form compatible with the language model's attention mechanism via cross-attention and prompted for text generation. At each timestep t, the LLM produces the probability distribution over the next token y_t based on previous tokens and the encoded visual embedding from 1:

$$P(y_t \mid \{y_{<t}\}, \mathbf{E}_{\text{img}}) = \text{LLM}(\{y_{<t}\}, \mathbf{E}_{\text{img}}),$$

yielding an initial unstructured text base $\mathcal{R}_{\text{base}} = \{y_1, y_2, \ldots, y_T\}$. We train this module with a language modeling objective:

$$\mathcal{L}_{\text{gen}} = -\frac{1}{T} \sum_{t=1}^{T} \log P(\hat{y}_t \mid \{y_{<t}\}, \mathbf{E}_{\text{img}}). \tag{2}$$

2.3　Pathology Classification Path

In the pathology classification path, the vision embedding $\mathbf{E}_{\text{img}}$ is fed into a classification head, which outputs logits $\mathbf{p} \in \mathbb{R}^{B \times K}$ for $K - 1$ pathologies with one additional normal class. Training uses a standard multi-label binary cross-entropy objective:

$$\mathcal{L}_{\text{cls}} = \frac{1}{B} \sum_{b=1}^{B} \sum_{k=1}^{K} \text{BCE}(p_{bk}, y_{bk}), \tag{3}$$

where $y_{bk} \in \{0, 1\}$ is the ground-truth label for the k-th pathology in the b-th sample. At inference time, each probability p_k determines the likelihood of the presence or absence of a specific pathology.

2.4 Pathology Classification-Informed Report Generation

Template Conversion. At inference time, for each pathology probability p_i, we define a template function $T(p_i)$ that converts the probability into a structured statement based on confidence levels:

$$
T(p_i) = \begin{cases}
\text{"highly likely } \mathcal{D}_i \text{ ("}\lfloor 100p_i \rfloor\text{"\%)"}, & \text{if } p_i \geq \tau_4 \\
\text{"likely } \mathcal{D}_i \text{ ("}\lfloor 100p_i \rfloor\text{"\%)"}, & \text{if } \tau_3 \leq p_i < \tau_4 \\
\text{"possible } \mathcal{D}_i \text{ ("}\lfloor 100p_i \rfloor\text{"\%)"}, & \text{if } \tau_2 \leq p_i < \tau_3 \\
\text{"low possibility of } \mathcal{D}_i \text{ ("}\lfloor 100p_i \rfloor\text{"\%)"}, & \text{if } \tau_1 \leq p_i < \tau_2 \\
\emptyset, & \text{if } p_i < \tau_1
\end{cases}
\tag{4}
$$

where $\mathcal{D}_i$ is the i-th pathology class ($i = 1, \ldots, K$) and $\tau_1 < \tau_2 < \tau_3 < \tau_4$ are confidence thresholds. For the special case of "No Finding" (NF), we use:

$$
T(p_{\text{NF}}) = \begin{cases}
\text{"Normal study ("}\lfloor 100p_{\text{NF}} \rfloor\text{"\%} confidence\text{)"}, & \text{if } p_{\text{NF}} \geq \tau_{\text{NF}} \\
\emptyset, & \text{otherwise}
\end{cases}
\tag{5}
$$

Then, we form the set of structured findings $\mathcal{R}_{\text{cls}} = \{T(p_i)\} \cup T(p_{\text{NF}})$.

LLM Fusion. The final report is obtained by fusing the initial free-text report $\mathcal{R}_{\text{base}}$ with the structured findings $\mathcal{R}_{\text{cls}}$. This step is performed by a separate, powerful language model, Gemini 2.5 Flash Lite [17], which is prompted to integrate the information:

$$
\mathcal{R}_{\text{final}} = \text{Gemini}\big(\mathcal{R}_{\text{base}}, \mathcal{R}_{\text{cls}}\big).
\tag{6}
$$

This fusion step effectively refines the base report by incorporating explicit, structured pathology classification information.

2.5 Model Training and Implementation Details

Our system employs a *multi-stage* training strategy, leveraging both large-scale pretraining and task-specific fine-tuning. We use MambaXray-VL both as a baseline representative of current SOTA vision-language approaches and as an ablation module to isolate the impact of the added classification pathway in the DuEL-Med architecture. The results reported for the *MambaXray-VL* baseline in our experiments (Table 1) correspond to the performance of this language generation path alone.

1. **Vision Encoder (MambaXray-VL).** We adopted the mamba-based vision encoder from [9], which has been pretrained on large-scale chest X-ray data using autoregressive objectives over visual patches and refined via image-text contrastive learning. We further fine-tune this encoder with LoRA [18], updating only a small fraction of the parameters for efficiency.

2. **Language Model (LLAMA-2).** We initialize the autoregressive text decoder from the pretrained LLaMA-2 model [19]. To reduce the memory footprint, we apply LoRA, refining only the attention-weight matrices.
3. **DuEL-Med Fine-Tuning** With both the vision encoder and LLM initialized, we *jointly* train them on the radiology report generation task and the multi-label pathology classification w.r.t. the combined loss

$$\mathcal{L}_{\text{total}} = \mathcal{L}_{\text{gen}} + \lambda \mathcal{L}_{\text{cls}},$$

where $\mathcal{L}_{\text{gen}}$ is the language generation loss (2), $\mathcal{L}_{\text{cls}}$ is the classification loss (3), and $\lambda = 10.0$ is a hyperparameter balancing text fluency and classification accuracy.

Implementation Setup. We implemented the framework in `PyTorch` and trained on up to 8 NVIDIA RTX A8000 GPUs. The LoRA adapters use a rank of 8 for both the vision encoder and the LLM. The classification head's linear parameters are optimized from scratch. We use the AdamW optimizer [20] with a base learning rate of 2×10^{-5}, a weight decay of 0.01, and train for 5 epochs. We monitor both the language modeling loss and classification performance on a held-out validation set.

Confidence thresholds were set as $[\tau_1, \tau_2, \tau_3, \tau_4] = [0.3, 0.5, 0.7, 0.9]$ in (4) and $\tau_{\text{NF}} = 0.7$ in (5) based on preliminary experiments and visual inspection of the generated reports.

3 Experiments

Our models were evaluated with both data-rich and data-scarce datasets. For each dataset, we fine-tuned the model on its respective training set, making all evaluations **in-domain**.

- **MIMIC-CXR-JPG** is a large repository of $\sim$377k chest radiographs with pre-computed structured labels for 14 CheXpert pathologies. We use the splits of training, validation, and test from the official website.
- **IU X-Ray** [21,22] serves as a "stress-test" dataset with its small size of $\sim$7.5k data. We evaluate on its complete test set of approximately 1,400 images.

For model evaluation, we report the METEOR [23] to assess lexical overlap with references, incorporating synonyms and stemming. We further report the GREEN metric [24] to measure clinically relevant text agreement facts as it inspects key pathologies, findings, and the internal consistency of descriptions.

4 Results

Figure 2 illustrates a typical example where the classification path helps capture crucial pathologies that may otherwise be omitted. In this example, the classifier correctly detects both *Lung Opacity* and *Pneumonia*, guiding the report towards a statement about "a subtle opacity in the right lower lobe." This additional detail improved clinical accuracy, with the GREEN score increasing to 0.600 from 0.429 in the MambaXray-VL report. There was a moderate compromise in text quality, as the METEOR score decreased from 0.140 to 0.132.

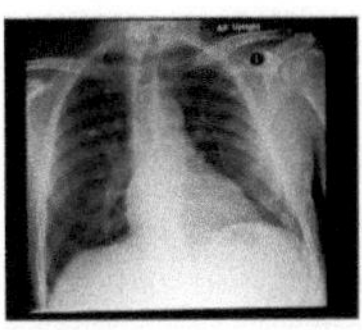

Classifier: Lung Opacity, Pneumonia
GT: Lung Opacity, Pneumonia
GREEN Score:
Report from MambaXray-VL: 0.429, Report from DuEL-Med: 0.600
METEOR Score:
Report from MambaXray-VL: 0.140, Report from DuEL-Med: 0.132

Ground-Truth Report

again seen are nonspecific bibasilar opacities which have increased from. the apices of lungs are clear. there is no evidence of pulmonary edema pleural effusion or pneumothorax. cardiomediastinal and hilar contours are unremarkable. no acute displaced rib fracture identified.

Report from MambaXray-VL

frontal and lateral views of the chest were obtained. there is no focal consolidation pleural effusion or pneumothorax. the cardiomediastinal silhouette is within normal limits. imaged osseous structures are intact. no free air below the right hemidiaphragm is seen. surgical clips project over the left upper abdomen.

Report from DuEL-Med

frontal and lateral views of the chest were obtained. there is no focal consolidation, pleural effusion, or pneumothorax. the cardiomediastinal silhouette is normal. imaged osseous structures are intact. no free air is seen below the right hemidiaphragm. surgical clips project over the left upper abdomen. there is a subtle opacity in the right lower lobe that may represent early pneumonia, clinical correlation is recommended.

Fig. 2. Example outputs from our report generation system. The chest X-ray and classifier results are shown at the top, followed by the Ground Truth (green box), the MambaXray-VL report (red box), and our DuEL-Med model's report (blue box). Color-coded text highlights corresponding findings, with red indicating details inserted by the classification path. (Color figure online)

Overall Results. A comprehensive performance evaluation against state-of-the-art methods is presented in Table 1.

On the large-scale **MIMIC-CXR** benchmark, *DuEL-Med* demonstrates its primary strength: enhancing clinical accuracy. It achieves a state-of-the-art GREEN score of **0.273**, surpassing all other methods, including its own baseline. This significant improvement in clinical correctness involves a deliberate trade-off with traditional NLP metrics. The baseline *MambaXray-VL* holds the top scores in METEOR, BLEU-4 [25], and ROUGE-L [26], showcasing its powerful generative fluency. This highlights a known trade-off: DuEL-Med's modifications, aimed at inserting or correcting specific clinical facts, can cause the

Table 1. Performance comparison on the MIMIC-CXR and IU X-Ray test sets. The best results are in **bold** and the second-best results are underlined. F-L: Flash Lite, B-4: BLEU-4, R-L: ROUGE-L.

Model	MIMIC-CXR Dataset			
	NLP Metrics			Clinical Metric
	METEOR	B-4	R-L	GREEN
Gemini 2.5 F-L [17]	0.052 ± 0.048	0.006 ± 0.009	0.083 ± 0.055	0.138 ± 0.214
RaDialog [15]	0.095 ± 0.063	0.023 ± 0.068	0.172 ± 0.093	0.253 ± 0.264
PromptMRG [14]	0.098 ± 0.054	0.016 ± 0.023	0.156 ± 0.000	0.009 ± 0.041
MSMedCap [27]	0.032 ± 0.027	0.000 ± 0.000	0.098 ± 0.061	0.164 ± 0.246
MedGemma [28]	0.125 ± 0.061	0.007 ± 0.008	0.096 ± 0.060	0.204 ± 0.275
LLaVA-Med [29]	0.072 ± 0.031	0.003 ± 0.004	0.074 ± 0.047	0.186 ± 0.367
MambaXray-VL [9]	**0.162 ± 0.090**	**0.099 ± 0.104**	**0.231 ± 0.138**	0.262 ± 0.248
DuEL-Med (Ours)	0.160 ± 0.086	0.046 ± 0.049	0.139 ± 0.097	**0.273 ± 0.258**
Model	IU X-Ray Dataset			
	NLP Metrics			Clinical Metric
	METEOR	B-4	R-L	GREEN
Gemini 2.5 F-L [17]	0.092 ± 0.096	0.008 ± 0.010	0.115 ± 0.094	0.411 ± 0.453
RaDialog [15]	0.185 ± 0.071	0.076 ± 0.072	0.200 ± 0.083	0.583 ± 0.256
PromptMRG [14]	0.086 ± 0.038	0.023 ± 0.000	0.169 ± 0.000	0.493 ± 0.281
MSMedCap [27]	0.174 ± 0.085	0.000 ± 0.000	0.027 ± 0.047	0.472 ± 0.297
MedGemma [28]	0.154 ± 0.039	0.007 ± 0.006	0.127 ± 0.036	**0.700 ± 0.337**
LLaVA-Med [29]	0.065 ± 0.040	0.000 ± 0.000	0.029 ± 0.039	0.652 ± 0.449
MambaXray-VL [9]	**0.199 ± 0.108**	**0.153 ± 0.160**	**0.351 ± 0.141**	0.616 ± 0.275
DuEL-Med (Ours)	0.196 ± 0.086	0.069 ± 0.088	0.215 ± 0.109	0.633 ± 0.274

generated text to diverge from the ground-truth's exact phrasing, thereby lowering performance on n-gram-based metrics like BLEU and recall-oriented metrics like **ROUGE-L**. However, *DuEL-Med* remains highly competitive, with strong scores for METEOR (0.160) and BLEU-4 (0.046).

This pattern of balancing clinical accuracy with fluency is further validated on the smaller **IU X-Ray** dataset. While *MedGemma* achieves the highest clinical score and *MambaXray-VL* again leads in all NLP metrics, *DuEL-Med* distinguishes itself as a strong all-around performer, achieving strong scores in METEOR (0.196) and ROUGE-L (0.215) while delivering a highly competitive GREEN score of 0.633. The model's clinical score improves upon its own powerful baseline (*MambaXray-VL*, 0.616). This consistently demonstrates that the *DuEL-Med* architecture successfully refines a strong generative model to be more clinically reliable across different data abundance conditions.

The high standard deviation in GREEN scores is attributable to the inherent variability of clinical language, where multiple descriptions can be valid for one image.

Impact of Classification Certainty. To analyze the mechanism behind DuEL-Med's performance, we evaluated the net improvement in the GREEN score based on the classifier's confidence categories, as shown in Fig. 3. This analysis reveals that the benefits of our architecture are adaptive and particularly impactful in different data contexts.

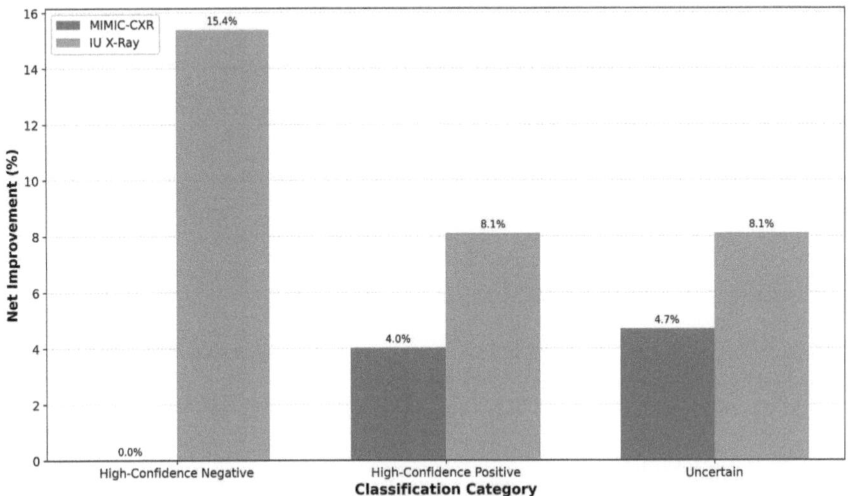

Fig. 3. DuEL-Med's Performance by Classification Confidence Category.

On the large-scale MIMIC-CXR dataset, DuEL-Med provides consistent, positive improvements for cases classified as 'High-Confidence Positive' (+4.0%) and 'Uncertain' (+4.7%). The 0.0% improvement for the 'High-Confidence Negative' category was caused by the performance saturation at the baseline level when well trained. On the smaller IU X-Ray dataset, it highlights the robustness of our architecture. DuEL-Med demonstrates an advantage in two aspects: it improves reports when adding correct findings ('High-Confidence Positive': +8.1%) and it provides a notable 15.4% improvement for 'High-Confidence Negative' cases.

5   Discussion and Conclusion

Our experiments on both large (MIMIC-CXR) and small (IU X-Ray) datasets show DuEL-Med adaptively enhances clinical correctness by correcting high-impact diagnostic errors in varied data settings. This improvement in clinical accuracy, reflected in the GREEN score, comes at an expected cost to traditional NLP metrics like ROUGE-L. This trade-off underscores a key challenge in medical report generation: improving diagnostic precision often requires deviating from the exact wording of ground-truth reports.

We acknowledge limitations, such as relying on empirically set confidence thresholds. Future work will focus on three key areas: First, developing more dynamic methods for integrating classifier probabilities. Second, pursuing rigorous clinical validation and reliability assessment through radiologist reader studies and classifier calibration metrics. Third, generalizing the DuEL-Med architecture to more complex modalities like CT and MRI, including a systematic characterization of its failure modes.

In summary, the dual-path architecture in DuEL-Med provides an effective and adaptive method for improving clinical accuracy and robustness in automated radiology reporting, offering a promising path toward more reliable AI in clinical practice.

Disclosure of Interests. The authors have no competing interests to declare that are relevant to the content of this article.

References

1. Irvin, J., Rajpurkar, P., Ko, M., Yu, Y., et al.: CheXpert: a large chest radiograph dataset with uncertainty labels and expert comparison. Nat. Mach. Intell. (2019)
2. Jing, B., Xie, P., Meng, E.: On the automatic generation of medical imaging reports. In: Proceedings of IEEE Conference Computer Vision and Pattern Recognition (CVPR) (2018)
3. Suganyadevi, S., Seethalakshmi, V., Balasamy, K.: A review on deep learning in medical image analysis. Int. J. Multimedia Inf. Retrieval **11**, 19–38 (2022)
4. Chen, H., Zhang, Y., Liao, G., et al.: Generative adversarial networks in medical image analysis: a survey. Med. Image Anal. (2020)
5. Radford, A., Kim, J., Hallacy, C., Ramesh, A., et al.: Learning transferable visual models from natural language supervision. arXiv:2103.00020 (2021)
6. Vaswani, A., Shazeer, N., Parmar, N., Uszkoreit, J., et al.: Attention is all you need. In: Advances in Neural Information Processing Systems (NeurIPS) (2017)
7. Devlin, J., Chang, M.W., Lee, K., Toutanova, K.: BERT: pre-training of deep bidirectional transformers for language understanding. arXiv:1810.04805 (2018)
8. Xu, K., Ba, J., Kiros, R., Cho, K., et al.: Show, attend and tell: neural image caption generation with visual attention. In: Proceedings of International Conference Machine Learning (ICML) (2015)
9. Wang, X., Wang, F., Li, Y., Ma, Q., et al.: CXPMRG-bench: pre-training and benchmarking for x-ray medical report generation on CheXpert plus. arXiv:2410.00379 (2024)
10. Hyland, S., Bannur, S., Bouzid, K., Castro, D., et al.: MAIRA-1: a specialised large multimodal model for radiology report generation. arXiv:2311.13668 (2023)
11. Rajpurkar, P., Irvin, J., Zhu, K., Yang, B., et al.: CheXNeXt: radiologist-level pneumonia detection on chest x-rays with deep learning. JAMA (2018)
12. Zhang, Y., Liu, X., Chen, S.: Uncertainty-aware deep learning for radiology. IEEE J. Biomed. Health Inf. (2022)
13. Chen, Z., Shen, Y., Song, Y., Wan, X.: Cross-modal memory networks for radiology report generation. arXiv:2204.13258 (2022)
14. Jin, H., Che, H., Lin, Y., Chen, H.: PromptMRG: diagnosis-driven prompts for medical report generation. In: Proceedings of AAAI Conference Artificial Intelligence, vol. 38, pp. 2607–2615 (2024)

15. Pellegrini, C., Özsoy, E., Busam, B., Wiestler, B., et al.: RaDialog: large vision-language models for x-ray reporting and dialog-driven assistance. In: Medical Imaging with Deep Learning (2025)
16. Gu, A., Dao, T.: Mamba: linear-time sequence modeling with selective state spaces. arXiv:2312.00752 (2023)
17. Google DeepMind: Gemini 2.5 Flash-Lite Preview [Model Card]. https://deepmind.google/models/gemini/flash-lite/ updated 30 July 2025; Accessed 17 June 2025
18. Hu, E., Shen, Y., Wallis, P., Allen-Zhu, Z., et al.: LoRA: low-rank adaptation of large language models. arXiv:2106.09685 (2021)
19. Touvron, H., Martin, L., Stone, K., Albert, P., et al.: Llama 2: open foundation and fine-tuned chat models. arXiv:2307.09288 (2023)
20. Loshchilov, I., Hutter, F.: Decoupled weight decay regularization. arXiv:1711.05101 (2017)
21. Demner-Fushman, D., Kohli, M., Rosenman, M., Shooshan, S., et al.: Preparing a collection of radiology examinations for distribution and retrieval. J. Am. Med. Inf. Assoc. (2016)
22. Alfarghaly, O., Khaled, R., Elkorany, A., Helal, M., Fahmy, A.: Automated radiology report generation using conditioned transformers. Inf. Med. Unlocked **24**, 100557 (2021)
23. Banerjee, S., Lavie, A.: METEOR: an automatic metric for MT evaluation with improved correlation with human judgments. In: Proceedings of ACL Workshop on Intrinsic and Extrinsic Evaluation Measures, pp. 65–72 (2005)
24. Ostmeier, S., Xu, J., Chen, Z., Varma, M., et al.: GREEN: generative radiology report evaluation and error notation. arXiv:2405.03595 (2024)
25. Papineni, K., Roukos, S., Ward, T., Zhu, W.: BLEU: a method for automatic evaluation of machine translation. In: Proceedings of 40th Annual Meeting, Association for Computational Linguistics, pp. 311–318 (2002)
26. Lin, C.: ROUGE: a package for automatic evaluation of summaries. In: Text Summarization Branches Out, pp. 74–81 (2004)
27. Zhang, Z., Wang, B., Liang, W., Li, Y., et al.: SAM-guided enhanced fine-grained encoding with mixed semantic learning for medical image captioning. In: ICASSP 2024, pp. 1731–1735 (2024)
28. Google: MedGemma – model collection on Hugging Face. https://huggingface.co/collections/google/medgemma-release-680aade845f90bec6a3f60c4 (2025)
29. Li, C., Wong, C., Zhang, S., Usuyama, N., et al.: LLaVA-Med: training a large language-and-vision assistant for biomedicine in one day. In: Advances in Neural Information Processing Systems, vol. 36, pp. 28541–28564 (2023)

On the Risk of Misleading Reports: Diagnosing Textual Biases in Multimodal Clinical AI

David Restrepo[1]([✉]) [iD], Ira Ktena[2] [iD], Maria Vakalopoulou[1] [iD], Stergios Christodoulidis[1] [iD], and Enzo Ferrante[3] [iD]

[1] MICS, CentraleSupélec - Université Paris-Saclay, Paris-Saclay, France
{david.restrepo,maria.vakalopoulou,
stergios.christodoulidis}@centralesupelec.fr
[2] Google DeepMind, London, UK
iraktena@google.com
[3] CONICET, Universidad de Buenos Aires, Buenos Aires, Argentina
eferrante@dc.uba.ar

Abstract. Clinical decision-making relies on the integrated analysis of medical images and the associated clinical reports. While Vision-Language Models (VLMs) can offer a unified framework for such tasks, they can exhibit strong biases toward one modality, frequently overlooking critical visual cues in favor of textual information. In this work, we introduce Selective Modality Shifting (SMS), a perturbation-based approach to quantify a model's reliance on each modality in binary classification tasks. By systematically swapping images or text between samples with opposing labels, we expose modality-specific biases. We assess six open-source VLMs–four generalist models and two fine-tuned for medical data– on two medical imaging datasets with distinct modalities: MIMIC-CXR (chest X-ray) and FairVLMed (scanning laser ophthalmoscopy). By assessing model performance and the calibration of every model in both unperturbed and perturbed settings, we reveal a marked dependency on text input, which persists despite the presence of complementary visual information. We also perform a qualitative attention-based analysis which further confirms that image content is often overshadowed by text details. Our findings highlight the importance of designing and evaluating multimodal medical models that genuinely integrate visual and textual cues, rather than relying on single-modality signals.

Keywords: VisionLanguage Models · Modality Bias · Calibration · Multimodal LLMs

1 Introduction

Clinical decision-making relies on the integrated analysis of medical images and associated textual or metadata information [9]. However, Vision-Language Models (VLMs) often exhibit modality bias on favor of language priors over visual inputs, which can lead to unsafe or misleading predictions. Recent work has

© The Author(s), under exclusive license to Springer Nature Switzerland AG 2026
J. Qiu et al. (Eds.): Agentic AI 2025/CMLLMs 2025/CREATE 2025, LNCS 16147, pp. 320–330, 2026.
https://doi.org/10.1007/978-3-032-06004-4_32

shown the limitations of large language models in medical domains such as oph-thalmology [17] and radiology [25], highlighting challenges like hallucinations, shortcut reasoning, and embedded biases. Even advanced multimodal decoders frequently overlook image content in favor of text, sometimes generating plausible yet hallucinated outputs [14,15]. These biases can come from the LLM backbone itself, with models confidently predicting outcomes without relying on image features [26]. Similar issues have been observed in unimodal and multimodal settings, such as in counterfactual text edits [2] and biasing feature interventions [19], both revealing a tendency to ignore visual evidence.

In medical applications, such biases are particularly concerning. A misleading textual report paired with a pathologic image may cause the model to ignore visual signs of disease. Beyond medicine, multimodal transformers trained on general-domain tasks have been shown to develop shortcut learning behaviors-e.g., answering "two" to all "how many" questions-without analyzing the visual scene [1,22]. This kind of spurious correlation, often caused by data imbalance or domain mismatch, undermines trust in model outputs. Recent studies also highlight that VLMs can produce high-confidence answers in the absence of meaningful visual content [26], suggesting that these models may not be grounded in the image modality even when such grounding is critical.

To address this, we propose a perturbation-based framework-Selective Modality Shifting (SMS)-that systematically swaps image or text components across samples with opposing labels, enabling us to quantify a model's reliance on each modality. We apply SMS to six VLMs (four general-purpose [5,7,11,12,21] and two medically fine-tuned [6,10]) on MIMIC-CXR [8] and FairVLMed [13] datasets. Our evaluation combines standard performance metrics (such as accuracy, precision, recall, and F1), as well as attention-map inspection [3,4,18], and first-token calibration via Expected Calibration Error (ECE) [16]. These experiments allows us to capture not only what the models predict but also where they attend and how confident they are. Results show that text bias persists even when visual cues contradict it, with miscalibration emerging as a secondary symptom of this over-reliance on language.

Our contributions can be summarized as: 1) we introduce a novel perturbation-based technique, *Selective Modality Shifting (SMS)*, to diagnose modality bias in VLMs for medical classification; 2) we conduct extensive tests with multimodal medical data, exposing how strong textual signals can overshadow image-based pathologies; 3) we employ attention-based explainability to show how attention patterns shift under SMS, providing qualitative evidence of this text dominance; and 4) we extend bias analysis to model uncertainty by computing the ECE of first-token probabilities in both clean and SMS conditions, revealing that modality bias is accompanied by systematic over-confidence.

2   Methodology

2.1   Problem Statement

Let $\mathcal{D}$ be a dataset comprising samples (I, T, y), where $I \in \mathcal{I}$ is a medical image (e.g., a chest X-ray from MIMIC-CXR [8] or a scanning laser ophthalmoscopy image from FairVLMed [13]), $T \in \mathcal{T}$ is an associated textual description (such as radiology reports or clinical notes), and $y \in \{0, 1\}$ is the ground-truth binary diagnosis label (e.g., abnormal vs. normal, or glaucoma vs. healthy). We define a function f_{VLM}:

$$\hat{y} = f_{\mathrm{VLM}}(I, T),$$

where f_{VLM} generates a binary diagnosis label $\hat{y} \in \{0, 1\}$ by prompting a VLM with an instruction prompt describing the diagnostic tasks, followed by a medical image I and a text description T. f_{VLM} maps the generated answer to a binary label by applying regular expressions (RE). The use of RE instead of the direct model answer is required as, sometimes, models do not answer "Yes" or "No" even if they are prompted to do so, but instead they provide more comprehensive answers likes "Yes, the patient presents abnormal findings" (as exemplified in Fig. 1) which require some post-processing.

The central challenge addressed in this work is to evaluate the extent to which these models genuinely integrate visual and textual cues rather than relying on textual priors- even when critical visual information is available. To this end, we propose the *Selective Modality Shifting* (SMS) framework. For a given sample (I, T, y), we generate counterfactual examples by selectively perturbing one modality. Specifically, for a sample with $y = 1$, we construct a perturbed sample (I, T', y) where T' is taken from a sample with $y = 0$, and similarly, a perturbed sample (I', T, y) by replacing I with an image I' from a sample with $y = 0$. The difference in the model's output between the original and perturbed inputs provides a direct measure of the contribution of each modality to the diagnostic decision, offering insights into how effectively these models balance and integrate both image and textual information in clinical settings.

2.2   Selective Modality Shifting

Figure 1 provides an illustrative example of SMS. Given a VLM-based function f_{VLM} that predicts a diagnosis $\hat{y} \in \{0, 1\}$ from an image I and text T, we systematically replace either the image or the text with a mismatched component from an opposing class. By comparing performance metrics (e.g., Accuracy, F1-score) between original and perturbed samples, we quantify the extent of single-modality bias. We hypothesize that extreme sensitivity to either text or image swapping implies over-reliance on that modality.

Text Swap. Let (I, T, y) be a sample with $y = 1$. We select another sample (I', T', y') where $y' \neq y$ (i.e., $y' = 0$), and replace only the text to construct a

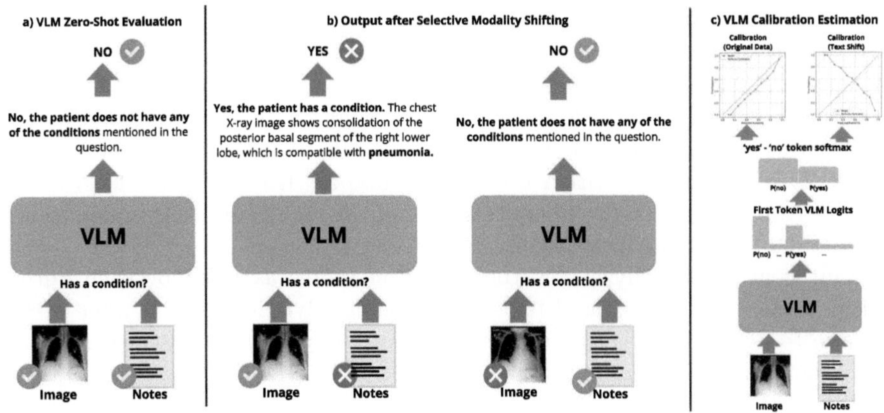

Fig. 1. Example of Selective Modality Shifting (SMS) and calibration evaluation on chest X-ray: (a) Original inputs yield correct "No condition" output. (b) Text swap induces false "Pneumonia" prediction, unlike image swap-indicating strong textual bias. (c) Calibration curves are computed over the probabilities of the tokens "yes" and "no".

perturbed sample (I, T', y). The model's prediction for this new sample is

$$\hat{y}_I^{(T')} = f_{\mathrm{VLM}}(I, T').$$

We perform this swapping for a number of data samples and measure how large the deviation from the originally predicted labels is by measuring changes in model performance. A large deviation in the predicted labels when only T is replaced indicates that the model relies heavily on the text modality. Conversely, minimal change suggests that visual features dominate-or that the text is not critical to the model's reasoning.

Image Swap. Similarly, for the same original sample (I, T, y) with $y = 1$, we select a different sample (I', T', y') where $y' = 0$ and replace only the image to obtain (I', T, y). The model's prediction here is

$$\hat{y}_T^{(I')} = f_{\mathrm{VLM}}(I', T).$$

If altering the image alone significantly affects the outcome, it implies that the model is integrating visual cues. Otherwise, the model may be underutilizing image information in favor of text.

2.3   Modality Attention

In order to provide a more mechanistic explanation of the model's behavior, we proposed to inspect the attention scores by modality. Modality attention was calculated for every (image or text) input token. Let $A_t \in \mathbb{R}^n$ be the attention vector for the t-th generated token over $n = n_T + n_I$ input tokens, where n_T and n_I denote the number of text and image tokens, respectively. We decompose A_t into modality-specific components: $A_t = \left[A_t^{(T)}, A_t^{(I)}\right]$.

In addition, following [20], the attention weight for the beginning-of-sequence (BOS) token is set to zero, i.e., $A_t(\text{BOS}) = 0$, to mitigate its confounding effect. This modality attention framework, adapted from [18] for Med-LLaVA, provides a precise token-level quantification of the contributions from image and text modalities during generation.

Our goal is to analyze how the attention scores assigned by each output token to the input image and text-tokens change dynamically. In other words, we aim to determine whether different output tokens exhibit distinct attention score patterns for images and text.

2.4 Calibration Evaluation

As illustrated in Fig. 1, we assess the calibration of first-token probabilities under both original and perturbed conditions. For each imagetext pair, we extract the model's logits for the "yes" and "no" tokens and apply softmax to compute predicted probabilities. While models do not always generate "yes" or "no" as the first token during free-form generation, we find that the probability assigned to these first-token choices remains highly consistent with the model's full answer. This justifies the use of first-token probabilities as a reliable proxy for binary decisions. We then compute the ECE by binning predictions into 10 uniform intervals and averaging the weighted absolute difference between predicted confidence and empirical accuracy within each bin following the equation:

$$\text{ECE} = \sum_{m=1}^{M} \frac{|B_m|}{n} |\text{acc}(B_m) - \text{conf}(B_m)|$$

3 Experimental Setup

3.1 Datasets

We analyze two medical imaging datasets, each suited for binary classification ("Yes" vs. "No") based on textual notes and corresponding images. Throughout our experiments, no fine-tuning or additional training is performed. Instead, each model is prompted in a zero-shot manner using images and/or medical reports, relying solely on its existing capabilities. Notably, medical records in both datasets do not always provide a clear diagnosis. In some cases, the text is inconclusive, requiring image analysis to arrive at a final diagnosis.

MIMIC-CXR [8]. We used 10k random chest radiographs from the test set of MIMIC-CXR repository. Its associated radiology reports are preprocessed to remove uncertain labels and included into a text prompt that asks if the patient's X-ray is normal or abnormal.

FairVLMed [13]. This dataset targets glaucoma detection via scanning laser ophthalmoscopy (SLO) images. We used the full test set comprised by 2k images. Metadata and clinical notes are concatenated into a textual prompt that queries for "glaucoma" vs. "healthy."

Prompting and Tokenization. In both datasets, we generate short textual templates instructing the model to consider the image and/or text and to provide a diagnosis depending on the dataset. See Fig. 4 for an example where both text and image are considered (the full set of instruction templates is included in the source code repository available at https://github.com/dsrestrepo/Selective-Modality-Shifting-SMS-. Tokenization and normalization for the text and image follow the format of each VLM. This setup ensures that all models receive a standardized format, enabling direct comparisons of zero-shot performance.

3.2 Models and Metrics

We evaluate several open-source VLMs from general domain: *LLaVA 1.5*(7B), *Qwen-2 VL*, *Llama 3.2 10B*, *Janus-Pro* (7B), and two from medical domain: *Med-LLaVA*, and *MedGemma*. The first four are primarily trained on natural-image tasks, whereas the latter two incorporate clinical/biomedical training data. All experiments are conducted on a single A100 GPU with PyTorch, using 16-bit precision (fp16), a temperature of 0, and no sampling for reproducibility.

We measure the classification performance using *Accuracy* (correct predictions over all samples), *Precision* (the proportion of true positive predictions among all predicted positives), *Recall* (the proportion of true positive predictions among all actual positives) and the *F1-score* (the harmonic mean of precision and recall). These metrics are computed in two settings: (i) *Unperturbed,* using each dataset as-is, and (ii) *Perturbed,* where we apply the proposed modality swaps to assess the model's reliance on text vs. image. We also include results for ablations where *only* text or image are fed into the model, directly removing the other modality.

To assess robustness, we adopt the *Negative Flip Rate (NFR)* from [24], which quantifies the fraction of correct predictions under the base condition $\hat{y}_i = y_i$ that flip to incorrect under a modality shift $\hat{y}_i^{(\text{shift})} \neq y_i$. It ranges from 0 (no flips) to 1 (all flipped). Formally:

$$\text{NFR} = \frac{1}{N} \sum_{i=1}^{N} \mathbb{1}\left(\hat{y}_i^{(\text{shift})} \neq y_i,\ \hat{y}_i = y_i \right)$$

4 Results and Discussion

Figure 2a,b shows results when applying the SMS framework to MIMIC and FairVLMed. The blue bars ("No Shift") summarize each model's performance in the unperturbed scenario. For MIMIC-CXR we observe that the generalist Llama 3 model achieves better accuracy and F1 score, while for FairVLMed, the specialist Med-LLava outperforms overall with a small margin. As a reference, we include results for the same datasets obtained by a SOTA zero-shot model as reported in [23]. We also compute the NFR (Fig. 2c,d), to quantify how many originally correct predictions are flipped by each perturbation.

Text Shift induces behavior akin to an inverse classifier, where models not only degrade but often flip correct predictions into confident errors, leading to performance far worse than random guessing. For example, in Fig. 2a,c Qwen-2 VL and LLaVA 1.5 lose over 20 points in performance with an NFR of above 0.60 when text alone is perturbed. The *Only Text* and *Only Image* columns further illustrate modality dominance: many models remain functional with text alone yet degrade severely with images alone, confirming the minimal role of visual features in certain predictions. Although domain-specific models like Med-LLaVA display a smaller gap, the overall findings underscore that even clinically tuned VLMs can be susceptible to shortcuts. Additionally, Fig. 3 shows ECE scores and calibration curves for both datasets. Under text shift, calibration degrades indicating that overconfident errors often coincide with modality-conflicting inputs, supporting our earlier claims on shortcut reliance. This phenomenon is especially pronounced in models like LLaVA-Med and LLaMA 3.2, as illustrated in Fig. 3-c,d, where calibration curves are inverted under the text-shift perturbation.

To assess reliance on a specific modality, we analyze how performance drops when either image (Image Shift) or text (Text Shift) is swapped. We also compute the NFR (see Fig. 2c,d), quantifying how many originally correct predictions are flipped by each perturbation. High NFR values-especially under Text Shift-confirm that models heavily depend on textual input. For example, in Fig. 2-c Qwen-2 VL has an NFR above 0.75 for text shift, indicating that text mismatches can mislead the prediction.

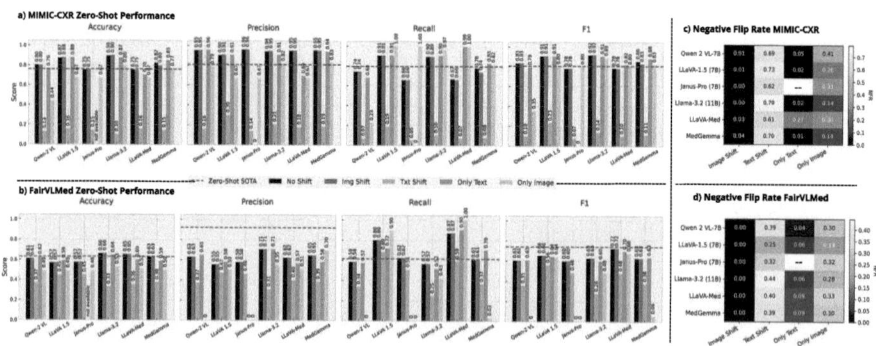

Fig. 2. Zero-shot classification results on (a) MIMIC-CXR and (b) FairVLMed under the Selective Modality Shifting framework. Panels (c) and (d) show the Negative Flip Rate (NFR), measuring how often correct predictions flip to incorrect ones after a perturbation. Janus-Pro does not support text-only input, so the corresponding ablation is omitted. Baseline results from a prior SOTA model are included for reference [23].

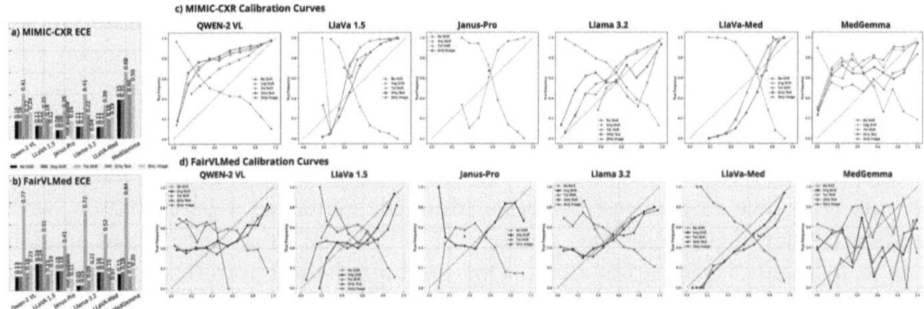

Fig. 3. (a) ECE values across shift conditions on MIMIC-CXR and (b) FairVLMed. (c,d) Calibration curves for all models. A strong misalignment between predicted probabilities and actual frequencies appears under text shifts, reflecting overconfidence.

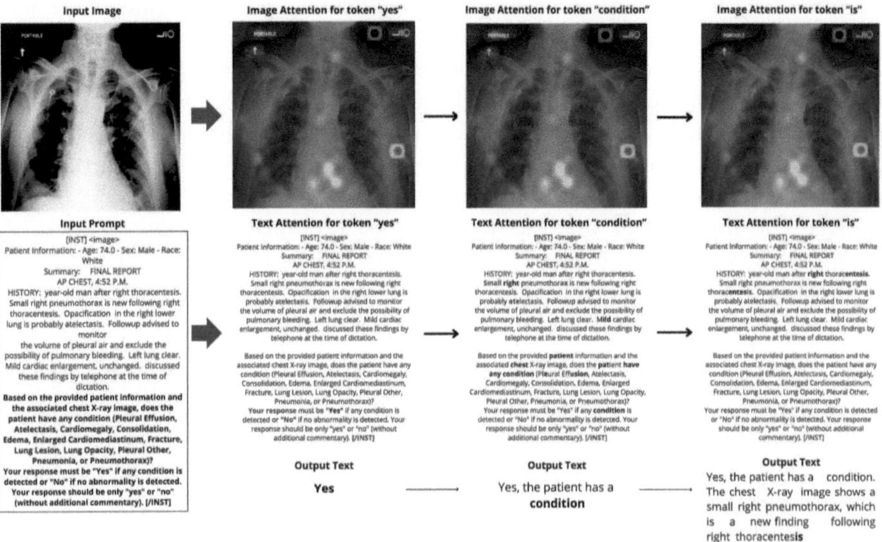

Fig. 4. Visualization of token-level attention during the generation of an X-ray classification output. The text input prompt is color-coded: violet for dataset metadata, green for clinical notes, and red for model instructions. Each panel shows how the model distributes attention across input image and text tokens for specific generated tokens ("yes" "condition", and "is"). The blue intensity of the highlighted text indicates the magnitude of attention directed to each token. (Color figure online)

We also performed a qualitative analysis of the attention patterns as discussed in Sect. 2.3. Due to space limitations, here we present an exemplar case that reflects the tendency observed in our experiments. As shown in Fig. 4, the attention patterns for input image tokens remain relatively stable across different output tokens, while for input text tokens vary as new tokens are generated.

This observation aligns with our hypothesis that images have minimal influence on the decoding process, whereas text plays a significantly more dominant role.

5   Conclusions

In this work, we introduced Selective Modality Shifting, a perturbation-based framework for evaluating the reliance of VLMs on text and image modalities in medical classification tasks. Our findings demonstrate a consistent bias toward textual information, with models frequently disregarding critical visual cues even when presented with complementary image data. This over-reliance on text persists across both general-domain and medically fine-tuned VLMs, raising concerns about the robustness of current multimodal architectures in clinical decision-making. Through experiments on MIMIC-CXR and FairVLMed, we observed that performance dropped significantly when textual notes were swapped, whereas image perturbations had minimal impact on predictions. Notably, we also found that predictions in these cases were not only less accurate but also poorly calibrated-highlighting overconfidence as a secondary failure mode. These results highlight the need for more rigorous evaluation methodologies to ensure that medical VLMs truly integrate both visual and textual data rather than exploiting shortcut patterns.

Here we limited our analysis to zero-shot prompting in open models. As future work, we plan to explore if similar conclusions hold for other prompting techniques such as few-shot prompting and more powerful closed models like the GPT or Gemini families.

Acknowledgments. This work was partially funded by the European Union's Horizon Europe programme through the Marie Skłodowska-Curie COFUND grant No. 101127936 (DeMythif.AI). This work was performed using HPC resources from the Mesocentre computing center of CentraleSupelec. EF was supported by the Google Award for Inclusion Research and a Googler Initiated Grant. EF and MV are supported by the STIC-AmSud CGFLRVE project.

Disclosure of Interests. The authors have no competing interests to declare that are relevant to the content of this article.

References

1. Ambsdorf, J.: Benchmarking faithfulness: towards accurate natural language explanations in vision-language tasks. arXiv preprint arXiv:2304.08174 (2023)
2. Atanasova, P., Camburu, O.M., Lioma, C., Lukasiewicz, T., Simonsen, J.G., Augenstein, I.: Faithfulness tests for natural language explanations. arXiv preprint arXiv:2305.18029 (2023)
3. Chefer, H., Gur, S., Wolf, L.: Generic attention-model explainability for interpreting bi-modal and encoder-decoder transformers. In: Proceedings of the IEEE/CVF International Conference on Computer Vision, pp. 397–406 (2021)

4. Chefer, H., Gur, S., Wolf, L.: Transformer interpretability beyond attention visualization. In: Proceedings of the IEEE/CVF Conference on Computer Vision and Pattern Recognition, pp. 782–791 (2021)
5. Chen, X., et al.: Janus-pro: unified multimodal understanding and generation with data and model scaling. arXiv preprint arXiv:2501.17811 (2025)
6. Google: Medgemma hugging face. https://huggingface.co/collections/google/medgemma-release-680aade845f90bec6a3f60c4 (2025), Accessed: [Insert Date Accessed 20 May 2025]
7. Grattafiori, A., et al.: The llama 3 herd of models. arXiv preprint arXiv:2407.21783 (2024)
8. Johnson, A.E., et al.: Mimic-cxr, a de-identified publicly available database of chest radiographs with free-text reports. Sci. Data **6**(1), 317 (2019)
9. Kline, A., et al.: Multimodal machine learning in precision health: a scoping review. npj Digit. Med. **5**(1), 171 (2022)
10. Li, C., et al.: Llava-med: training a large language-and-vision assistant for biomedicine in one day. Adv. Neural. Inf. Process. Syst. **36**, 28541–28564 (2023)
11. Liu, H., Li, C., Li, Y., Lee, Y.J.: Improved baselines with visual instruction tuning. In: Proceedings of the IEEE/CVF Conference on Computer Vision and Pattern Recognition, pp. 26296–26306 (2024)
12. Liu, H., Li, C., Wu, Q., Lee, Y.J.: Visual instruction tuning (2023)
13. Luo, Y., et al.: Fairclip: harnessing fairness in vision-language learning. In: Proceedings of the IEEE/CVF Conference on Computer Vision and Pattern Recognition, pp. 12289–12301 (2024)
14. Parcalabescu, L., Frank, A.: Mm-shap: a performance-agnostic metric for measuring multimodal contributions in vision and language models & tasks. arXiv preprint arXiv:2212.08158 (2022)
15. Parcalabescu, L., Frank, A.: Do vision & language decoders use images and text equally? how self-consistent are their explanations? arXiv preprint arXiv:2404.18624 (2024)
16. Posocco, N., Bonnefoy, A.: Estimating expected calibration errors. In: International Conference on Artificial Neural Networks, pp. 139–150. Springer (2021)
17. Restrepo, D., et al.: Multi-ophthalingua: a multilingual benchmark for assessing and debiasing llm ophthalmological qa in lmics. In: Proceedings of the AAAI Conference on Artificial Intelligence, vol. 39, pp. 28321–28330 (2025)
18. Stan, G.B.M., et al.: Lvlm-interpret: an interpretability tool for large vision-language models. arXiv preprint arXiv:2404.03118 (2024)
19. Turpin, M., Michael, J., Perez, E., Bowman, S.: Language models don't always say what they think: unfaithful explanations in chain-of-thought prompting. Adv. Neural. Inf. Process. Syst. **36**, 74952–74965 (2023)
20. Vig, J., Belinkov, Y.: Analyzing the structure of attention in a transformer language model. arXiv preprint arXiv:1906.04284 (2019)
21. Wang, P., et al.: Qwen2-vl: enhancing vision-language model's perception of the world at any resolution. arXiv preprint arXiv:2409.12191 (2024)
22. Wu, J., Mooney, R.J.: Faithful multimodal explanation for visual question answering. arXiv preprint arXiv:1809.02805 (2018)
23. Xia, P., et al.: RULE: reliable multimodal rag for factuality in medical vision language models. In: Al-Onaizan, Y., Bansal, M., Chen, Y.N. (eds.) Proceedings of the 2024 Conference on Empirical Methods in Natural Language Processing, pp. 1081–1093. Association for Computational Linguistics, Miami, Florida, USA, November 2024. https://doi.org/10.18653/v1/2024.emnlp-main.62, https://aclanthology.org/2024.emnlp-main.62/

24. Yan, S., et al.: Positive-congruent training: towards regression-free model updates. In: Proceedings of the IEEE/CVF Conference on Computer Vision and Pattern Recognition, pp. 14299–14308 (2021)
25. Yang, Y., Liu, X., Jin, Q., Huang, F., Lu, Z.: Unmasking and quantifying racial bias of large language models in medical report generation. Commun. Med. **4**(1), 176 (2024)
26. Zhang, Y.F., et al.: Debiasing multimodal large language models. arXiv preprint arXiv:2403.05262 (2024)

Aligning Multimodal Large Language Models with Patient-Physician Dialogues for AI-Assisted Clinical Support

Junyong Lee, Jeihee Cho, Jiwon Ryu, and Shiho Kim[✉]

School of Integrated Technology, Incheon, South Korea
{jjunilee,shiho}@yonsei.ac.kr, jwdylan@khu.ac.kr

Abstract. Recent advancements in Multimodal Large Language Models (MLLMs) have created new opportunities for medical AI, particularly through the integration of Medical Vision-Language Models (VLMs) into clinical applications. While these models exhibit robust capabilities in processing multimodal medical data, their practical adoption remains challenging due to the complexity of patient-physician interactions and clinical workflows. Understanding how to effectively adapt these models for real-world use is therefore essential. We propose a practical-oriented method that aligns multimodal medical data with synthetic patient-physician dialogues. Rather than relying on real-world dialogue datasets, we generate structured medical conversations using VLM-based techniques. This study aims to bridge the gap between Medical VLM research and clinical applicability by examining whether MLLMs trained on synthetic dialogues can effectively support AI-assisted medical decision-making. We expect our findings to contribute to the development of safer, more reliable AI-driven clinical assistants and to facilitate their successful implementation in healthcare settings.

Keywords: Medical Vision-Language Models (VLMs) · Multimodal Large Language Models (MLLMs) · AI-Assisted Medical Decision-Making

1 Introduction

The integration of Multimodal Large Language Models (MLLMs) into clinical practice represents a significant advancement in medical artificial intelligence (AI) [15], offering new opportunities for patient care. Recent advances, particularly through Medical Vision-Language Models (VLMs), have demonstrated their capability to process diverse forms of medical data, including radiological images [25], clinical notes [19], and laboratory reports [13]. However, the practical adoption of these models remains limited due to the inherent complexity of patient-physician interactions [16] and the nuanced nature of clinical workflows [28].

J. Qiu et al. (Eds.): Agentic AI 2025/CMLLMs 2025/CREATE 2025, LNCS 16147, pp. 331–340, 2026.
https://doi.org/10.1007/978-3-032-06004-4_33

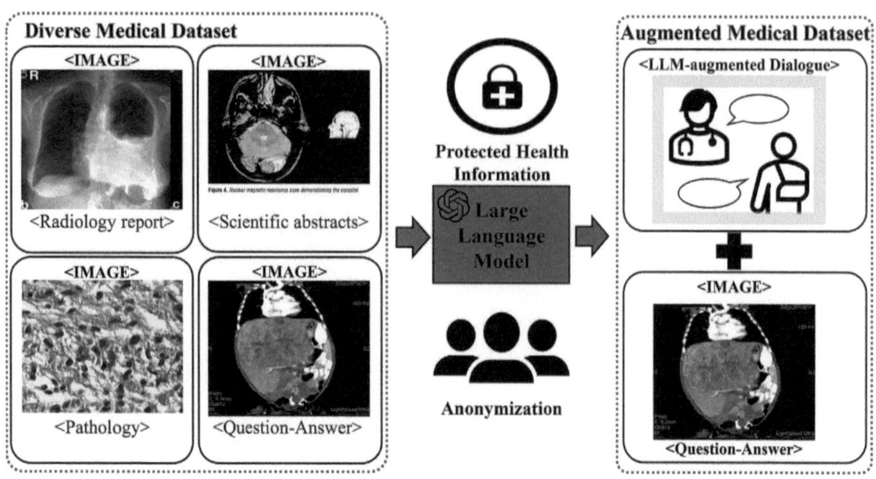

Fig. 1. Overview of proposed framework of generating augmented dataset using diverse medical datasets from different sources. Dataset is processed through large language models to generate dialogue between patient and physician, which can further assist clinical process.

Traditional approaches to fine-tuning MLLMs typically rely on large-scale, real-world datasets of medical dialogues [9]. However, these datasets [11,17, 29] are challenging to obtain due to privacy concerns, data scarcity, and the heterogeneity of clinical conversations. As a result, many models face challenges in adapting to the dynamic and context-specific nature of medical discussions, which are crucial for effective AI-assisted clinical support.

This study addresses these challenges by proposing a practical approach that aligns medical data with synthetically generated patient-physician dialogues as shown in Fig. 1. By leveraging LLM-based techniques, we generate structured medical conversations that emulate real-world clinical exchanges while preserving patient privacy. Our method focuses on the following core objectives:

1. Synthetic Dialogue Generation: Utilize LLMs to generate realistic, diverse, and context-aware patient-physician dialogues across various medical domains.
2. Clinical Utility Assessment: Evaluate the performance of the fine-tuned MLLM on tasks such as medical question answering, diagnostic assistance, and clinical reasoning.
3. Label Smoothing: Applying a label smoothing [18] method suitable for natural language. This approach enables the model to better handle noisy or ambiguous clinical labels, thereby improving the robustness of predictions across different modalities.

We hypothesize that synthetically generated patient-physician dialogues, when aligned with multimodal data, can provide the necessary contextual grounding for MLLMs to serve as effective AI assistants in real-world clinical environments. This study contributes to the ongoing effort to leverage AI in healthcare by proposing a method that balances data availability, model performance, and clinical relevance. Our approach establishes a connection between research on Medical VLMs and real-world clinical applications, supporting the development of reliable AI-driven assistants.

2   Related Works

2.1   MLLMs in Healthcare

Multimodal Large Language Models (MLLMs) integrate textual and visual medical data to support clinical decision-making [1]. Models like Med-PaLM2 [23] and GatorTronGPT [27] advance medical NLP through large-scale biomedical corpora, but their reliance on structured datasets limits adaptability to dynamic clinical settings.

2.2   Medical VLMs and Multimodal Alignment

Medical Vision-Language Models (VLMs) bridge textual and visual medical information [6]. Models such as BiomedCLIP [30], LLaVA-Med [10], and Med-Flamingo [16] improve multimodal reasoning but struggle with generalization due to supervised dataset constraints. Robust contextual adaptation and fusion methods remain necessary.

2.3   Synthetic Medical Dialogue Generation

Synthetic dialogues enhance AI-assisted clinical support by simulating patient-physician interactions [26]. While resources like MedDialog [29], ChatDoctor [11], and MTsample [17] diversify conversation formats, challenges persist in replicating real-world complexity. Adaptive frameworks are needed for broader clinical generalization.

2.4   Privacy Issues in Medical Data

Strict regulations limit access to real clinical data, especially patient-physician dialogues [4,31]. PHI removal and anonymization reduce utility [20,21]. This study instead uses LLMs to generate privacy-free synthetic dialogues that preserve clinical relevance.

2.5   Label Smoothing and Generalization in Medical AI

Label smoothing improves generalization in ambiguous medical contexts [12,18]. Prior work using meta-learning [22] or contrastive learning [3] lacks adaptability. Our approach synthesizes diverse conversations to enhance robustness in complex settings.

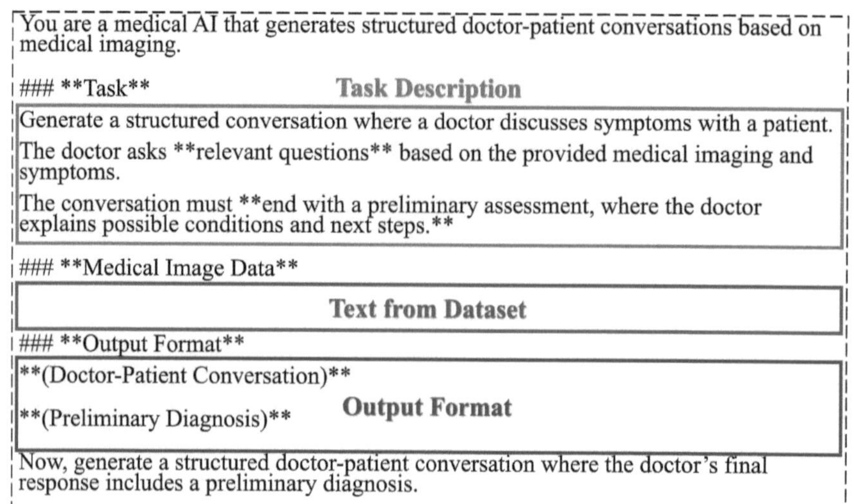

Fig. 2. Prompt used for generating conversation based on the existing medical dataset. The generated text from the language model is then processed as generated conversation and preliminary diagnosis.

3   Method

We propose a framework of generating augmented dataset to narrow the gap between the clinical usage and the training of VLM. To facilitate the understanding of clinical interactions, we generate the plausible conversation between the patient and the doctor along with the preliminary diagnosis.

To generate doctor-patient dialogues, we use three medical datasets: MIMIC-CXR [8], MedICaT [24], and VQA-Med 2021 [7]. These datasets vary in textual structure and serve distinct roles in medical AI research. Due to their diversity, direct adaptation for conversational modeling is impractical. To address this, we design dataset-specific structured prompts to generate clinically relevant interactions while preserving domain-specific language and contextual integrity.

Our approach extends beyond conventional label smoothing by leveraging LLMs to synthesize diverse conversational representations of medical conditions. This method captures the nuances of both patient expressions and doctor responses, enabling a more adaptable model that reflects the variability inherent in clinical settings.

3.1   Prompt Design

To ensure realism, consistency, and clinical relevance, we developed dataset-specific prompts that guide the model in generating structured doctor-patient conversations.

The prompt for our research follows a structured format consisting of four components as shown in Fig. 2: image of the patient, a [**Task Description**] that outlines the objective, [**Text from Dataset**] that provides data information, and an [**Output Format**] that defines the expected response. This design ensures that LLMs generate contextually rich patient-physician dialogues.

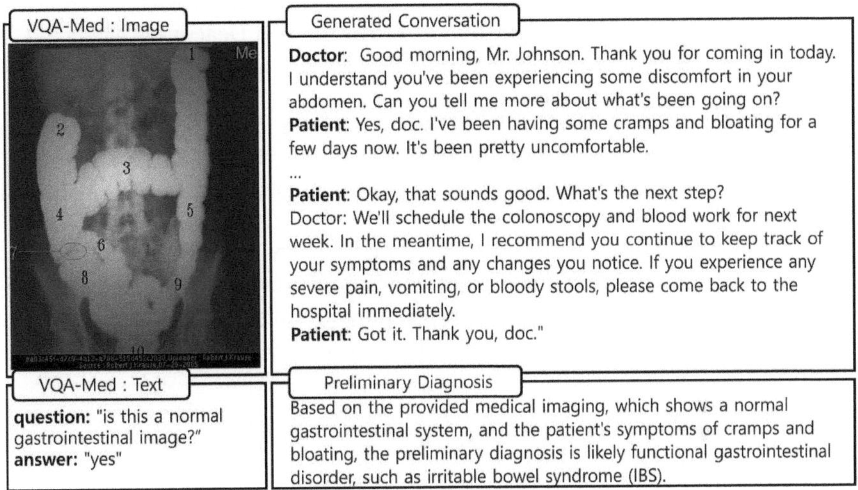

Fig. 3. Example dataset augmented with conversation generation.

3.2 Structured Dialogue Generation with Segmentation

To facilitate fine-tuning in later stages, we structured the generated dialogues into two distinct parts:

- **Generated Conversation (X):** General medical discussion where the doctor gathers symptoms, explains medical findings, and discusses potential next steps.
- **Preliminary Diagnosis (Y):** A structured segment where the doctor provides a diagnosis and suggests clinical decisions.

This segmentation improves adaptability during fine-tuning, enabling structured training with specific objectives. By designing structured prompts, we ensure that the generated dialogues maintain medical accuracy and align with clinical communication protocols. The structured nature of these interactions supports informed patient decision-making while maintaining clinical integrity. Since all dialogues are fully synthetic, they eliminate the use of patient-identifiable data, thereby ensuring full compliance with privacy regulations. This prompt-based approach enables scalable, privacy-free medical dialogue generation, making it suitable for training and evaluating medical VLMs without ethical or legal constraints. The example of augmented dataset using proposed framework is provided in Fig. 3, which is generated based on sample data from VQA-Med 2021.

3.3 Medical Datasets Used for Generation

The three datasets used in our approach provide distinct types of medical knowledge.

- **MIMIC-CXR**: Consists of 377,110 chest X-rays paired with 227,835 free-text radiology reports, de-identified from Beth Israel Deaconess Medical Center [8]. The dataset primarily contains posteroanterior (PA) and lateral view radiographs in DICOM format.
- **MedICaT**: Comprises 217,060 figures extracted from 131,410 open-access PMC papers, focusing on radiology and biomedical imaging [24]. It includes structured metadata and inline references, allowing the extraction of clinically relevant figure captions and descriptions.
- **VQA-Med 2021**: Provides 5,500 radiology images with structured question answering (QA) pairs sourced from MedPix, an open-access medical image database [7]. Designed for visual question answering (VQA) tasks, it includes structured Q&A related to abnormalities in medical images (Table 1).

Table 1. Meta Data Statistic of Generated Dialogues

Attribute	MIMIC-CXR	MedICaT	VQA-Med 2021
Domain	Radiology	Clinical Reports	General Medical
Modality	Image + Text	Figure + Text	Image + QA
# of Total Data	227,835	217,060	5000
Success rate	96.1%	93.3%	100%
Conversation Avg. Length	207.1 words	280.2 words	267.8 words
Conversation Avg. Turns	5.7	5.3	5.0
Preliminary diagnosis Avg. length	74.0 words	66.2 words	45.9 words

4 Experiment

4.1 Experiment Setup

In order to check the direct impact of augmented dataset on medical contexts, we conduct experiment on VQA tasks. Augmented dataset is generated using LLaMa 3.2 Vision Instruct model, and two different VLMs, LLaVA 1.5 [14] and InstructBLIP [5], are trained based on the augmented dataset. The performance of clinical assistant of the models the evaluation set from VQA-Med 2021 [2] is used for VQA task. The evaluation metric strictly follows an accuracy-based approach, where a model-generated response is considered correct only if it *exactly* matches the ground-truth answer. This stringent evaluation method ensures a reliable comparison between models without relying on partial credit or semantic similarity scoring.

We fine-tune the model in three strategies:

- **Base**: The pretrained model without any fine-tuning.
- **Fine-Tuning on Original Data (FT-Ori)**: The model is fine-tuned on image-question pairs from the training dataset, generating an answer without any additional context.
- **Fine-Tuning with Augmented Unlabeled Data (FT-Unlabeled)**: The training dataset is augmented using **generated dialogue without ground-truth answers**, allowing the model to learn from expanded conversational context but without explicitly seeing ground-truth answers during training.
- **Fine-Tuning with Augmented Labeled Data (FT-Labeled)**: Similar to FT-Unlabeled, but during augmentation, the model is also provided with **ground-truth answers as part of the generated dialogue** to reinforce learning.

For efficient fine-tuning, we utilize Low-rank-adaptation (LoRA) instead of full fine-tuning. We provide hyperparameters used for the experiments in Table 2.

Table 2. Hyperparameter configuration used throughout the experiments.

Hyperparameter	Value	Description
LoRA Rank (r)	8	Rank of LoRA adaptation matrices
LoRA Alpha	16	Scaling factor for LoRA adaptation
LoRA Target Modules	{q_proj, v_proj}	Layers where LoRA is applied
LoRA Dropout	0.1	Dropout probability for LoRA layers
LoRA Task Type	SEQ_2_SEQ_LM	Task type for LoRA adaptation
Batch Size (Train)	8	Training batch size per device
Number of Epochs	10	Total training epochs
Learning Rate	2×10^{-4}	Initial learning rate

Additionally, during validation, the augmented dataset does not include labels, ensuring that the models rely on learned representations rather than direct supervision during inference (Table 3).

4.2   Experiment Result

Among the evaluated models, fine-tuning with augmented dataset demonstrated superior performance, with InstructBLIP-13B achieving the highest accuracy (25.0%) in the FT-Labeled setting. The results suggest that augmenting training data with structured doctor-patient interactions improves model understanding of medical image contexts, making it a promising direction for vision-language adaptation in clinical AI. Moreover, generating dialogues based on available information proves beneficial even without ground-truth labels, highlighting its utility for data augmentation in scenarios where labeled data is scarce.

Table 3. Accuracy on VQA task with diverse vision-LLM Model.

	Accuracy (%)			
	Base	FT-Ori	FT-Unlabeled	FT-Labeled
LLaVA1.5-7B	15.4	18.4	23.4	**23.8**
LLaVA1.5-13B	14.8	18.0	**23.6**	23.4
InstructBLIP-7B	7.6	6.0	**24.0**	23.8
InstructBLIP-13B	8.2	6.0	24.4	**25.0**

5 Conclusion

In this study, we proposed a novel approach for aligning **MLLMs** with **patient-physician dialogues** to enhance AI-assisted clinical decision-making. By integrating **structured medical conversations** into the training process, we demonstrated that VLMs can be effectively fine-tuned to process medical data. Our key contribution is the demonstration that different modalities of medical data can be aligned, enabling AI models to bridge textual and visual information in a clinically relevant manner.

Future research could explore pretraining on multimodal clinical datasets such as MiMIC-CXR [8] or MedICaT [24] to assess generalizability and cross-modal robustness, developing interactive AI models that dynamically adapt to physician feedback, and deploying them in real-world clinical workflows to assess practical impact. Furthermore, investigating generlization performance of synthetic dialogue on other clinical domains beyond radiology, such as pathology or oncology can validate the applicability to other medical specialties.

We acknowledge the risk of hallucination in synthetic doctor-patient dialogues. Although structured prompts help reduce this risk, it remains a concern for clinical deployment. To mitigate this, we plan to evaluate the generated dialogues with medical professionals to ensure factual accuracy and clinical relevance.

Future work will explore the use of synthetic dialogues in more complex clinical decision support scenarios beyond single-turn VQA. By grounding conversations in EMR-derived content such as SOAP notes, lab results, and radiology summaries, we aim to better reflect real-world clinical reasoning and enhance applicability to AI-assisted diagnosis and management.

By aligning diverse medical data modalities with structured doctor-patient interactions, we pave the way for more interpretable, reliable, and clinically useful AI systems. Our findings highlight the potential of multimodal AI in healthcare and provide a foundation for future advancements in context-aware medical assistants.

References

1. AlSaad, R., et al.: Multimodal large language models in health care: applications, challenges, and future outlook. J. Med. Internet Res. **26**, e59505 (2024)
2. Ben Abacha, A., Sarrouti, M., Demner-Fushman, D., Hasan, S.A., Müller, H.: Overview of the vqa-med task at imageclef 2021: visual question answering and generation in the medical domain. In: CLEF 2021 Working Notes. CEUR Workshop Proceedings, CEUR-WS.org, Bucharest, Romania, 21–24 September 2021
3. Chaitanya, K., Erdil, E., Karani, N., Konukoglu, E.: Contrastive learning of global and local features for medical image segmentation with limited annotations. Adv. Neural. Inf. Process. Syst. **33**, 12546–12558 (2020)
4. Clunie, D., et al.: Summary of the national cancer institute 2023 virtual workshop on medical image de-identification-part 2: pathology whole slide image de-identification, de-facing, the role of ai in image de-identification, and the nci midi datasets and pipeline. J. Imaging Inf. Med. 1–15 (2024)
5. Dai, W., et al.: InstructBLIP: towards general-purpose vision-language models with instruction tuning. In: Thirty-seventh Conference on Neural Information Processing Systems (2023), https://openreview.net/forum?id=vvoWPYqZJA
6. Hartsock, I., Rasool, G.: Vision-language models for medical report generation and visual question answering: a review. Front. Artif. Intell. **7**, 1430984 (2024)
7. Ionescu, B., et al.: Overview of the ImageCLEF 2021: multimedia retrieval in medical, nature, internet and social media applications. In: Candan, K.S., et al. (eds.) CLEF 2021. LNCS, vol. 12880, pp. 345–370. Springer, Cham (2021). https://doi.org/10.1007/978-3-030-85251-1_23
8. Johnson, A.E., et al.: Mimic-cxr, a de-identified publicly available database of chest radiographs with free-text reports. Sci. Data **6**(1), 317 (2019)
9. Leong, H.Y., Gao, Y.F., Shuai, J., Zhang, Y., Pamuksuz, U.: Efficient fine-tuning of large language models for automated medical documentation. arXiv preprint arXiv:2409.09324 (2024)
10. Li, C., et al.: Llava-med: training a large language-and-vision assistant for biomedicine in one day. In: Advances in Neural Information Processing Systems, vol. 36 (2024)
11. Li, Y., Li, Z., Zhang, K., Dan, R., Jiang, S., Zhang, Y.: Chatdoctor: a medical chat model fine-tuned on a large language model meta-ai (llama) using medical domain knowledge. Cureus **15**(6) (2023)
12. Liao, Z., Xie, Y., Hu, S., Xia, Y.: Learning from ambiguous labels for lung nodule malignancy prediction. IEEE Trans. Med. Imaging **41**(7), 1874–1884 (2022)
13. Liu, B., Zhan, L.M., Xu, L., Wu, X.M.: Medical visual question answering via conditional reasoning and contrastive learning. IEEE Trans. Med. Imaging **42**(5), 1532–1545 (2022)
14. Liu, H., Li, C., Wu, Q., Lee, Y.J.: Visual instruction tuning. In: NeurIPS (2023)
15. Meskó, B.: The impact of multimodal large language models on health care's future. J. Med. Internet Res. **25**, e52865 (2023)
16. Moor, M., et al.: Med-flamingo: a multimodal medical few-shot learner. In: Machine Learning for Health (ML4H), pp. 353–367. PMLR (2023)
17. MTSamples: Mtsamples: medical transcription samples and example reports (2025), https://mtsamples.com, Accessed 19 Feb 2025
18. Müller, R., Kornblith, S., Hinton, G.E.: When does label smoothing help? In: Advances in Neural Information Processing Systems, vol. 32 (2019)

19. Naseem, U., Khushi, M., Kim, J.: Vision-language transformer for interpretable pathology visual question answering. IEEE J. Biomed. Health Inform. **27**(4), 1681–1690 (2022)

20. Norgeot, B., et al.: Protected health information filter (philter): accurately and securely de-identifying free-text clinical notes. NPJ Digit. Med. **3**(1), 57 (2020)

21. Olatunji, I.E., Rauch, J., Katzensteiner, M., Khosla, M.: A review of anonymization for healthcare data. Big Data **12**(6), 538–555 (2024)

22. Singh, R., Bharti, V., Purohit, V., Kumar, A., Singh, A.K., Singh, S.K.: Metamed: few-shot medical image classification using gradient-based meta-learning. Pattern Recogn. **120**, 108111 (2021)

23. Singhal, K., et al.: Toward expert-level medical question answering with large language models. Nat. Med. 1–8 (2025)

24. Subramanian, S., et al.: Medicat: a dataset of medical images, captions, and textual references. arXiv preprint arXiv:2010.06000 (2020)

25. Thawkar, O., et al.: Xraygpt: chest radiographs summarization using medical vision-language models. arXiv preprint arXiv:2306.07971 (2023)

26. Wang, J., et al.: Notechat: a dataset of synthetic doctor-patient conversations conditioned on clinical notes. arXiv preprint arXiv:2310.15959 (2023)

27. Yang, X., et al.: Gatortron: a large clinical language model to unlock patient information from unstructured electronic health records. arXiv preprint arXiv:2203.03540 (2022)

28. Yu, F., et al.: Evaluating progress in automatic chest x-ray radiology report generation. Patterns **4**(9) (2023)

29. Zeng, G., et al.: Meddialog: large-scale medical dialogue datasets. In: Proceedings of the 2020 Conference on Empirical Methods in Natural Language Processing (EMNLP), pp. 9241–9250 (2020)

30. Zhang, S., et al.: Biomedclip: a multimodal biomedical foundation model pretrained from fifteen million scientific image-text pairs. arXiv preprint arXiv:2303.00915 (2023)

31. Zhang, X., Ding, J., Wu, M., Wong, S.T., Van Nguyen, H., Pan, M.: Adaptive privacy preserving deep learning algorithms for medical data. In: Proceedings of the IEEE/CVF Winter Conference on Applications of Computer Vision, pp. 1169–1178 (2021)

Author Index

© The Editor(s) (if applicable) and The Author(s), under exclusive license
to Springer Nature Switzerland AG 2026
J. Qiu et al. (Eds.): Agentic AI 2025/CMLLMs 2025/CREATE 2025, LNCS 16147, pp. 341–343, 2026.
https://doi.org/10.1007/978-3-032-06004-4

MIX
Papier aus verantwortungsvollen Quellen
Paper from responsible sources
FSC® C105338

If you have any concerns about our products,
you can contact us on
ProductSafety@springernature.com

In case Publisher is established outside the EU,
the EU authorized representative is:
Springer Nature Customer Service Center GmbH
Europaplatz 3, 69115 Heidelberg, Germany

Printed by Libri Plureos GmbH
in Hamburg, Germany